PROTECTION OF ELECTRONIC CIRCUITS FROM OVERVOLTAGES

Ronald B. Standler, Ph.D.

DOVER PUBLICATIONS, INC.
Mineola, New York

Bibliographical Note

This Dover edition, first published in 2002, is an unabridged republication of
the work published by John Wiley & Sons, Inc., New York, in 1989.

The citation of trade names and names of manufacturers in this book is not
to be construed as approval of commercial products, nor is the absence of a particular manufacturer's name to be construed as an adverse judgment by the
author. The schematic circuit diagrams, description of circuits, specifications, or
other data in this book are not to be regarded (by implication or otherwise) as in
any manner licensing the holder or any other person or corporation, or convey-
ing any rights or permission to manufacture, use, or sell any patented invention.

Library of Congress Cataloging-in-Publication Data

Standler, Ronald B. (Ronald Bruce), 1949–
    Protection of electronic circuits from overvoltages / Ronald B. Standler.
        p. cm.
    Originally published: New York : John Wiley & Sons, 1989.
    Includes bibliographical references and index.
        ISBN-13: 978-0-486-42552-8
        ISBN-10: 0-486-42552-5
        1. Electronic apparatus and appliances—Protection. 2. Electronic cir-
cuits—Protection. 3. Overvoltage. 4. Transients (Electricity) I. Title.

    TK7870 .S77 2002
    621.381'044—dc21

                                                        2002067356

www.doverpublications.com

Preface

Electrical overstresses, such as from lightning, electromagnetic pulses from nuclear weapons, and switching of reactive loads, can cause failure, permanent degradation, or temporary malfunction (upset) of electronic devices and systems. Cost-effective protection of industrial, military, and consumer electronic systems from these overstresses is of great importance. However, this subject is ignored as an applications detail in engineering education. The literature on this subject is scattered throughout professional journals, patents, conference proceedings, military technical reports, and standards. It is no wonder that protection from electrical overstresses has a reputation as being a "black art" instead of a science. One of the goals of this book is to index and organize information on protection from overvoltages in one convenient place.

Another goal is to present practical rules and strategies for the design of circuits to protect electronic systems from damage by transient overvoltages. Whenever possible, these rules have been related to the physics of the situation rather than given as the author's opinion or a "rule of thumb." Much needs to be done to place this subject on a firm scientific basis; however, engineers who need to solve problems now cannot wait for experts to find a more complete understanding of overvoltage protection from fundamental principles.

Information is included for users of electronic equipment who wish to make knowledgeable decisions about specifying or purchasing overvoltage protection, pariculary in Chapters 19 and 20.

The scope of this book is limited to protection of electronic circuits and systems. There are four parts to this book:

1. discussion of various threat waveforms
2. properties of nonlinear protection components
3. applications of these components to form protection circuits
4. validating protective measures, including laboratory techniques

Because many of these circuits are operated from the ac supply mains, protection of equipment operating from mains with nominal voltages up to 1 kV rms is also discussed in this book.

As would be expected from the title, the following topics are excluded from this book:

1. protection of electric power transmission and distribution equipment
2. details of the physics of generation of overvoltages (e.g., lightning, nuclear electromagnetic pulse, and electrostatic discharge), although this subject is reviewed in Chapter 2

In addition, the following topics are excluded:

3. specific examples of protection of telephone equipment
4. specific details of protecting three-phase power circuits
5. details of shielding of conductors and equipment from transient electromagnetic fields
6. testing of devices and equipment with potential differences greater than about 6 kV
7. recitation of the complete set of military specifications, commercial standards, and government regulations

Readers interested in these topics will find some related material and references cited in this book but must seek detailed treatment elsewhere.

Professor F. Reif in the preface of his book[1] stated that "an author never finishes a book, he merely abandons it." Although this project has been my principal activity for 2 years, I keep finding better ways to explain this material. Furthermore, research and development by many engineers continue to develop new knowledge and products faster than can be appreciated. I hope that this book is useful, and I look forward (I think!) to preparing a second edition in the future.

RONALD B. STANDLER

State College, Pennsylvania
June 1988

[1] *Fundamentals of Statistical and Thermal Physics*, McGraw-Hill, New York, 1965.

Acknowledgments

A draft of part of this book was done in 1983–84 when the author was a Research Scholar at the U.S. Air Force Weapons Laboratory under Maj. James Kee. Messrs. Dennis Andersh, Mark Snyder, and David Hilland provided constructive criticism of that first draft. Mrs. Joy Bemesderfer, Mrs. Virginia King, Ms. Ann Kloss, and Mr. Keith Newson of the AFWL Technical Library assisted with literature searches and obtained more than a hundred documents from the Defense Technical Information Center, National Technical Information Center (NTIS), and interlibrary loans during 1983–84. Ms. Elaine Lethbridge of The Pennsylvania State University has obtained copies of articles for me since 1984.

The author thanks the U.S. Air Force Weapons Laboratory, the U.S. Army Harry Diamond Laboratory, and the Allegheny Power System for support of his research projects. The generous support of these organizations made it possible for this book to be written. The author also thanks Messrs. François D. Martzloff and Peter Richman for many interesting discussions and critical review of a draft of this book.

R. B. S.

Contents

Appendices **402**

INDEX **431**

Protection of
Electronic Circuits
from Overvoltages

Part 1

Symptoms and Threats

1

Damage and Upset

A. NATURE OF ELECTRICAL OVERSTRESS PROBLEM

Electrical overstresses (e.g. from lightning, electromagnetic pulses from nuclear weapons, and switching of reactive loads) can cause failure, permanent degradation, or temporary malfunction of electronic devices and systems. The characterization of these overstresses and the design of effective protection from them is of great importance to manufacturers and users of industrial, military, and consumer electronic equipment.

Electrical overstresses have received increasing attention during the period between 1960 and the present (1988). This trend can be expected to continue. There are several reasons for this trend: (1) devices are becoming more vulnerable; (2) vulnerable systems are becoming more common; and (3) awareness of the existence of overstresses has increased.

Modern semiconductor integrated circuits are much more vulnerable to damage by overstresses than earlier electronic circuits, which used vacuum tubes and relays. Progress in developing faster and denser integrated circuits has been accompanied by a general increase in vulnerability. At the same time that electronic circuits were becoming more vulnerable, they were also becoming more widely used. (As an example, consider desktop computers and videotape recorders: these were nonexistent items in 1960 but are quite common now.) Therefore, there are now more systems to protect from overstresses. Finally, as awareness of overstresses increases, users of vulnerable systems request appropriate protective measures.

In general, techniques for protection against transient overvoltages can be divided into three classes:

1. shielding and grounding
2. application of filters
3. application of nonlinear devices

Shielding, while important, is not sufficient protection against electromagnetic fields from lightning or nuclear weapons, because compromises in the integrity of the shield must be made (e.g., windows in aircraft; long lines must enter the shielded volume to supply electric power and carry communication signals). Various shielding and grounding techniques are covered in detail in books by Ott (1976), Morrison (1977), Ricketts et al. (1976), and Lee (1986) and in government reports by Lasitter and Clark (1970) and Sandia Laboratories (1972). The design of filters is covered in many electrical engineering text and reference books. The emphasis in this book is on the third class of techniques, nonlinear transient protection devices, although some information on filters is included in Chapters 13 and 19.

B. ORGANIZATION OF THIS BOOK

This book is divided into four parts:

1. symptoms and threats
2. nonlinear protection components
3. applications of protection components
4. validating protective measures

1. Symptoms and Threats

Transient overvoltages in electronic circuits can arise from any of the following causes: lightning, electromagnetic pulse produced by nuclear weapons (NEMP), high-power microwave weapons (HPM), electrostatic discharge (ESD), and switching of reactive loads. These sources are described in Chapter 2. These transient overvoltages can be coupled to vulnerable circuits in several different ways:

1. direct injection of current—for example, a lightning strike to an overhead conductor
2. effects of rapidly changing magnetic fields—for example, induced voltage in a conducting loop from changing magnetic fields owing to nearby lightning or NEMP

3. effects of rapidly changing electric field—for example, charging by induction from ESD
4. changes in reference ("ground") potential due to injection of large currents in a grounding conductor that has nonzero values of resistance and inductance

A discussion of surveys of transient overvoltages in specific environments is given in Chapter 3. The propagation of transient overvoltages from their source to the vulnerable equipment is discussed in Chapter 4. Chapter 5 discusses standard overstress test waveforms that are simplifications of the environment. Chapter 6 gives a brief sketch of protective circuits and devices, which is useful in preparing the reader for the following two parts of the book.

2. Nonlinear Protection Components

Chapters 7 to 14 contain a discussion of properties of various components that are useful to protect circuits and systems from overvoltages. Spark gaps, metal oxide varistors, and avalanche diodes are emphasized. However, other components, such as semiconductor rectifier diodes, thyristors, resistors, inductors, filters, and optoisolators are also discussed. Chapter 15 explains why minimization of parasitic inductance is critical in practical transient overvoltage clamping circuits.

3. Applications of Protective Devices

Chapters 16 to 20 are concerned with application techniques to protect circuits and systems from damage by transient overvoltages. Specific applications of the components in Part 3 are discussed in the context of signal lines, dc power supplies, and low-voltage mains. Protection of circuits and systems from upset is covered in Chapter 21.

4. Validating Protective Measures

Chapters 22 to 24 discuss how to validate protective measures against damage by overstresses. Preparation of a test plan, high-voltage laboratory procedures, and safety are discussed.

C. NOMENCLATURE

There is no general agreement on a name for electrical overstress. The Institute of Electrical and Electronics Engineers (IEEE) in the United States has adopted the word *surge* to denote an overstress condition that has a duration of less than a few milliseconds. American engineers also use the word surge to mean something quite different: an increase in the rms voltage for a few cycles, which is called a *swell* by Martzloff and Gruzs (1987). To avoid misunderstanding, the author favors the use of *overvoltage,* which is translated from the German *Überspannung.* To emphasize the brief nature of the event, one may say *transient overvoltage.* The term *electrical overstress* is more general, because it includes excessive current or energy as well as voltage. To be precise, it is necessary to say *electrical* overstress, because there are other kinds of adverse environments—for example, extreme temperature. Because this book is only concerned with electrical overstresses, the modifier "electrical" is omitted.

Overstresses can cause two different kinds of adverse outcomes in sensitive electronic circuits and systems: damage or upset. *Damage* is a permanent failure of hardware. A damaged system may fail completely or partially. The only way to recover from damage is to replace defective components. *Upset* is a temporary malfunction of a system. Recovery from upset does not require any repair or replacement of hardware. An example of damage is a charred printed circuit board after a lightning strike. An example of upset is loss of the contents of the volatile memory in a computer when there is a brief interruption of power. A system or circuit is said to be *vulnerable* to damage but *susceptible* to upset.

Components and circuits that protect vulnerable devices and systems from damage by electrical overstresses are members of a class of devices called *terminal protection devices* (TPDs) or *surge protective devices* (SPDs). The term TPD is used by the U.S. military; SPD is used by the engineers in commercial practice. A surge protective device that is intended for electrical power systems is called an *arrester.*

The word *mains* is used in this book to refer to low-voltage ac power distribution circuits inside of buildings. In this context, *low-voltage* means less than 1 kV rms, and ac means a sinusoidal waveform with a frequency between 50 and 400 Hz.

Definitions of these and other specialized terms are contained in the glossary in Appendix A.

D. DAMAGE AND UPSET THRESHOLDS

Because many modern semiconductor devices (small signal transistors, integrated circuits) can be damaged by potential differences that exceed 10 V, the survivability of modern electronics is limited when exposed to transient overvoltages. Modern electronic technology has tended to produce smaller and faster semiconductor devices, particularly high-speed digital logic, microprocessors, metal oxide semiconductor (MOS) memories for computers, and GaAs FETs for microwave use. This progress has led to an *increased* vulnerability of modern circuits to damage by transient overvoltages, owing to the inability of small components to conduct large currents and to breakdown at smaller voltages.

Smaller devices make a more economical use of area on silicon wafers and decrease components cost. These smaller devices also have less parasitic capacitance and are therefore faster. However, devices often fail when the current per unit area becomes too large. The magnitude of transient currents is determined principally by external circuit parameters (e.g., nature of the source, characteristic impedance of transmission lines, resistance and inductance between the source of the transient and the vulnerable circuit, etc.). Smaller devices obviously have less area and are thus more vulnerable to damage from a current of a given level. When breakdown is considered, smaller devices that have less spacing between conductors will break down at lower voltages.

A logical approach to transient protection would be (1) to determine the threshold at which damage occurred, (2) to determine the worst-case electrical overstress that could arrive at a particular device, and (3) to design and install a protective circuit that would limit the worst-case overstress to less than the damage threshold. This simple, scientific approach has become a practical nightmare. First, as is described below there are apparently no simple criteria for determining the maximum overstress that a part can withstand without being damaged. Second, as described throughout this book, a protective circuit that can survive a worst-case overstress is often extremely expensive, massive, and bulky.

1. Damage Threshold

Failure of transformers and motors is caused by breakdown of insulation. The most important parameter in insulation breakdown is the magnitude of the peak voltage, although the rise time may also be important. Once the insulation has broken down, some of the winding is shunted with a low-impedance arc. Transformers and motors can withstand voltages that are much greater than those that cause failure in semiconductor devices. Therefore, recent concerns about damage caused by overvoltages has focused on semiconductors and has tended to ignore damage to transformers and motors.

The value of voltage, current, or power that is necessary to cause permanent damage to semiconductor devices is known as the *damage threshold* or *failure threshold*. In general, the value of the threshold is a function of the duration of the overstress. Such information is essential during the design of protective circuits, since the protection must attenuate overstresses to below the damage threshold.

The damage threshold is defined as the minimum power transfer through a terminal such that the device's characteristics are significantly and irreversibly altered. The damage threshold is a function of the waveshape and is particularly sensitive to the duration of the transient.

The most widely used model for damage threshold was presented by Wunsch and Bell (1968). They showed that the maximum power, P, that could be safely dissipated in a semiconductor junction was given by

$$P = kt^{-1/2}$$

where t is the time duration of a pulse and k is the *damage constant*. Devices with a larger value of k are able to withstand larger transients. The inverse square-root dependence on the pulse duration was derived by Wunsch and Bell (1968) for adiabatic heating of the junction. This relation is approximately valid for pulse durations that satisfy

$$0.1 \, \mu s < t < 20 \, \mu s$$

This simple model is known as a *thermal model*, since the mechanism for damage is melting of the semiconductor by excessive energy deposited in the bulk semiconductor.

Ideally, the value of k would be a constant for a device with a particular model number. Much effort has been devoted to finding the proper value of the damage constant, k, for hundreds of different silicon diodes and transistors. Several conclusions are clear from this effort.

A relation of the form

$$P = At^{-B}$$

fits the empirical failure data better than the form where B is one-half. Efforts to predict the values of k, A, or B from parameters on the specification sheet (e.g., thermal resistance) have not been particularly successful, so the value of k or the values of A and B must be determined by experiment.

Enlow (1981) described variations in the mean failure threshold for samples of 100 transistors from each of five manufacturers for four different 2N part numbers. Even for devices of the *same* model number and *same* manufacturer, the standard deviation for the failure threshold was often about 25% of the mean value. When failure thresholds for devices of the same model number but *different* manufacturers were compared, it was clear that specifying the same model number was not adequate to ensure that the two lots of devices came from the same statistical population for failure thresholds. For example, 2N718 transistors from Texas Instruments had a failure threshold of 452 ± 73 W, whereas 2N718 transistors from IDI had a failure threshold of 97 ± 7.2 W (these data are written as $\bar{P} \pm \sigma$). The mean for the IDI devices is $4.86 \, \sigma$ from the mean of the TI devices; the mean for the TI devices is $49.3 \, \sigma$ from the mean of the IDI devices:

$$97 = 452 - (4.86 \times 73)$$

$$452 = 97 + (49.3 \times 7.2)$$

These two distributions of failure thresholds are clearly distinct. Measurement of Texas Instruments 2N718 transistors tells nothing about IDI 2N718 transistors.

Kalab (1982) investigated failure thresholds for 2N1613 and 2N4237 transistors from 30 and 29 manufacturers, respectively. One transistor of each model from each manufacturer was opened and inspected with a microscope to determine the geometry. Sixteen different geometries were used for each model—evidently some manufacturers used the same pattern as other manufacturers. For the 2N4237 transistor, the ratio of smallest to largest chip area was a factor of 20. Clearly the model number does not specify how these transistors are fabricated. Twenty samples of each model from each manufacturer were pulsed with a rectangular waveform with a duration of 1 μs and a polarity that reverse-biased one junction. The ratio of minimum to maximum failure power for the collector-base junction for these 1800 transistors of each model was about 3×10^4. Clearly the collector-base failure power cannot be specified by model number.

These damage thresholds are a statistical concept, not precise numbers that are applicable to a particular piece part. When damage thresholds are being determined, the device either fails or it does not fail. If it does fail, one has an upper bound for the damage threshold but no information about the effect of slightly smaller stresses. After the device fails, the experiment cannot be repeated for that particular piece part.

By testing a large number of devices, one can obtain a statistical distribution of damage thresholds and fit various models to these data. Such effort is expensive, and the results are not applicable to components of the same model number from a different manufacturer. Worse yet, there are apparently no discussions in the literature about whether such statistical distributions are applicable to different production lots of the *same* model number *and same* manufacturer.

One of the major reasons for this variation in damage thresholds is inherent in the device specifications. Specifications for electronic components give maximum or minimum values for various parameters that are important in the original application of the device, but they do *not* specify how the device is to be constructed. Therefore, parts from different manufacturers with the same model number will probably have different electrode geometries and different compositions. Also, manufacturers will revise their process for a particular model from time to time without changing the model number of the parts. Such unspecified features can be important in determining transient performance and damage thresholds.

The simple thermal model for damage in semiconductors ignores effects caused by different failure mechanisms such as second breakdown, metalization failure, and breakdown of gate oxide in MOSFETs.

Chowdhuri (1965) showed that diodes that were conducting when a transient voltage was applied had a breakdown voltage that was about half that for a diode that had no initial bias. This would be expected to affect the failure thresholds for these devices. However, most empirical studies of failure thresholds are made with an initially unbiased device. These thresholds are not necessarily valid for devices that are conducting when the transient occurs. This is an important point, since during normal operation of a system, many vulnerable devices may be conducting when the transient occurs and thus have different failure thresholds.

Moreover, most empirical studies of failure thresholds of transistors apply the transient pulse to the base-emitter junction (with a polarity that will reverse-bias this junction) while the collector terminal is an open circuit *and* there is no initial current in any of the transistor's terminals. This is certainly not the usual way to operate a transistor, and one would expect that failure thresholds obtained in this way would not be representative of failure thresholds during normal operation.

Devices that are initially conducting may enter the *second breakdown* region of operation as a result of transient overstress. Once the transient pushes the device into second breakdown, the dc power supply in the system may kill the semiconductor. In this scenario, the transient does not necessarily need a large energy content for the device to fail, since the transient only initiates the failure process.

An insidious possibility is that a stressed part may suffer an adverse change of parameters (called *degradation*) but still be capable of performing its intended function. A degraded part may have values of its parameters that are either within or outside the minimum and maximum limits, although it may be of more concern when the values of the parameters are outside the limits on the specification sheet. A degraded component may fail later when subjected to some small stress that a virgin component would survive. This late failure is, at least in part, due to *latent damage* caused by an earlier overstress.

In 1984 a committee of the U.S. National Academy of Science examined the methodologies of *calculating* the hardness of electronic systems. The committee was "skeptical of the assurance one can have in the hardness of systems protected" by designs based on mathematical models of a system (Pierce, 1984, p. 21). The lack of reliable information about the damage threshold of individual transistors in an electronic system was part of the reason for this lack of confidence. The committee specifically noted that "uncertainties in the damage thresholds of military specification components persist" (Pierce, 1984, p. 45).

The author wonders if determinations of damage thresholds are useful for hardening. Certainly one can spend large sums of money testing components and calculating statistical parameters such as mean and standard deviation. Research cited above shows that there are large variations in damage thresholds for a specific model of transistor. Large, complicated systems have many component parts. If damage thresholds are to be used to design overvoltage protection circuits, one must concentrate on devices with extremely low damage thresholds, because these devices are the "weak link in the chain" that composes the system. It is difficult to characterize extreme values with statistical analysis, because these "outlying" data do not fit standard statistical distributions. Indeed, the outlying points may not be members of the same population as the data nearer the mean value.

One attack on the problem of determining damage thresholds involves more research in pure and applied solid-state physics regarding failure mechanisms in devices. With more knowledge one might be able (1) to design devices with greater damage thresholds or (2) to control production parameters to eliminate devices with low damage thresholds without testing every piece part. Although such research is certainly worthwhile, it is of no help to today's circuit designer.

Another attack on the problem of determining damage thresholds is to use the *absolute maximum ratings* given in the specification sheet. These parameters are usually steady-state ratings. It is well known that components can tolerate, for a few microseconds, values of current and power that are factors of 10^2 (or more) greater than their maximum steady-state ratings (Wunsch and Bell, 1968; Alexander et al., 1975). Therefore this approach is very conservative. *If* protecting a circuit to the absolute maximum ratings imposes a considerable hardship on the designer, stresses of a factor of 2 above these steady-state maximum ratings are easily tolerable for a suitably brief time (e.g., a few microseconds). Such an approach also avoids the problem of specifying in advance the transient waveforms to which the component will be exposed, although it will be necessary to have an estimate of the worst-case peak current and a few other parameters.

Such an approach has been endorsed by U.S. Department of Defense Military Handbook 419 (p. 1-50, 1982). In the absence of data from manufacturers or laboratory testing, the following surge withstand levels were given as typical:

Integrated circuits: 1.5 times normal rated junction and supply voltage

Discrete transistors: 2 times normal rated junction voltage

Diodes: 1.5 times peak inverse voltage

Under this approach, components should be protected against the following conditions:

1. rise time on the order of few nanoseconds
2. continuation of stress at reduced level for 0.5 second
3. polarity reversals
4. estimate of worst-case peak current and total energy in transient

The brief rise time is typical of the threat from NEMP and ESD. The continuation of stress for 0.5 second is typical of continuing currents in cloud-to-ground lightning, as explained in Chapter 2.

Some will object that using the absolute maximum ratings as an estimate of the damage threshold is much too conservative, since components are known to withstand greater stresses for brief periods of time. However, *it is often possible* to design economical protective circuits that can protect components from stresses that are greater than the absolute maximum ratings. The use of these circuits can provide substantial assurance that equipment will survive exposure to adverse electrical environments.

2. Upset Threshold

Nearly all circuits that protect vulnerable devices from damage allow a small fraction of the incident transient, called the *remnant,* to propagate to the protected devices. In a properly designed protection circuit, the remnant will have insufficient energy, current, or voltage to damage protected devices. However, there is still concern that the remnant could be misinterpreted by the system as valid data. Such misinterpretation may cause upset. The threshold for upset is usually within the normal range of input voltages to the system. Therefore, one cannot discriminate against upset on the basis of voltage levels alone. Some ways of dealing with the problem of upset are discussed in Chapter 21.

Some engineers associate large magnitudes of voltage with damage and large magnitudes of dV/dt with upset. This simple association can be misleading. Overvoltages corrupt data, which can lead to upset. Large magnitudes of dV/dt can propagate through the parasitic capacitance between primary and secondary windings of a transformer and damage components in a power supply. One of the most common causes of upset is an outage on the mains, which has a *zero* value of dV/dt.

2

Threats

There are several different sources of overvoltages that are of interest to electronic circuit designers: lightning, electromagnetic effects of detonation of nuclear weapons, high-power microwave weapons, electrostatic discharge ("static electricity"), and switching reactive loads. Properties of these sources are described in this chapter. This discussion is deliberately concise: interested readers will find a more detailed account in the references cited.

A. LIGHTNING

The physics of the lightning discharge have been reivewed by Uman (1969, 1987), Golde et al. (1977), and Uman and Kreider (1982). There are two common forms of lightning: cloud-to-ground lightning and intracloud lightning.

Cloud-to-ground lightning begins when a highly ionized plasma, called the *stepped leader*, propagates from a thundercloud toward the ground. When the leader is within about 50 m of the ground, another electrical discharge, called a *streamer*, propagates upward from the ground and establishes a highly conducting path between the ground and the stepped leader. At this time an arc current, called the *return stroke*, flows from the ground up the ionized channel into the thundercloud. The return stroke produces the intense luminosity that is seen as lightning. At the end of a return stroke, there is often a *continuing current* on the order of 100 A in the channel. This continuing current can have durations between a few milliseconds and a half-second. After a few tens of milliseconds or more, another leader can travel down the same ionized channel toward the ground

TABLE 2-1. Major Parameters in Cloud-to-Ground Lightning Flash

Parameter	Typical value	Worst-case Value
Peak return stroke current, *I*	20 kA	200 kA
Total charge transfer	20 C	300 C
Rise time of return stroke	0.2 μs (?)	
Maximum *dI/dt* of return stroke	10^{11} A/s	

and produce a second return stroke. This process can repeat itself several times. The entire event is called a *lightning flash*. One flash typically contains between three and five leader-return stroke sequences. Most lightning research has been directed at understanding the return stroke process in a cloud-to-ground lightning flash. Some of the major parameters in cloud-to-ground lightning flashes are summarized in Table 2-1.

Use of the term "worst case" to describe natural phenomena is risky. Since only a trivial fraction of lightning strokes are measured by scientists, it is quite possible that rare events may have more severe properties. Events with large magnitudes of currents or charge transfer often exceed full-scale limits of electronic measuring and recording instruments. Some reports of large magnitudes in the literature were obtained by inference from indirect techniques.

Rise times of the order of 1 μs are commonly reported in the older literature. These values for the rise time are too large, owing to inadequate bandwidth of recording devices (e.g., tape recorders, oscilloscopes) and electronic signal processing circuits. Even when oscilloscopes with adequate bandwidth were used, the sweep rate was usually set to a relatively slow rate in order to capture most of the return stroke waveform. Therefore, data on submicrosecond rise times could not be obtained. Recent measurements with faster electronics, rapid analog-to-digital data conversion, and storage in semiconductor digital memories have revealed rise times on the order of 0.1 μs. These data may still suffer from limited bandwidth.

For practical reasons, direct measurements of lightning currents are usually done with instruments located on tall towers or tall buildings that are struck by lightning many times during each year. However, the presence of the tall object may alter the properties of the lightning. In particular, many of the lightning events are upward-propagating discharges that are not preceded by a stepped leader.

The combination of a 20 kA peak current and a 0.2 μs rise time implies a value of *dI/dt* of 10^{11} A s^{-1}. This large value of *dI/dt* implies that transient protection circuits must use radio frequency design techniques, particularly considerations of parasitic inductance and capacitance of conductors, which are discussed in Chapter 15.

Although peak currents of the order of 10 kA in cloud-to-ground lightning are certainly impressive, one should recognize that the bulk of the charge transferred by lightning flashes occurs during the continuing currents, which are usually between 50 and 500 A, for a duration between about 0.04 and 0.5 seconds. The continuing current is responsible for much of the damage by direct strikes, including arc burns on conductors and forest fires (Fuquay et al., 1972; Brook et al., 1962; Williams and Brook, 1963).

Kreider et al. (1977) have measured the electric field from stepped leaders and report a mean rise time on the order of 0.3 μs and a full width at half maximum of about 0.5 μs. They estimate that the peak leader currents are typically between 2 and 8 kA when the leader is near the ground. This is of particular importance to aircraft and missiles in flight, since these vehicles may be near a leader.

In contrast to cloud-to-ground lightning, rather little is known about the properties of intracloud lightning. Measurements of lightning strikes to instrumented aircraft often show peak currents of a few kiloamperes (Thomas and Pitts, 1983). The presence of the aircraft probably "triggers" many lightning events that would otherwise not occur. Nevertheless, data obtained in this way are certainly relevant to assessments of lightning hazards to aircraft.

B. ELECTROMAGNETIC PULSE (EMP) FROM NUCLEAR WEAPONS

Detonation of nuclear weapons produces an intense electromagnetic pulse (EMP) which is a threat to many electronic systems, both military and civilian. There are three different types of EMP, which depend on the location of the weapon and observer:

1. *h*igh-altitude *e*lectro*m*agnetic *p*ulse (HEMP): the weapon is detonated above the atmosphere, and the observer is on the surface of the earth or in the lower part of the atmosphere within line of sight of the detonation
2. surface burst EMP: the weapon is detonated at ground level, and the observer is on the ground within a few kilometers of ground zero.
3. *s*ystem-*g*enerated *e*lectro*m*agnetic *p*ulse (SGEMP): the weapon is detonated above the atmosphere, and the observer is also above the atmosphere and within line of sight of the detonation

When a nuclear weapon is detonated, a very large flux of photons (gamma rays) is produced. The interaction of these photons produces the various types of EMP.

1. High-Altitude Electromagnetic Pulse

Detonation of nuclear weapons above the atmosphere (altitude greater than 30 km, typically between 100 and 500 km) produces HEMP that illuminates all objects within line of sight of the weapon on the surface of the earth or in the lower atmosphere of the earth. A burst 300 to 500 km above Kansas would illuminate most of the continental United States.

Gamma rays from the weapon interact with air molecules through the Compton effect to produce pairs of electrons and positive ions. The electrons are ejected with large speeds and are turned by the earth's magnetic field to produce HEMP. The volume where the charge is separated and the electromagnetic field is produced is known as the *source region*. For HEMP, the source region is in the upper atmosphere above the observer on the ground or in the lower atmosphere, as shown in Fig. 2-2. The physics of the generation of HEMP has been reviewed in a number of references (Sherman, 1975; Longmire, 1978, 1985a; Glasstone and Dolan, 1977, pp. 514–540; Lee, 1986; Longmire et al., 1987).

The analytical expression for the unclassified HEMP waveform is given by Eq. 1, where E is the electric field in units of volts per meter, t is time in seconds, and H is the magnetic field in amperes per meter (Sherman, 1975; Stansberry, 1977).

$$E(t) = 5.25 \times 10^4 [\exp(-4 \times 10^6\, t) - \exp(-4.76 \times 10^8\, t)] \tag{1}$$

$$H(t) = E(t)/377 \tag{2}$$

This HEMP waveform given in Eq. 1 is plotted on two different time scales in Fig. 2-1. As shown in Fig. 2-1, the HEMP waveform is a pulse with a rise time (10–90% of peak) of about 5 ns, a full width at half-maximum of about $0.2\,\mu s$, and a peak electric field of 50 kV/m.

It is important to recognize that Eq. 1 is a simple model that is representative of some worst-case features of HEMP: it has the minimum rise time and maximum duration that are likely to be observed. It is not an accurate description of the waveform at any particular location. For example, the $0.2\,\mu s$ duration does not occur at the same location where the 5 ns rise time is observed (Sherman, 1975, pp. 13, 22; Longmire et al., 1987). The details of the HEMP waveform vary depending on type of weapon, explosive yield, altitude of burst, and location of the burst and observer. The use of Eq. 1 is justified as a threat waveform for use in designing protection, because the location of the detonation is unknown in advance.

The energy density, w, is given by Eq. 3:

$$w = \int_0^\infty E \times H\, dt \tag{3}$$

where E and H are given by Eqs. 1 and 2. The value of w is $0.9\,J/m^2$. The physical interpretation of w is that a loop antenna with a cross-sectional area of $1\,m^2$ oriented perpendicular to the direction of propagation of the EMP

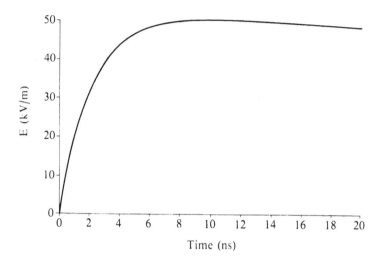

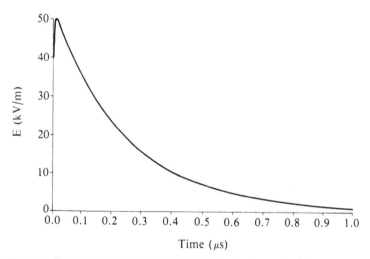

Fig. 2-1. Free space electric field of HEMP waveform as a function of time.

wave would provide a pulse with an energy of 0.9 J. This is sufficient energy to destroy most small-signal high-frequency transistors or integrated circuits. This computation is not representative of the energy delivered by HEMP at any particular location, because a simple threat waveform was used for E and H rather than actual values of fields at a specific location. The value of 0.9 J/m^2 may be considered an approximate upper bound for actual values of w.

Equation 1 describes the electromagnetic field in free space. It does *not* describe the current or voltage waveform at the end of a long cable that is

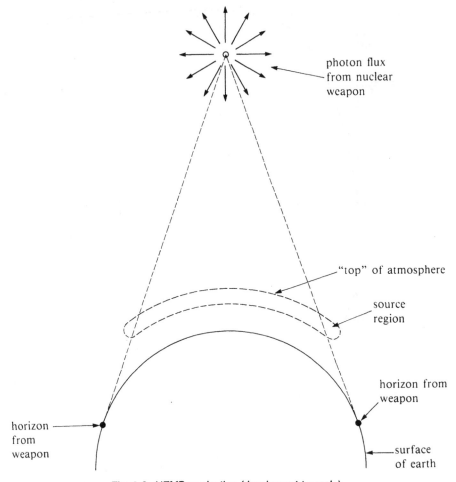

Fig. 2-2. *HEMP production (drawing not to scale).*

illuminated by HEMP. Klein et al. (1985) have calculated the current in an overhead high-voltage transmission line that would be caused by HEMP. They found that the initial peak current occurs at about 0.03 μs, which is much longer than the peak of the free-space HEMP waveform.

2. Effects of HEMP

Little unclassified information is available on the effects of HEMP on modern electronic systems. Weapons tests in the 1950s and 1960s were mostly concerned with damage from blast and thermal radiation from detonations in the lower atmosphere or at the earth's surface. At long ranges, the magnitude of EMP from such tests is small.

Most of the unclassified data on HEMP comes from the STARFISH PRIME test, 8 July 1962 at about 23 h Hawaii local time. A 1.4 megaton device was detonated at an altitude of 400 km above Johnson Island (Glasstone and Dolan, 1977, pp. 45, 523). Some disruption occurred in civilian systems in Honolulu, Hawaii as a result of HEMP from this burst. In particular, "hundreds" of burglar alarms were activated by HEMP, and some fuses were opened in long series strings of street lamps. Longmire (1985b) has calculated the expected magnitude of HEMP at Honolulu. The peak electric field was calculated to be 5.6 kV/m, and the energy density about 0.01 J/m^2. The peak electric field was about 10% of the value given in Eq. 1. The implication is that STARFISH PRIME at Honolulu was far from a worst-case event.

Modern systems may be much more vulnerable to damage by HEMP than circuits in 1962, owing to the increasing use of integrated circuits. One anecdotal report* states that waves from an EMP simulator damaged electronic ignition systems in automobiles, so that the cars that were parked near the simulator would not start after the tests. Modern telephone systems use semiconductor switches instead of electromechanical relays; there is no doubt that the semiconductor switches are more vulnerable to damage.

HEMP may be a particular threat to systems that depend on electrical power or data that are carried on long lines. This includes nearly every civilian system and many military systems. The peak current in overhead lines may be on the order of 10 kA (Vance, 1975; Glasstone and Dolan, 1977, p. 530). Considerable shielding from HEMP can be obtained by burying the cable in the ground. However, depending on soil conductivity and depth of burial, peak currents are predicted to be between a few hundred amperes and a few kiloamperes. Such currents are not negligible.

There are two principal areas of concern about HEMP and the electric power transmission network. The abnormal voltages and currents caused by HEMP may trip circuit breakers on these lines, causing a momentary interruption of power. Some of the circuit breakers will probably automatically reclose, so that the power will be restored. Still this momentary outage could produce malfunction (upset) in computers and electronic systems. One must be concerned not only with abnormal currents and voltages on the transmission lines themselves, but also with the effects of HEMP on electronic control circuits for electric power transmission. It is possible that HEMP could destabilize the electric power grid and cause a nationwide outage. It is difficult to ascertain the effects of HEMP on large systems, such as the electric power grid. Large systems cannot be tested in a laboratory simulator; computer calculations that are unsupported by experimental confirmation are notoriously unreliable.

HEMP may also adversely affect aircraft and missiles in flight. The electromagnetic field changes produce skin currents that excite resonances

* *Science News*, 119:344, 30 May 1981.

in the aircraft. The resonance frequencies are generally between 1 and 20 MHz. The electromagnetic field from skin currents, in turn, induces currents in cables inside the wing and fuselage. In an independent process, transducers and antennae on the exterior of the aircraft are illuminated directly by the electromagnetic field from HEMP. Electronic circuits that are connected to these transducers or antennae may be exposed to relatively large transient currents.

Nuclear warheads in antiballistic missile (ABM) systems pose a special threat. Detonation of the ABM's warhead above the atmosphere will produce HEMP that may adversely affect operation of electronic systems at the launch site on the ground.

3. Magnetohydrodynamic HEMP

Nearly all discussions of HEMP concentrate on overvoltages that have durations of the order of no more than a few microseconds. However, there is a late-time effect, known as *magnetohydrodynamic* (MHD) EMP, that persists for hundreds of seconds after the detonation. The MHD effect is generated by distortion of the earth's geomagnetic field by slowly moving high-conductivity material in two places: (1) the source region in the upper atmosphere and (2) near the point where the weapon was detonated. The magnitude of MHD electric field is on the order of 10 V/km, which is much smaller than the submicrosecond HEMP described above. The component of the MHD electric field that is parallel to the earth's surface causes abnormal currents in long conductors, such as the electric power transmission system (Legro et al., 1986). These currents provide a dc bias in transformers and produce saturation of the magnetic field inside the core of transformers.

Electric fields of similar magnitudes and frequencies are produced by severe solar magnetic activity at higher latitudes, which also produce aurora. There have been reports of upset of power transmission by tripping of circuit breakers and saturation of transformer cores during geomagnetic storms (Albertson et al., 1973; Albertson and Thorson, 1973; Kappenman et al., 1983). Saturation of the magnetic field inside the core of transformers can result in severe harmonic distortion of the power waveform and possible overheating of the transformer. Albertson et al. (1973) report reductions in rms voltage to as low as 45% of normal during geomagnetic storms.

A power system can be protected from some of the effects of geomagnetic currents, for example by connecting a capacitor in series with the neutral of a three-phase system to block dc current. This might be done for power systems at extreme latitudes, where severe geomagnetic storms are more common. It will not be economical to install this kind of protection on power systems at lower latitudes, so these systems will remain susceptible to upset by MHD EMP.

4. Surface Burst EMP

The EMP *from surface bursts,* in contrast to high-altitude bursts described above, is confined to a relatively small region, about 3 to 8 km in radius centered about ground zero (Glasstone and Dolan, 1977, pp. 517–518). This region will also be affected by blast and thermal radiation. However, surface burst EMP is still an important problem for hardened targets (e.g., underground missile silos and command bunkers). Combat troops who are far enough from ground zero to survive the blast may have their electronic equipment destroyed or degraded by EMP. Furthermore, EMP from surface bursts will produce large transient currents on both overhead and buried cables such as power lines and telephone lines. These transient currents could travel far from the region of the burst and cause damage to facilities that would not be affected by the blast or heat from the burst. There is no standard waveform for surface burst EMP, because this phenomenon is strongly dependent on the type of weapon, explosive yield, altitude of burst, and distance between the burst and observer. In general the electric field has a rise time of a few nanoseconds to a positive peak, followed by a negative peak after a few tens of microseconds. Surface burst EMP has relatively more energy at frequencies below 100 kHz than does HEMP.

The physics of the generation of surface burst EMP has been reviewed by Longmire (1978), Glasstone and Dolan (1977, pp. 517–518, 535–536), Gilbert and Longmire (1985), and Lee (1986, pp. 33–40). A schematic drawing of the source region is shown in Fig. 2-3. Because most of the large electromagnetic fields are confined to the source region, EMP from surface bursts is often called *source region EMP* (SREMP).

The electric field at a distance of about 1 km from ground zero of a 10 megaton detonation is sufficient to trigger lightning that travels from the ground upward (Uman et al., 1972). One lightning stroke had a continuing current that provided enough illumination to be recorded on photographic film for 75 ms.

French researchers measured the current in a field of wires placed on the sand beneath a surface burst in the Sahara Desert and connected to earth about 3 km from ground zero (Ferrieu and Rocard, 1961). The current had a peak value of 150 kA, which decayed to zero at 150 μs. The current had a second peak of 56 kA of opposite polarity to the first. No other results were given.

Little unclassified information is available on the effects of surface burst EMP on modern electronic systems. If any electronic systems were exposed to surface burst EMP before the test ban treaty of 1963, those systems contained vacuum tubes and not semiconductor integrated circuits. It is now known that integrated circuits are many orders of magnitude more vulnerable to damage by EMP than vacuum tubes.

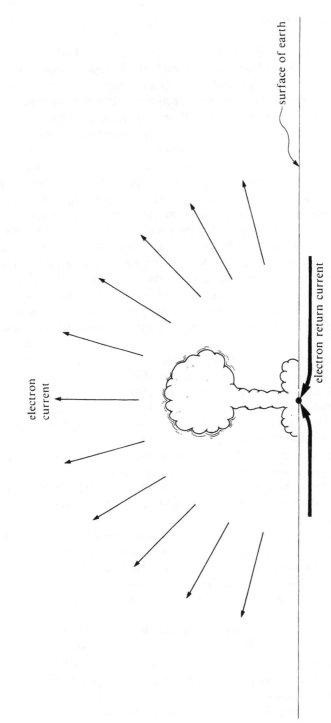

electron
current

electron return current

surface of earth

Fig. 2-3. Surface-burst EMP production mechanism.

5. System-Generated Electromagnetic Pulse

When nuclear weapons are detonated in space, gamma rays and subatomic particles travel away from the explosion. When these particles impinge on the metal container of spacecraft, an intense electromagnetic pulse is produced inside the space vehicle. This pulse is known as *system-generated electromagnetic pulse* (SGEMP), because the pulse is produced by the interaction of the system and the incident particles.

The most important part of SGEMP is produced when the photons (X-rays and gamma rays) from a nuclear detonation impinge on materials in a spacecraft (Higgins et al., 1978; Lee, 1986, p. 42). The interaction between the photons and atoms in the spacecraft occurs through three different processes—the photoelectric effect, the Compton effect, and pair production, all of which produce electrons. There are two effects that occur as electrons are ejected from the conducting walls of the spacecraft: a dipole electric field is created by the ejected electrons and the positive ions that remain in the conducting material; a magnetic field is created by the current in the conducting material as an equilibrium charge distribution is sought. Unclassified estimates are that the magnitude of the electric field are between 100 kV/m and 1 MV/m (Glasstone and Dolan, 1977, p. 522).

The incident particles from the detonation and electrons produced by interactions can directly effect electronic circuits inside the spacecraft. These effects are often known as TREE, an acronym for *t*ransient *r*adiation *e*ffects on *e*lectronics. In particular, each electron can ionize many atoms of material inside the system and produce either damage or upset of the electronic system. The flux of neutrons from a nuclear detonation can also damage microelectronic circuits. Because the current is injected throughout the system, there is no input port for this overstress. Therefore, conventional circuits for protection against overvoltages are of no use in TREE.

One cannot rely on a Faraday Cage to shield the interior of the spacecraft from SGEMP, since time will be required for an equilibrium charge distribution to occur in the conducting shield. Most of the time delay will be caused by the finite speed of light in the conducting material. Furthermore, a perfect Faraday Cage would make a space vehicle useless: one must have penetrations for antennae, optical sensors, and possibly panels of solar cells for electric power.

Mitigation of SGEMP could be done in several different ways. One way is to attenuate the incident photon flux with a thick barrier of copper, lead, or iron. This requires massive shielding that is exorbitantly expensive to place in orbit above the earth. Furthermore, electrons produced in the outer layer of the shield will produce an intense electric field perpendicular to the surface of the shield.

Another way to mitigate SGEMP would be to make the spacecraft and the electronic system essentially transparent to the photons. Also one could try to harden the electronic circuit against damage by ionizing radiation, possibly by using vacuum tubes.

SGEMP is a threat to artificial satellites for communications, navigation, reconnaissance, and Strategic Defense Initiative battle stations. Space mines that produce SGEMP in spacecraft near the detonation will also produce HEMP for systems on the ground.

C. COMPARISON OF LIGHTNING AND HEMP

Several authors (Uman et al., 1982; Vance and Uman, 1988) have compared HEMP and lightning. HEMP has more energy at high frequencies than lightning, whereas lightning has more energy at lower frequencies. This is easy to understand: the HEMP waveform (Eq. 1) has a duration of about $0.2\,\mu s$, whereas return stroke currents have a duration of a few tens of microseconds, followed by continuing currents that persist for tens of milliseconds.

There are major differences between electromagnetic fields from lightning and HEMP. Over a small region on the surface of the earth the fields from HEMP approximate a plane wave with an amplitude that is independent of location. Near the lightning stroke, the fields vary approximately inversely with distance. Far from the lightning stroke, the fields are weak and unlikely to cause damage (although they can be measured).

Consider systems that are spread over a very large area, such as the electric power grid and telephone communications. Most of the effects of lightning are confined to a local region, which includes the object struck by lightning and other objects that may be a few kilometers from the lightning channel. Therefore, upset caused by lightning may be avoided, in part, by redundancy in the system. HEMP illuminates the entire system almost simultaneously, so redundancy may be ineffective to prevent system upset by HEMP.

Some engineers have been concerned with deciding which is worse, lightning or HEMP. This is not a particularly productive exercise, since electronic systems should be designed to withstand *both* threats. Moreover, the details of the system/threat interaction are often sensitive to particular features of a specific system. This tends to preclude general conclusions from being valid for a specific system.

If hardening a system against lightning ensured that the system was also hardened against HEMP, then there would be no need for research and development to provide hardening against HEMP. This hypothesis seems to have been the basis for various attempts to decrease funding for projects related to HEMP. The author believes that measures that protect against lightning *can* also provide some protection against HEMP. However, there are protective devices (e.g., the silicon carbide varistor in series with an air gap) that may be effective against lightning but ineffective against HEMP.

Both lightning and HEMP should be considered when designing protection, and further research is needed on both topics.

D. SIGNIFICANCE OF EMP

Although this is not the place for an examination of the threat of EMP and its effects on military strategy, a few short paragraphs may demonstrate the grave importance of protection against EMP.

Modern defense policy is based on the doctrine of "mutually assured destruction." During the first strike, the aggressor could destroy a fraction of the offensive weapons of his opponent. However, the opponent would have enough weapons surviving the first strike to enable the aggressor to be destroyed in the retaliation. There is no victor in this exchange. By waiting until after many of the aggressor's weapons have detonated before the opponent launches his retaliation, there is then no doubt about the aggressor's intentions. In other words, there is no "accidental" retaliation.

Suppose that HEMP and SGEMP from the aggressor's first strike disable the command, control, and communications (C^3) of the opponent. Then there could be no large-scale retaliation, and the aggressor would be the victor.

To make nuclear war unwinnable, the opponent must then launch his retaliatory strike *before* the first strike weapons have detonated and produced EMP (Broad, 1981; Steinbruner, 1984). This is called *launch on warning,* or *use it or lose it.* Since long-range missiles have a travel time of only a few tens of minutes, the decision to launch on warning will probably be reached with many unanswered questions. This is a dangerous policy, but it appears to be the only one that makes war unwinnable given the uncertain ability of the C^3 system to survive EMP.

In this view, engineers who work to harden electrical and electronic systems to withstand EMP effects are working toward a safer policy, that of abandoning launch on warning. The end result is much more significant than survival of mere electronic circuits.

E. HIGH-POWER MICROWAVE

It is possible to upset, or even damage, electronic circuits by illuminating them with a beam of intense electromagnetic radiation. When this effect is exploited as a weapon, it is known as *high-power microwave* (HPM). One can envision sweeping a battlefield with HPM, disabling the C^3 systems, and interfering with electronic systems in aircraft, including helicopters.

The HPM threat is a burst of microwaves. The microwaves themselves have a frequency between 500 MHz and 100 GHz. The duration of the burst is of the order of a microsecond or less (Florig, 1988). HPM can deliver more energy at high frequencies than lightning or the various EMP phenomena.

F. ELECTROSTATIC DISCHARGE

Electrostatic discharge (ESD) occurs when two conducting, charged objects are brought near each other. The mechanism for the separation of charge is the triboelectric effect, which involves motion of dissimilar materials. Common examples of triboelectric charging occur when people walk across an insulating carpet, or when layers of clothing rub against skin. People are not the only source of ESD. Sheets of insulating materials (e.g., plastic or paper) moving on rollers can become charged by triboelectricity. Aerosol sprays can produce highly charged droplets.

Early measurements of the ESD voltage waveform (Tucker, 1968) showed a rise time of a few nanoseconds and a duration of 0.1 to 0.3 μs. The peak potential difference between a person and earth can be as large as 40 kV but is more commonly 0.5 to 5 kV. A person charged to a potential of 40 kV can produce ESD with a peak current of about 70 A (Tucker, 1968). Recent measurements show that the ESD current waveform is a brief pulse with a duration between 2 and 4 ns and a peak value of about 20 to 30 A (for a 5 kV open-circuit voltage), followed by a slower exponential decay (Tasker, 1985). The rise time of this initial pulse is less than 1 ns. Early measurements missed the initial pulse, probably owing to inadequate bandwidth in the oscilloscope and probe.

From analysis of the discharge waveform, the person can be modeled as a RLC circuit, as shown in Fig. 2-4 (Richman and Tasker, 1985). The values of R_B and C_B for the human body vary from person to person but are generally between 100 Ω–1.5 kΩ and 50 pF–300 pF. One must also model

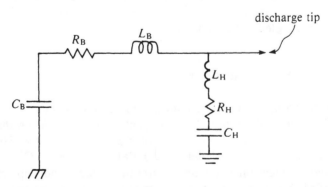

Fig. 2-4. *ESD circuit model for discharge from human hand (Richman and Tasker, 1985). Typical values for C_B, R_B, and L_B are 150 pF, 500 Ω, and 1 μH. Typical values for C_H, R_H, and L_H are 8 pF, 60 Ω, and 0.1 μH.*

the circuit parameters, R_H and C_H, of the human hand in order to duplicate actual ESD events from people touching an object. The circuit elements that simulate the hand account for the initial pulse, while the circuit elements that simulate the body account for the slower, exponentially decaying waveform.

ESD is a widely recognized hazard during shipping and handling of many semiconductor devices. Semiconductors that are especially sensitive include those that contain unprotected metal oxide semiconductor field effect transistors (MOSFETs), semiconductors for use at microwave frequencies, and very high speed logic with switching times on the order of 2 ns or less. In response to the ESD threat, most semiconductors are routinely shipped in containers that are made of conductive material. A less suitable but commonly used method of protection is plastic containers that have been treated to reduce triboelectric charging, the so-called antistatic treatment. In addition to these shipping precautions, assembly line workers are grounded (through a 1 MΩ resistance to prevent accidental electric shock), and they use grounded soldering irons, ionized air blowers, and other techniques to avoid applying large potential differences to semiconductors during assembly. ESD continues to be a problem for many circuits after assembly: these completed circuit boards are shipped in conductive bags to avoid damage.

Most of the anti-ESD technology has been concerned with damage that occurs during shipping and assembly of vulnerable devices. However, ESD is still a hazard to many electronic systems during normal use. Any person who walks across a carpet and touches a keyboard of a computer terminal may damage the electronics inside the keyboard or the interface electronics inside the computer. Conventional techniques to avoid this hazard include (1) shielding vulnerable circuits inside conducting enclosures, and (2) installing conductive carpets or mats near computer terminals or routinely spraying regular carpets with antistatic chemicals.

G. SWITCHING OF REACTIVE LOADS

Switching reactive loads is probably the most common cause of overvoltages on ac supply mains. There are several specific situations that can cause problems.:

1. interruption of current in an inductive load (motor, coil of solenoid or relay, lightly loaded transformer)

2. switching power factor correction capacitors on line and off line

3. "showering arc" produced by interrupting current in an inductive load with a mechanical switch

4. interruption of current by a fuse or circuit breaker may produce an
   overvoltage in the parasitic inductance of the wiring upstream from
   the fuse

Each of these situations is discussed below. Many of these transient
overvoltages are produced on high-voltage transmission lines, but the
overvoltages propagate through step-down transformers and appear on the
low-voltage ac supply mains.

1. Interruption of Current in Inductive Load

Interrupting the current in an inductive load is well known to generate a
high-voltage pulse that can cause dielectric breakdown of insulation. This is
likely to be the most common cause of transient overvoltages in power
circuits, such as the ac supply mains in a building. A simple demonstration
is to unplug a vacuum cleaner in a dark room while the motor is running.
The sudden interruption of current in the motor will create an arc in air at
the socket where the connection is broken. Martzloff and Hahn (1970)
found that a transient overvoltage with a peak value between 1.4 and 2.5 kV
was injected into the 120 V rms mains when the ignition system of an oil
furnace created a spark to ignite the oil.

A special case of a transient produced by interrupting current in an
inductor occurs in a transformer. When the magnetizing current in the
primary coil is interrupted, a transient overvoltage is produced across the
secondary winding (Liao and Lee, 1966; Smith and McCormick, 1982, p.
10). The overvoltage across the secondary winding can destroy rectifier
diodes in a dc power supply.

Connection of large inductive loads to ac supply mains produces a large
phase angle between the voltage and current, which decreases the efficiency
of the electric power distribution system by increasing losses in conductors
and transformers. The inductive loads can be balanced by installation of
capacitor banks to correct the power factor (make the current and voltage in
phase). When the inductive loads are switched on and off line, it is
necessary to also switch the power factor correction capacitors, because a
large capacitive load is just as undesirable as a large inductive load. When
the power factor correction capacitors are switched, substantial problems
may be produced (Cuffman et al., 1976). Oscillations can be produced with
amplitudes of several times the normal mains voltage, frequencies between
about 300 Hz and 25 kHz, and durations of several milliseconds. While the
amplitude of these oscillations is much smaller than that produced by
lightning or by interrupting current in an inductive load, the long duration
of overvoltages from switching power factor correction capacitors can
transfer a large amount of energy to overvoltage protective devices or
vulnerable equipment.

Switching power factor correction capacitors does not always produce an overvoltage. The magnitude of any overvoltage will depend on the amount of inductance and resistance present in the circuit as well as the instantaneous values of current and voltage when the switching occurs.

Johnson (1961) reviewed the literature about overvoltages on power transmission lines (nominal voltage of at least 100 kV rms) and concluded that switching power factor correction capacitors and interrupting magnetizing current in transformers produced overvoltages of up to three times the amplitude of the normal system voltage. Such overvoltages propagate through step-down transformers and appear on the low-voltage supply mains.

2. Propagation of Surges Through Transformers

The details of the propagation of surges through transformers were explained by Palueff and Hagenguth (1932) and Bellaschi (1943). There are four mechanisms for coupling a transient overvoltage through a transformer: parasitic capacitance between the coils, mutual inductance, and oscillations in each of the two coils.

An incident surge with a large value of dV/dt is coupled through the transformer by the parasitic capacitance between the primary and secondary coils (see Chapters 14 and 18). This coupling is independent of the ratio of turns in the two coils.

Each of the two coils has some inductance, series resistance, and shunt capacitance, which forms an *LRC* circuit. This circuit will respond with a damped oscillation when excited with a brief pulse. The oscillation of the current in the primary coil will be coupled to the secondary by the mutual inductance between the coils. The oscillation of the current in the secondary coil produces a voltage directly across the secondary. Additional oscillations can occur owing to reflections from nodes on cables attached to the secondary winding. These reflections can double the expected voltage across the secondary.

Martzloff (1986) reported one case where an overvoltage created by switching power factor correction capacitors traveled a distance of 3 km and passed through two step-down transformers before damaging an overvoltage protection circuit.

While the waveshape of the transient voltage may change as it propagates through a transformer, it is important to realize that a transformer does *not* block transient voltages. Low-frequency oscillations produced by switching power factor correction capacitors will likely propagate through transformers with little change of waveshape, whereas submicrosecond features of other types of transient overvoltages may be altered.

3. Showering Arc

When a mechanical switch is opened in a circuit that has an appreciable inductive load, a phenomenon called a *showering arc* occurs at the switch

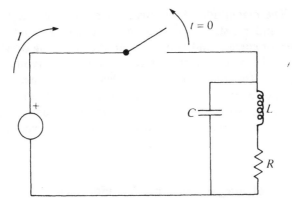

Fig. 2-5. *Generation of transient overvoltage, which is called a showering arc, by interrupting current in an inductor.*

contacts. Showering arcs can also be produced when a circuit is opened owing to switch bounce. The production of showering arcs have been described by many investigators (Mills, 1969; Mellitt, 1974; Howell, 1979; Shi and Showers, 1984). A simple circuit diagram for production of the showering arc is shown in Fig. 2-5. The voltage source at the left edge of Fig. 2-5 may be either ac or dc. The inductive load has a parasitic capacitance, *C,* and a resistance, *R.*

An example of the voltage waveform across the inductor during a showering arc is shown in Fig. 2-6. This example was produced with a

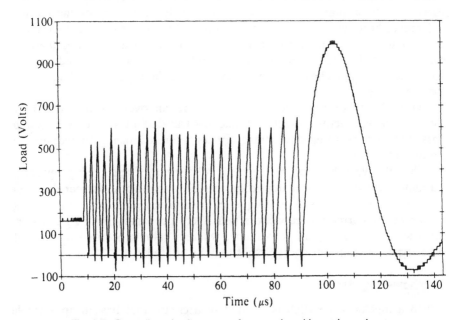

Fig. 2-6. *Recording of voltage waveform produced by a showering arc.*

mechanical switch, an inductance of about 25 mH, and a capacitance of about 2 nF. The switch and load were connected to the ac supply mains that had a nominal voltage of 120 V rms. A digital oscilloscope was used to record the voltage across the load; the small steps in voltage visible prior to 10 μs and after 90 μs in Fig. 2-6 are caused by the resolution of the digital sampling (about 20 V per least significant bit). At the left edge of Fig. 2-6 the mains voltage of about 165 V is shown, the switch opens at 8 μs, and a showering arc ensues. The peak voltage at the end of the showering arc is 1 kV. Even greater peak voltages are possible with different circuits.

A showering arc is produced by the following sequence of events:

1. The switch opens, which causes the magnetic field in the inductor to decay.
2. The inductor supplies a current to the shunt capacitance, C, so that the voltage across the load increases.
3. The voltage across the load increases until a spark occurs in the air between the switch contacts (dielectric breakdown); this is followed by the formation of a highly conducting arc between the switch contacts that restores the current in the inductor.
4. The voltage across the inductor suddenly collapses as the current is restored.
5. Just before the current reaches zero, the arc extinguishes, which causes the magnetic field in the inductor to decay.

The process in steps 2 to 5 repeats many times. Finally the contacts are separated by enough distance so that breakdown of the air gap will not occur. The LRC circuit then begins a smooth, damped oscillation.

If the current in the inductive load when the switch initially opens is I, then the collapse of the magnetic field in the inductor after the switch opens will maintain this same current. Therefore, the initial rate of change of voltage across the inductive load will be given by

$$dV/dt = I/C$$

The time interval between the successive arcs is determined by the values of I and C as well as the breakdown voltage and speed of the switch contacts. The time interval between successive arcs tend to increase with time. For example, the first pair of arcs in Fig. 2-6 are separated by about 2.3 μs, the last pair by 5 μs.

As the switch contacts move farther apart with time, one might expect that a larger voltage would be required to cause dielectric breakdown between the contacts. This would cause the amplitude of the showering arc to increase monotonically with time and the repetition rate to decrease, and sometimes it does. In other cases, such as in Fig. 2-6, the amplitude of the showering arc is almost constant with time. Perhaps the dielectric break-down was initiated by field emission from a small particle on one of the contacts. The showering arc is not an exactly reproducible phenomenon because of the statistical nature of breakdown between the contacts as well as variations in the speed of the contacts (Mills, 1969).

The sudden collapse of voltage caused by each restoration of current creates electromagnetic noise, some of which is conducted in the circuit, and some of which is radiated. The radiated noise can interfere with com-munications, particularly AM radio reception. The large magnitude of voltage across the inductive load can damage electronic devices that are connected to the same bus. The large values of dV/dt at the beginning of each arc can propagate through the parasitic capacitance between primary and secondary windings in transformers and may cause damage or upset of electronic equipment. The showering arc is also responsible for erosion of the metal contacts in the switch, which often limits the lifetime of the switch in an inductive circuit.

There are many ways to prevent a showering arc. A metal oxide varistor can be shunted across the inductive load. If the source is dc, a rectifier diode can be shunted across the inductive load. These techniques are discussed further in Chapters 18 and 19. There are other suppression techniques that tend to be more expensive, such as connecting a large capacitance (of the order of 1 μF) or an avalanche diode in shunt with the inductive load.

A showering arc requires mechanical switch contacts that move apart; it will not occur with semiconductor switches (e.g., SCR, triac, transistor). However, unprotected semiconductor switches can be destroyed by the transient overvoltage caused by interrupting the current in an inductive load.

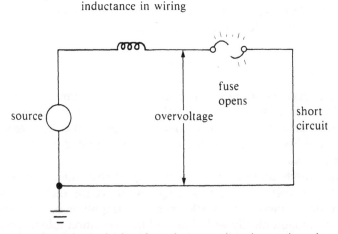

inductance in wiring

Fig. 2-7. *Circuit for production of transient overvoltage by opening a fuse.*

4. Overvoltage Produced by Fuse Opening

Inductance in the conductors in the power transmission and distribution network can also cause problems. Figure 2-7 shows a simple schematic diagram for discussion of this type of overvoltage. When a fault (short circuit) occurs, a large current, I, passes down the line. A fuse or circuit breaker opens to isolate the fault and interrupts the current, which produces a large magnitude of dI/dt. The large value of dI/dt produces a transient overvoltage in the inductance of the wiring upstream from the point where the current was interrupted.

Meissen (1983) showed that the duration of the surge (measured at the time the voltage was half the peak value) was often between about 150 and 1000 μs. This is an unusually long duration compared with many other types of overvoltages. Most of the surges had a peak value that was less than 3 times the amplitude of the normal mains voltage, but a few were as large as 10 times the amplitude of the normal mains voltage.

3

Surveys of Threats in Specific Environments

This chapter contains a brief review of surveys of overvoltages that have been conducted in various environments. The best-characterized environment, although many questions remain unanswered, is the low-voltage ac supply mains. There have been several studies of transients on telephone lines. Characterization of other environments has been largely neglected.

A. OVERVOLTAGES ON TELEPHONE LINES

Many telephone lines are routed on overhead poles and can be exposed to direct lightning strikes. Transient current can also be induced in overhead wires by electromagnetic field changes during nearby lighting strikes or high-altitude EMP.

Another source of transients on telephone lines is induction between other conductors that contain transients and the telephone lines. Telephone lines are customarily installed immediately below overhead power distribution lines. The mains and telephone lines are often routed together inside buildings. Transients that occur on the mains can be coupled by electric or magnetic fields into nearby parallel telephone lines.

Bodle et al. (1976) reviewed the effects of lightning and damaging interference from ac supply mains on telephone systems. Bodle and Gresh (1961) did the pioneering survey of voltage waveforms on telephone lines caused by lightning. In summarizing their data, they proposed a "10/1000 μs" test waveform, which is discussed in Chapter 5. Bennison et al. (1973) monitored overvoltages on telephone lines in Canada and validated the 10/1000 μs waveform.

Carroll and Miller (1980) measured transients on a telephone loop with a length of 16.6 km. The far end of the line was terminated with a 550 Ω resistance between each wire and ground. More than 1230 surges occurred in 116 days; for a few days there were more than 100 surges per day. Current transformers were inserted in the grounded "ring" lead. The largest peak current that was observed was 59 A. Most of the transients had damped oscillatory waveforms.

Carroll (1980, p. 1650) showed an example of transients in a telephone cable that was caused by a faulty switch in power factor correction capacitors on the mains. Although these transient overvoltages in the telephone conductors had a rather small amplitude (the largest peak potential and current were 836 V and 3.1 A), there were 5480 transients from this cause. In fact, 820 of these events were recorded during a single day! Transient overvoltages on the mains are discussed later in this chapter.

Yet another source of electrical overstress is the accidental connection of the mains to a signal line (e.g., telephone, cable television), which is called a *power cross*. Strictly speaking, a power cross is a continuous phenomenon, not a transient. However, the techniques for assuring the survival of electronic circuits after a power cross are similar to techniques that are used for protection against transient overvoltages.

B. OVERVOLTAGES ON COMPUTER DATA LINES

Data transmission between two computers or between one computer and a peripheral (e.g., a terminal, printer, or plotter) is done by sending digital data over either a serial or parallel interface. The serial interface requires fewer conductors (minimum of 3) but can only transfer one binary digit (bit) at a time. The parallel interface typically requires 10 to 20 conductors but can transfer either 8 or 16 bits simultaneously. For short distances, where the increased expense of multiconductor cable is not prohibitive, parallel interfaces are preferred. For longer distances, serial interfaces are common.

Only one study of transients on computer data lines has been published. Tetreault and Martzloff (1985) reported measurements of the voltage on an RS-232 computer data line that was connected between two buildings. The line had a length of 650 m, of which 225 m was "strung overhead on utility poles." They measured the voltage between one data line (pin 2 of the standard 25 pin connector) and building ground with a digital oscilloscope. They simultaneously measured the voltage between the common conductor of the data line (pin 7 of the 25 pin connector) and building ground. During local thunderstorms, they measured transient overvoltages that resembled a damped sinusoidal oscillation. Tetreault and Martzloff (1985) gave the following description of a typical transient waveform: rise time, about 0.6 μs (range <0.1 μs to 5 μs); duration, about 50 μs (range 5 μs to >100 μs); and an oscillation frequency of about 250 kHz. Full scale on their digital oscilloscope was set at ±200 V. The largest transient observed during their measurements had a peak value of about 300 V, which was determined by extrapolation from the data. The computed differential-mode voltage between pins 2 and 7 was no larger than 48 V for one transient with a common-mode voltage of about 200 V. This suggests that transients on RS-232 computer data lines are principally common-mode events.

Computer data are commonly sent in serial format on twisted pair cable according to the RS-232C protocol. This protocol requires that one of the two wires be connected to local ground at each end of the cable. This makes RS-232C communications susceptible to damage from changes in ground potential. Consider the case when an RS-232 cable connects a computer in one building with a terminal in another building, and lightning strikes one of the buildings, as illustrated in Fig. 3-1. The peak current in a typical lightning stroke is of the order of 20 kA. When this large current flows to earth through the building, the potential of earth will rise owing to the finite conductivity of the earth. Some of the lightning current will flow out of the building that was struck on the various electrical cables. The current on these cables will flow to earth in other buildings. In this way, lightning injects a transient common-mode current and voltage into RS-232 computer data cables. Transient overvoltages can get into computer data lines through the same mechanisms discussed above for telephone lines.

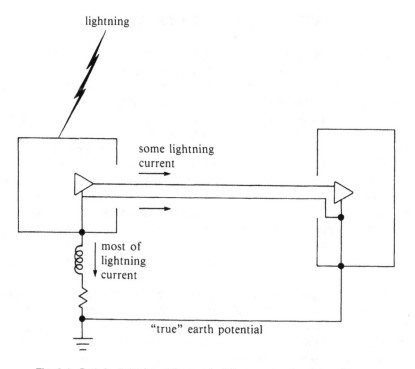

Fig. 3-1. *Path for lightning strike to a building may involve data cables.*

C. DEFINITIONS OF DISTURBANCES ON THE MAINS

More material has been published about disturbances on the ac supply mains than any other environment. Before this literature is reviewed, precise definitions of the disturbances are necessary.

There are many terms in use by engineers for disturbances of the mains waveform. Some of these words are accepted engineering terms. Others, such as "spikes" and "dips," are descriptive terms that have no precise meaning. The word "surge" is used to indicate either a high-voltage transient (e.g., 1 kV peak on the 120 V rms mains) or a momentary increase in rms voltage (e.g., five cycles of 135 V rms on a nominal 120 V rms mains). These are two very different events, and to use the same word for both of them causes confusion (Martzloff and Gruzs, 1987).

The following discussion proposes precise definitions of various disturbances. Some of these definitions are supported by citations to the literature; others are simply the author's personal preference. It is hoped that various standards working groups will incorporate mathematical definitions of disturbances into future editions of appropriate standards.

For the purpose of defining disturbances it is convenient to consider an ideal sinusoidal waveform given by Eq. 1:

$$f(t) = V_p \sin(\omega t + \phi) \tag{1}$$

We will not pretend that the actual mains waveform is this idealized waveform, but the nominal amplitude, V_p, will be used to define magnitudes of events, and the nominal angular frequency, ω, to define durations of events. Making parameters in the definitions relative to nominal parameters of the mains voltage is better than using numerical values, since normalized parameters are independent of the nominal voltage and frequency of the ac supply mains, and there are many different nominal values throughout the world.

1. Disturbance

If the instantaneous mains voltage, $V(t)$, satisfies Eq. 2

$$|V(t) - f(t)| > V_p/4 \tag{2}$$

then $(V(t) - f(t))$ is a *disturbance* (Standler, 1987). This definition is illustrated in Fig. 3-2. The solid line in Fig. 3-2 is the ideal mains waveform given by Eq. 1; the two dashed lines are the boundaries that are 0.75 and 1.25 times the idealized waveform. A disturbance-free mains waveform would remain in the region between the two dashed lines. Any event that causes the mains waveform either to go above the upper dashed line or to go below the lower dashed line is a disturbance. (In Fig. 3-2, it appears that

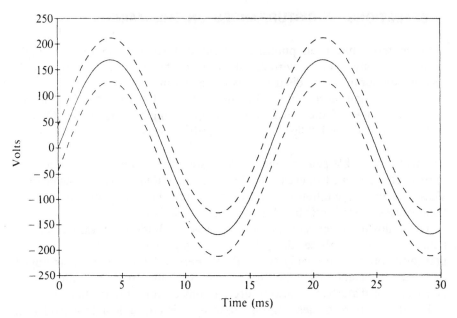

Fig. 3-2. *The solid line is a plot of f(t) versus time for mains with a voltage of 120 V rms and a frequency of 60 Hz. The two dashed lines represent the upper and lower boundaries for a disturbance according to the definition in Eq. 2.*

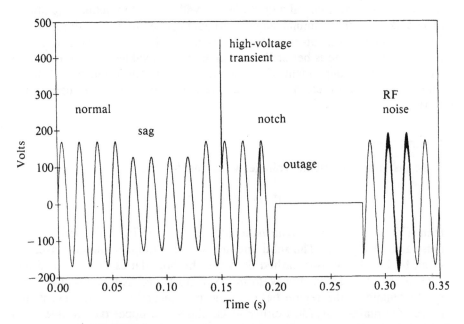

Fig. 3-3. *Artificially generated examples of various disturbances superimposed on the nominal mains voltage, which is a sinusoid with an amplitude of 120 V rms.*

the voltage difference between the two dashed lines is larger at the peaks than at the zero crossing. This is an optical illusion.) This definition of a disturbance is a mathematical statement of the definition proposed by Peter Goodwin in 1985 to define a transient overvoltage or "surge" at a meeting of the Working Group that was revising ANSI C62.41-1980. The author believes that this is not a good way to define an overvoltage but that it is a good definition of a disturbance.

It should be noted that small changes in rms voltage (e.g., 135 V rms or 95 V rms on mains with a nominal voltage of 120 V rms) do not satisfy this definition of disturbance. This is not to say that high or low values of rms voltage do not cause problems. However, it is convenient to distinguish defects in the instantaneous voltage, $V(t)$, from defects in the time-averaged or rms voltage.

A plot of an ideal mains waveform with various types of disturbances superimposed is shown in Fig. 3-3. Each of these types of disturbances is defined in the following paragraphs.

2. Overvoltage

If $|V(t)| > 1.25V_p$, then $V(t)$ is an *overvoltage*. An overvoltage is a condition in which the magnitude of the voltage increases. A *transient* overvoltage has a duration of less than one-half cycle of the normal mains waveform. A *high-voltage transient* is defined as a waveform that satisfies $|V(t)| > 2V_p$ for less than one-half cycle of the normal mains waveform. This definition of "high-voltage transient" was used by Martzloff and Hahn (1970) for "surge" and incorporated into ANSI C62.41-1980.

3. Notch

A *notch* is a condition in which the magnitude of the voltage decreases toward zero for a fraction of a half-cycle of the nominal mains waveform, as shown in Fig. 3-4. A notch does not cross the voltage zero. Notches are

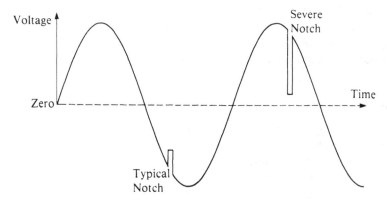

Fig. 3-4. *Artificial generated examples of notches in the mains voltage.*

usually caused by switching of thyristors in motor speed controls, light dimmers, etc. A nonrepetitive notch can be produced by switching on a load that initially has a small impedance, such as an incandescent lamp, uncharged capacitor (e.g., power factor correction, low-pass filter for attenuation of conducted electromagnetic interference).

4. Swell

A *swell* is defined by Martzloff and Gruzs (1987) as an abnormal increase in rms voltage on the mains. Because rms is a time-averaged measure, the minimum duration of a swell is at least a half-cycle of the nominal mains waveform, π/ω. The rms voltage of a swell is more than about 1.08 times the nominal rms voltage. Some engineers and technicians in the United States use "surge" to indicate "swell," a practice that should be avoided (Martzloff and Gruzs, 1987).

5. Sag

A *sag* is defined as an abnormal decrease in rms voltage on the mains. Because rms is a time-averaged measure, the minimum duration of a sag is at least a half-cycle, π/ω. The rms voltage of a sag is less than 0.87 times the nominal rms voltage.

The numerical values for these thresholds of a swell and a sag, 1.08 and 0.87 times the nominal rms voltage, have a simple basis. ANSI Standard C84.1-1982 recommends that "prompt corrective actions" be taken when the rms voltage at the wall outlet of buildings (the "utilization voltage") is below 0.88 or above 1.06 times the nominal system voltage. A margin of 1.5% of the nominal system voltage has been applied to the ANSI limits to obtain the threshold values given above for swell and sag.

6. Brownout

A sag has the connotation of lasting for only a few cycles, certainly not more than a few seconds. A *brownout* is an abnormal decrease in rms voltage on the mains that lasts for many minutes or hours. A brownout often occurs when demand for electric power exceeds the available supply.

7. Outage

The word "outage" does not have a consistent definition. One might define an outage as the absence of useful power. For example, if $|V(t| < V_p/2$ during any continuous interval, $t_1 < t < t_2$, where $(t_2 - t_1) > 3.8/\omega$, then there is an outage at time t. The minimum duration of an outage in this example is 60% of the period of the normal mains waveform. An outage with a lesser duration may not affect the output of a dc power supply, since current flows only during brief pulses twice per cycle.

An alternative way to detect outages might be to search for continuous intervals for which the magnitude of the instantaneous voltage is small. For example, if $|V(t)| < V_p/15$ for any continuous interval, $t_1 < t < t_2$, where $(t_2 - t_1) > 0.4/\omega$, an outage begins. An outage ends when the criterion for the beginning of an outage is not satisfied for a continuous interval of, for example, six cycles. This definition of an outage allows more precise measurement of brief interruptions of power than the definition in the preceding paragraph, and was used by Standler (1987) in a monitoring system.

8. Combinations of Disturbances

Some events are not easy to classify. Goldstein and Speranza (1982) discuss what happens when lightning strikes an overhead power conductor. Obviously, a transient overvoltage occurs. This is often followed by a brief outage or reduction in rms voltage (sag) due to tripping of circuit breakers to extinguish arcs caused by the lightning and follow current. Since the voltage is continuous, it is difficult to establish a precise time when one kind of disturbance ends and another kind of disturbance begins.

D. SURVEYS OF DISTURBANCES ON THE MAINS: REVIEW OF THE LITERATURE

More is known about disturbances on the mains than on any other electrical or electronic environment. Research on these disturbances began as an inquiry into transient overvoltages. Soon it was realized that other kinds of disturbances could also cause upset of electronic systems, so site surveys became broader as other disturbances were also studied.

This book is principally concerned with transient electrical overstresses and how to protect circuits and systems from them. In a larger sense, this book is concerned with how to make electronic circuits and systems reliable in adverse electric environments. Many of the protective measures that one uses against transient overvoltages on the mains also give protection against other disturbances. For these reasons, surveys of other kinds of disturbances on the mains are included in this review. Protective measures against these disturbances are discussed in Chapter 20.

Each of the major studies is summarized in the following paragraphs. The work is organized chronologically.

1. Bull and Nethercot (1964)

Bull and Nethercot (1964) reported results of a survey of 19 sites for several weeks that found 36,000 transient overvoltages. They passed the mains voltage, V_{HN}, through a high-pass filter with a cutoff frequency of 2 kHz to remove the normal mains waveform and its first few harmonics. They then counted excursions of only one polarity and assumed that excursions of the other polarity were equally probable. (This is probably not correct: later investigators found oscillatory overvoltages that contain both polarities.) There were two models of monitors: one was sensitive to pulses as short as 0.1 μs, and the other had limited response to pulses with a duration of less than 0.3 μs. The monitors were designed to sort the peak voltage into four to six different values, which were then recorded on mechanical counters.

They plotted the peak voltage versus the average number of overvoltages per day and found that as the peak voltage increased, the number of events decreased monotonically. They found that most of the transient overvoltages on the low-voltage mains were coming from load switching on the low-voltage side of the distribution transformer and not from the high-voltage distribution system.

2. Hayter (1964)

Hayter (1964) measured transient overvoltages in a laboratory environment. These overvoltages were caused by the interruption of current in various inductive loads. A line impedance stabilization network (LISN) of his own design was connected between the source of the overvoltages and the secondary of the distribution transformer. Details of the oscilloscope and probes were not given. These transient overvoltages enter digital computers through the power supply and can cause malfunction of the computer.

In Hayter's study 37% of the transients has a rise time of less than 2 ns, and 83% had a rise time of less than 10 ns. The peak amplitude was between 100 and 400 V in 65% of the transients; the largest amplitude seen during this study was 1 kV.

The duration of most of the individual transients was between 20 and 200 ns. The individual transients occurred in sub-bursts of 100 to 300 μs duration. Each sub-burst occurred in bursts with a duration of 1 to 3 ms. A description of the generation of showering arcs is given in Chapter 2.

Hayter mentioned that an electric clock produced transients that had an amplitude that was 10 times greater than transients from a 400 horsepower motor. This incidental finding is important, since it clearly shows that common objects can be a potent source of transient overvoltages.

3. Martzloff and Hahn (1970)

Martzloff and Hahn (1970) were the first to perform a large-scale study of transients on mains with a nominal voltage of 120 V rms. They used a specially modified Tektronix 515 oscilloscope with a camera that contained 100 feet of 35 mm film. The film automatically advanced after each sweep or after a 1 hour exposure, whichever came first. The oscilloscope was set to trigger on waveforms of either polarity with a level greater than 300 V.

They found that the electrical ignition of a oil-burning furnace could inject a pulse with an amplitude between 1500 and 2500 V on the 120 V rms mains. These transients occurred at an average rate of 1.4 to 9.6 times per day.

In other cases, transient overvoltages that were observed were due to operation of a fluorescent lamp or an electric motor in a water pump, refrigerator, or food mixer. Again, common objects were recognized as potent sources of transient overvoltages. Transients were also associated with lightning. Many transient overvoltages were of unknown origin.

The oscilloscope in Martzloff and Hahn's study completed its sweep in 100 μs. No analysis of rise time or waveform was presented, except for duration and type of waveform for the most severe and most frequent transient at each of 21 sites. Durations were between 5 and 30 μs. Most of the transients were damped oscillations with frequencies between 0.1 and 2 MHz. This waveshape has been found in subsequent investigations and is incorporated in IEEE Standard 587 (American National Standard C62.41-1980), a Guide to the surge voltage environment. This standard wave shape is discussed in detail in Chapter 5.

Although data from this oscilloscope was quite valuable, the equipment was expensive and could not be used to gather data from a large number of different geographical sites for statistical analysis. Martzloff and Hahn (1970) developed a simple instrument to detect transients that had a peak potential greater than 1200 V. The number of transients that exceeded this threshold was recorded on a mechanical counter. First 249 homes were surveyed for transients for 1 to 2 weeks each. Six homes (2.4%) showed repetitive transients. Then, a total of 91 homes were surveyed for transients during an average of 9.3 weeks each. Most homes had no recorded transients. Four homes had one transient overvoltage; two homes had two transients each. If one assumes that transients occur randomly at a constant rate throughout the year, the average rate of positive transients that exceeds 1200 V is 0.5 per year. The trigger circuit used by Martzloff and Hahn detected only positive overvoltages. Since many overvoltages on the mains have an oscillatory waveform, an overvoltage with an initial negative peak might be detected during the next half-cycle. Martzloff and Hahn's best guess is that the number of transients of both polarities that exceeds 1200 V is 1.6 times the number of positive transients given above.

4. Cannova (1973)

Cannova (1973) used a modified oscilloscope and a camera that contained 100 feet of 35 mm film to record mains transients aboard U.S. Navy ships. There is no mention of the bandwidth of the oscilloscope or its sweep rate. A total of 51 transients, each with a peak of at least 200 V above the amplitude of the normal waveform, were found during the monitoring period. Six of the 11 ships had no transients during the monitoring period.

One ship had 31 transients in a 47 hour period! Data on peak voltage of the transient on both 120 and 450 V rms services at 60 or 400 Hz were combined for analysis. The peak voltage was between 400 and 999 V above the amplitude of the normal power waveform in 78% of the observed transients. The largest transient had an 1800 V peak and was found on a 120 V rms 60 Hz system. Cannova did not present any waveforms. Cannova's only comment about waveshapes was that the "duration of most of the transient voltages recorded was 4 to 6 μs with a few at 19 μs."

5. Allen and Segall (1974)

Allen and Segall (1974) of IBM undertook a comprehensive monitoring of mains with a nominal voltage of either 208 or 240 V rms at computer sites. Like Martzloff and Hahn (1970), they also used two different detector systems in their investigation. One was a Tektronix 549 storage oscilloscope and a 35 mm camera with a 20 exposure roll of film. The oscilloscope was set to have one sweep at a relatively fast rate, followed by a second sweep at a slower rate. The film was automatically advanced after the second sweep. Three different sets of sweep rates were considered (Allen, 1971):

1. First sweep 0.5 μs/cm, second sweep 2 ms/cm. This "did not provide sufficient information about the long-term effects of disturbances."
2. First sweep 250 μs/cm, second sweep 15 ms/cm.
3. First sweep 1 ms/cm, second sweep 50 ms/cm. This "did not provide sufficient information about the recorded voltage spikes."

This oscilloscope system had three problems: the oscilloscope tube had a limited lifetime, the oscilloscope required daily adjustments, and the film was troublesome to develop and analyze. To supplement the oscilloscope system, Allen (1971) also designed and built a digital transient detector. This detector produced a printed record of the following three types of disturbances:

1. Transients with frequency components in the band from 3 to 30 kHz. Allen (1971) says that these transients are "caused by lightning and other noise sources."
2. Transients with frequency components in the band from 0.3 to 3 kHz. Allen (1971) says these transients are often caused by switching of power factor correction capacitors or network switching.
3. Undervoltages and overvoltages that persist for longer than 8 ms. The mains voltage was sent through an 80 Hz low-pass filter to obtain this information. The disturbances were classified in five levels of under-voltage (90%, 80%, 70%, 57%, and 20% of nominal) and one level of overvoltage (more than 110% of nominal). Duration was measured in 8 ms increments up to 1.65 seconds.

Allen and Segall reported their results at the 1974 Winter Meeting of the IEEE Power Engineering Society. Only 3 of 610 transient overvoltages had an amplitude of more than 100% of the steady-state 60 Hz mains waveform. Martzloff (personal communication, 1985) suspects that their analog storage oscilloscope had insufficient writing rate to record transient overvoltages. The 30 kHz limit of their digital transient counter was certainly inadequate to detect all transient overvoltages.

6. Bachman et al. (1981)

Bachman et al. (1981) reported the results of a comprehensive survey of transient overvoltages on the mains aboard U.S. Navy ships. About 80% of their data was taken with a Dranetz model 606-3 monitor, about 15% was taken with a Franklin Electric model 3500 monitor, and about 5% was taken with a Nicolet model 2090-3 digital storage oscilloscope. Data were taken aboard 13 different ships. Transient overvoltages were defined as more than 50 V above the steady-state amplitude on a 120 V rms mains or more than 200 V above the steady-state amplitude on a 460 V rms mains. The Dranetz monitor had a capacitor across its mains terminals so that it suppressed overvoltages that it was supposedly measuring.

A total of 2300 transients were found during 9400 hours of monitoring. As in Cannova's (1973) study, some ships had a much greater incidence of transients than other ships. When operating on shore power, about 1% of the transients on 120 V rms service had a peak of more than 250 V. Transients with large amplitudes were less common while at sea than while operating on shore power.

Data from the digital storage oscilloscope were analyzed to get information about transient waveforms. About 20% of the transients on 120 V rms service had indicated rise times of 0.05 μs or less. The average rise time for transients on 120 V rms mains was 0.25 μs. Bachman et al. (1981) state that it is likely that bandwidth limitations "reduced transient amplitudes." The average duration of transients on 120 V rms mains was about 14 μs. All transients were observed to have a duration of at least 0.3 μs. About 80% of the transient overvoltages had a duration of less than 100 μs. About 20% of the transients had durations between 100 μs and 1 ms.

Nine different types of transient waveforms were identified. About 75% of the waveforms were either a critically damped or overdamped unipolar waveform. Of these, about half had high-frequency noise superimposed. The classical ringwave observed by Martzloff and Hahn (1970) accounted for only 8% of the transients observed by Bachman et al.

Bachman et al. (1981) tried to measure current and voltage simultaneously and then calculate the energy in the transient. No large transient was observed during this part of the experiment. The most severe transient that was observed had an energy content of only 4 μJ. This type of measurement needs to be continued, since some transients are expected to have an energy content of at least 0.1 J. (A surge with a rectangular waveshape with an amplitude of 350 V and 30 A and a duration of 10 μs would contain 0.1 J of energy. This is a relatively mild surge.)

7. Goldstein and Speranza (1982)

Goldstein and Speranza (1982) and Speranza (1982) reported results of monitoring power disturbances at 24 Bell Telephone computer sites by using Dranetz model 606-3 analyzers. Monitoring equipment was operated at each site for at least 2.1 months; the average duration of monitoring at each site was 11.3 months. Their results are briefly summarized below.

1. 56% of the sites experienced at least one instance of the rms voltage (10 second average) outside the ANSI acceptable range from 106 to 127 V rms
2. The temporary disturbances had the following relative distribution:

Sags	87.0%	(<96 V rms, duration > 16 ms)
Transient overvoltages	7.5%	(> 200 V peak)
Outages	4.7%	
Swells	0.8%	(>130 V rms, duration > 16 ms)
Total	100.0%	

3. Half of the sags had durations of less than 0.12 seconds, 90% of the sags had a duration of less than 0.53 seconds
4. Half of the outages had durations of less than 38 seconds, 90% of the outages had durations of less than 4.2 hours
5. The average duration of a swell was 0.10 second

8. Wernström et al. (1984)

Wernström et al. (1984) measured transient overvoltages on a 220/380 V mains in Sweden. They used a Phillips PM3310 digital oscilloscope and a desktop computer to store their data. The oscilloscope was operated at 50 samples/μs for a 5 μs interval. They found rise times as small as 20 ns, which was the limit of their measuring instrumentation. The greatest rise time was 0.6 μs. The transient overvoltages observed in this study had durations (defined as full width at half-maximum) between 0.2 and 4 μs.

9. Odenberg and Braskich (1985)

Odenberg and Braskich (1985) measured the current and voltage during transients. They stored the peak value, rise time, and decay time parameters, as defined in ANSI Standard C62.4-1980. They did not record the waveforms. During a two-year period, they recorded about 2.8×10^5

transients. The average rise time for voltage was reported to be 1 μs, for current 60 μs. The average time to decay to half of the peak value was about 1000 μs for both current and voltage. This duration is remarkably long: the most commonly accepted standard test waveforms for transient overvoltages on the mains decay to half of the peak value in 50 μs for voltage, in 20 μs for current (see Chapter 5).

Odenberg and Braskich also report that the peak of transients on mains with a nominal voltage of 120 V rms had an average value of about 350 V and 40 A. They do not provide a distribution function for these peak values, and the pulse energy content was not estimated.

The results of Odenberg and Braskich have been questioned by Martzloff (1985), because they fitted diverse real-world waveforms to a simplified waveform that is described in an ANSI Standard, rather than collect and save data on the real-world waveforms. Further, about 90% of their surges had a duration between 900 and 1100 μs, which is remarkably consistent for surges from many different sources at 63 locations in 9 different cities (Richman, 1985). Odenberg and Braskich did not disclose the details of their instrumentation in their paper, nor is a plot of the frequency response of their measuring circuit given.

10. Goedbloed (1987)

Goedbloed (1987) reported a summary of almost 28,000 transient overvoltages recorded during 3400 hours of monitoring at 40 sites in Germany. The mains voltage was routed through a high-pass filter to remove the normal mains sinusoidal waveform, through an antialiasing filter, a logarithmic amplifier, and then recorded on two digitizers: one sampled every 0.01 μs for an interval of 20 μs, the other sampled every 1 μs for an interval of 2000 μs. The logarithmic amplifier was used to obtain good resolution of relatively small events *and* have full-scale limits of ±3 kV. The data were then processed in real time by a computer. The processing was limited in order to minimize the dead time. (The *dead time* is the amount of time that the system is busy and unable to capture additional data.) The processed results, but not the raw data, were stored on magnetic tape.

Goedbloed displayed the maximum magnitude of slope, $|dV/dt|$, as a function of the magnitude of the peak voltage. For transients with peak voltages of about 100 V, the median slope was about 2 kV/μs. For peak voltages of 1 kV, the median slope was about 7 kV/μs.

11. Standler (1989)

In 1986 Standler measured V_{HG} and V_{NG} simultaneously at a wall outlet with a digital oscilloscope and transferred the data to a computer for storage on a magnetic disk. Although the particular wall outlet that was monitored did not have overvoltage protection devices, several overvoltage protection devices were located elsewhere in the building.

Bursts of high-frequency noise with a peak-to-peak value between 40 and 100 V occurred at this residential site at an average rate of 0.5 per hour. The noise was mostly a common-mode phenomena (see Chapter 4 for definition of common mode). About half of the events had a maximum value of dV/dt of more than $0.7\,kV/\mu s$, and 10% had slopes greater than $1.3\,kV/\mu s$. The actual slope may have been even greater, since the sampling rate of 10 sample/μs is rather small for such rapid events. Indications were that approximately 60% of the transients had an origin inside the building.

Standler showed voltage waveforms of some naturally occurring transients, as well as events generated by deliberately switching motors in vacuum cleaners and fans. Switching motors were observed to produce magnitudes as large as 350 V and slopes as large as $17\,kV/\mu s$. Switching on a tungsten lamp or connecting a fully discharged shunt capacitor in an EMI low-pass filter to the mains caused a decrease in magnitude of V_{HG} with a value of $|dV_{HG}/dt|$ of more than $6\,kV/\mu s$. It is clear that routine operation of common loads can generate large values of dV/dt.

These measurements are being continued with a system that has a metal oxide varistor connected between the hot and ground conductors and a second varistor between the neutral and ground conductor (Standler, 1987). The voltage across each varistor and the current in each varistor are sampled 25 times per microsecond for 41 μs. The raw data are transferred to a computer for storage on a magnetic disk from each event that satisfies one of the following trigger criteria: (1) $|V_{HG}| > 210$ volts; (2) the portion of V_{HN} above 50 kHz exceeds 60 volts. In addition, the voltage across the varistors is measured once every 40 μs, and a record of 0.16 seconds is stored when there is either a disturbance or an outage (as defined earlier in this chapter). The raw data are stored for processing at a later time. When data have been collected for 2 years, a statistical summary will be published.

The varistor serves to compress the voltage data so that a measured voltage beyond the full-scale limits of the oscilloscope is highly improbable. Furthermore, there is no longer any doubt that transient overvoltages exist. The issue now is to characterize surge currents and energy deposited in representative protective devices, such as a varistor.

E. CONCLUSIONS FROM REVIEW OF THE LITERATURE

1. Transient Overvoltages

Buildings that are served by an overhead ac power distribution line have a major transient protection problem. The overhead line can be struck directly by lightning. The overhead line will also act as a very long antenna for electromagnetic field changes due to lightning and EMP from nuclear weapons. These transients can be attenuated (but not eliminated) by burying the conductors in moist soil. The most common method of getting transient overvoltages on the mains is by switching reactive loads.

If lightning should strike the overhead power line outside a building, not all of the lightning current will travel into the building (Martzloff, 1980). Part of the lightning current will travel down the conductors away from the building, part will flow to ground at the nearest distribution transformer, and part will travel down the pole that got struck. In addition, some of the lightning current may be shunted to ground by arcs between the overhead wires and adjacent trees. Despite this current division, a lightning strike on an overhead power line can be a disaster for electrical equipment inside nearby buildings.

In homes without transient protection devices on the mains, one may hear an occasional click or pop from the stereo loudspeaker when the stereo amplifier is on. This transient usually originates when an inductive load on the mains (e.g., motor in a refrigerator or an air conditioner) is switched off. While such a transient is a minor annoyance, it does illustrate two important points: (1) dissimilar systems can be coupled, and (2) transients are ubiquitous. The same transient that causes a click in a loudspeaker could alter the contents of a computer's memory and cause the computer to "crash."

Most of the studies cited above have concentrated on finding the probability that a transient overvoltage will occur. Although there is some disagreement among the results of these studies, all of them found disturbances on the mains. The average rate at which transient overvoltages occur varies greatly from one location to another. Some sites have several high-voltages transients per day; other sites have no such transients during several months of continuous monitoring. Furthermore, at each location the average rate is a function of time, as equipment that is connected to the mains is changed. These studies were important and useful, and it is now clear than there is not a typical environment with a well-defined average rate.

2. Questions to Be Answered by Future Surveys

Despite the large number of surveys of disturbances on the mains, many questions remain unanswered. Future surveys will need to answer these questions. Development of digital oscilloscopes, storage of digital waveforms on magnetic disks, and numerical analysis of waveforms on a digital computer will make it much easier to do comprehensive studies in the future.

More needs to be known about the waveshape of transient overvoltages. Martzloff and Hahn (1970) found that most transients have a damped oscillatory waveform, but only a few transients found by Bachman et al. (1981) had this waveshape.

Although Bachman et al. (1981) attempted to compute the energy in a transient overvoltage, they did not observe any large transients during that portion of their experiment. There have been no other published reports of energy deposited in loads by transient overvoltages. Because it is now well known that transient overvoltages exist and that protection is needed, measurements are needed of the energy deposited in protective devices. Standler (1987) has described an experiment, which is in progress at present, to do this.

3. Distribution of Peak Voltages

Martzloff and Hahn (1970) reported that failures of electric clock motors dropped to 1% of the previous value when the manufacturer increased the insulation level from 2 to 6 kV. This implies that there are a considerable number of transients on 120 V rms mains with peak voltages that are greater than 2 kV but less than 6 kV. In the absence of surge protective devices, the maximum transient voltage inside a building on 120 V rms branches is limited to between 6 and 10 kV by insulation breakdown in the wiring and devices (e.g., outlets, circuit breaker boxes, etc.).

Bull and Nethercot (1964) and, later, Martzloff have plotted on a log-log scale the cumulative number of transient overvoltages per year, $N(V)$, that have a maximum magnitude of voltage that is greater than V. The latest version of this plot is shown in Fig. 3-5, which is taken from a paper by Martzloff and Gruzs (1987). Several important conclusions can be drawn from the summary of data shown in Fig. 3-5. Transient overvoltages with large peak voltages are less common than overvoltages with small peak voltages, regardless of the location of the sites being monitored. Furthermore, the slope, $d \log(N)/d \log(V)$, is approximately constant for data from all of these different sites. Therefore, N and V are related by Eq. 3:

$$N = kV^a \tag{3}$$

where $-5 < a < -3$. The value of k, which is necessary to predict the rate of occurrence of transient overvoltages, varies markedly from one site to another. Goedbloed (1987) found that the value of the parameter a was approximately -3.

One might expect the amplitude of transient overvoltages on the mains to be proportional to the nominal rms mains service voltage. However, review of Fig. 3-5 shows no such relation. If there is a relationship, it is masked by the large statistical variation in amplitude of overvoltages for different sites with the same nominal rms mains voltage.

Although it is conventional to plot the magnitude of the peak voltage in a statistical distribution of overvoltages, one should ask whether peak voltage is the appropriate measure of the severity of a surge. This is a complicated question, and there is no simple answer, except perhaps "probably not." When the failure mechanism is insulation breakdown, the peak voltage is only one relevant parameter; the rise time is also important. When the failure mechanism is thermal damage, the *energy* delivered to a particular

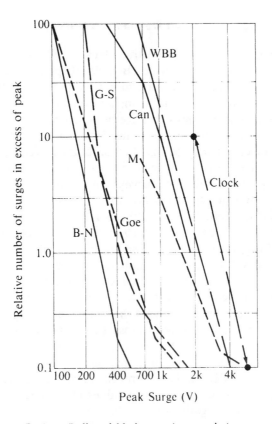

B-N   = Bull and Nethercot (composite)
M   = Martzloff and Hahn
Can   = Cannova
G-S   = Goldstein and Speranza
WBB   = Wernstrom *et al.* (upper limit)
Goe   = Goedbloed
Clock = 2 kV and 6 V points only

Fig. 3-5. *Cumulative distribution of number of surges with peak values in excess of the abscissa. This plot is reproduced from Martzloff and Gruzs (1987), with permission of the authors.*

component is certainly the relevant parameter. When upset is considered, the magnitude of the slope, $|dV/dt|$, may also be an important parameter. One reason that peak voltage is widely accepted as the only measure of the severity of a surge is that it is easy to determine, particularly from photographs of oscilloscope displays. Very little information is available for computation of other, more appropriate measures of severity of surges. There is probably not a single measure of the severity of a surge that is relevant to all types of vulnerable devices.

4. Effects of Surge Protective Devices on Surveys of Overvoltages

There are about 10^9 surge protective devices (SPDs) installed on the low-voltage supply mains worldwide (B.I. Wolff, personal communication, 1988). These SPDs, which are described in Chapters 8 and 19, will limit the magnitude of overvoltages on the mains and provide substantial protection to vulnerable equipment against damage by transient overvoltages. Because protective devices are often included inside a chassis by manufacturers of electronic equipment or may be installed on branch circuits, many environments will now have limited peak overvoltages. This poses a difficult problem for interpreting the results of surveys of transient overvoltages. The worst-case environment, which has no SPDs, has large overvoltages that are limited by insulation breakdown in the wiring and connectors. However, many environments may have much smaller overvoltages because of clamping by SPDs (Martzloff, 1985; Martzloff and Gruzs, 1987). This makes measurements of voltage alone of limited value, because the value of the voltage depends not only on the properties of the threat but also (1) the relative location of the source of the surge, the SPD, and the monitor, and (2) the V-I characteristics of the SPD that limits the voltage.

In sites with SPDs connected across the nodes where the voltage is monitored, one would expect the relation between N and V to agree with Fig. 3-5 for small values of V, for which the current in the SPD is small. However, when the SPD conducts appreciable current, there will be a knee in the log-log plot that corresponds to values of parameter a in Eq. 3 that will probably be more negative than -10.

5. Other Disturbances

Computers are particularly vulnerable to malfunctions caused by reductions in mains voltage. Key (1979) found that "every recorded severe sag was found to disturb the computers." Speranza (1982, p. 150) also found that sags caused computers to malfunction immediately. He found that when the rms voltage dropped below 96 V for as little as 16 ms, computers failed. Kania et al., (1980, p. 42) state that large mainframe computers will shut down during a sag to 80% of nominal mains voltage that persists for 33 ms or more. Whereas undervoltages are not as dramatic as transient overvoltages, they are probably more common disturbances at most sites for computers and other critical electronic systems.

Brownouts are particularly common near 6 PM on weekdays in the summer or winter: people come home from work and turn on the air conditioner or electric furnace as well as the oven to cook dinner. Brownouts are also common on unusually cold days, when people use supplemental electric heaters (even if their principal heating fuel is natural gas, oil, or coal).

F. SUGGESTIONS FOR MONITORING DISTURBANCES

1. What to Measure

Analog oscilloscopes are too inconvenient and too expensive to gather large amounts of overvoltage waveform data. Digital transient counters, although inexpensive and convenient, do not provide a picture of the waveform. These limitations can be avoided by using a high-speed digital oscilloscope to collect voltage and current data as a function of time during transient overvoltages. Several manufacturers of mains monitoring equipment have recently incorporated analog-to-digital converters in their instruments.

There are two classes of disturbances that should be recorded in different ways. Transient overvoltages often have a brief duration and sub-microsecond features; these should be recorded with a high-speed digitizer. Sags, flickers, and outages are disturbances of rms voltage and should be recorded by a slow-speed digitizer.

Hayter (1964) is the only investigator to publish descriptions of transient overvoltages with rise times of a few nanoseconds. Other investigators have used instruments with more limited bandwidths and have missed this feature. As discussed above, many sources of transients produce waveforms that have durations of tens of microseconds. It is difficult to record waveforms that have durations of at least tens of microseconds and have resolution adequate to discern features on a nanosecond time scale. However, it is unnecessary to have resolution of nanoseconds for all studies of transient overvoltages on mains. Such short-time phenomena correspond to very high frequencies. Since attenuation in mains wiring increases markedly as frequency increases above about 1 MHz, the rise time of such fast transients will be degraded by propagation, as explained in Chapter 4. Thus, nanosecond rise times will occur only when the measurement point is within about 10 m of the source. The mains conductors themselves will act like a low-pass filter and remove phenomena with a short rise time. In addition, passive low-pass filters are commonly installed in electronic equipment to meet government regulations for electromagnetic interference from such equipment. These filters may greatly attenuate submicrosecond features in transient overvoltages.

Although nanosecond resolution is not always necessary, most previous studies have been handicapped by inadequate bandwidth and time resolution. A time resolution between 10 and 40 ns is desirable. The duration of the high-speed record should be at least 40 μs, preferably at least 100 μs.

The slow-speed digitizer should record at least 25 samples per period of the nominal mains waveform. The duration of the slow-speed record should be at least 10 periods of the nominal mains waveform. Faster sampling rates and longer durations are desirable, subject to the amount of available memory and processing time.

Monitoring equipment should contain overvoltage protective devices to protect the instruments from damage and upset by the phenomena being monitored. However, this same protection can suppress transient overvoltages that are being monitored. When the results of the monitoring show no large overvoltages, it is natural to declare that overvoltages are not a problem at the site that was monitored. However, overvoltages may again be a problem when the monitor is disconnected, since the monitor was protecting the site from overvoltages. There are two solutions to the dilemma of how to protect the instruments without removing the phenomena being studied: (1) determine the *V-I* characteristics of the protective devices and infer the current and energy in them from measurements of voltage across them, and (2) use a ferroresonant line conditioner, which is described in Chapter 20, to isolate the instruments from the mains.

2. Duration of Monitoring at One Site

It is very difficult to predict how long monitoring should be done at a site. To characterize a site properly, measurements would need to be made continuously for at least 12 months to observe seasonal fluctuations. For example, oil-burning furnaces do not routinely ignite during the summer. However, few clients with possible mains disturbances would want to wait 12 months to take corrective action. It is clear that long-term monitoring is a research project and not a routine engineering evaluation.

If observation of all seasonal loads is excluded, a site might be characterized in a shorter interval—for example, 4 to 6 weeks. However, rare events will not be properly characterized. Consider, for example, a fictional site where exactly one outage occurs per year. If one outage is observed during a 4 week monitoring interval, one would conclude (erroneously) that about 13 outages would be expected each year. If no outage is observed during the 6 week interval, not much can be concluded except that the frequency of outages is small. One might use statistical analysis to assign an upper bound for the frequency of outages, but this analysis will need to *assume* that a particular probability distribution is applicable to outages at the site. This assumption may be difficult to defend.

There are two hazards of brief monitoring durations: (1) the frequency of rare events will be overestimated, or (2) no rare events will be observed. If there is an unusually large rate of disturbances at a site, then monitoring for a few days might help identify the problem. If a quick survey is desired, deliberate switching of loads in the building (one at a time) while someone watches the monitoring instruments is advisable. Offending equipment can be quickly identified in that way.

4

Propagation of Overvoltages

A. INTRODUCTION

The propagation of transient overvoltages on cables is discussed in this chapter. In a sense, there is nothing special about the propagation of transient overvoltages—they propagate in the same way as other electromagnetic waves. Therefore this chapter is a review of material from circuit theory and electromagnetic field theory, with emphasis on topics of particular importance to the propagation of transient overvoltages.

Many electrical and electronic systems have two or more nongrounded conductors in a cable (e.g., telephone cable, many computer data cables, some electric power cables, etc.). It is often convenient to analyze the propagation of signals on two conductors in terms of differential-mode and common-mode components. The reasons for these two modes and their usefulness in discussing propagation of transient overvoltages is discussed first in this chapter.

The discussion of propagation is continued by reviewing the theory of electrical transmission lines. Ideal transmission lines are reviewed, and then complications arising from inclusion of series resistance in the line are discussed.

In a practical application, the transient overvoltage that reaches the device under consideration usually has very different properties from the overvoltage at the source. This difference between overvoltages at the source and point of interest is due to various effects encountered during propagation along cable, which are discussed in detail.

The chapter ends with a discussion of the situation when the duration of the transient overvoltage is much longer than the time required for round-trip propagation of a wave on the longest transmission line in the system.

B. COMMON AND DIFFERENTIAL MODES

1. Voltages

The terms *differential mode* and *common mode* are often used to distinguish two kinds of signals that can propagate on a two-wire transmission line. These concepts are quite useful when the transmission line is balanced—for example, for telephone lines and some analog data lines in instrumentation. The voltages between each conductor (A, B) and ground (G) are denoted V_{AG} and V_{BG}, as shown in Fig. 4-1. The differential mode voltage, V_D, and the common-mode voltage, V_C, are given by Eq. 1.

$$V_C = (V_{AG} + V_{BG})/2 \qquad (1a)$$

$$V_D = V_{AG} - V_{BG} \qquad (1b)$$

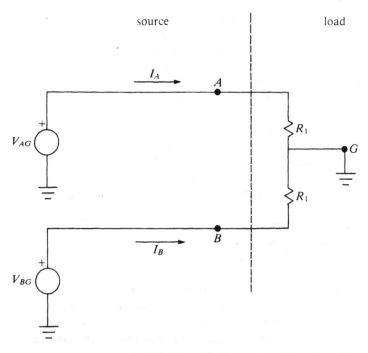

Fig. 4-1. *General schematic of two voltage sources, each referenced to earth ground. Reference will be made to the potential difference between the two points A and B and ground in succeeding figures.*

If these relations for V_{AG} and V_{BG} are solved as a function of V_D and V_C, Eq. 2 are obtained.

$$V_{AG} = V_C + (V_D/2) \tag{2a}$$

$$V_{BG} = V_C - (V_D/2) \tag{2b}$$

The equivalent representation shown in Fig. 4-2 is the schematic that describes Eq. 2. The branch currents and node voltages are identical in Figs. 4-1 and 4-2.

The common-mode voltage, V_C, is always common to *both* the A and B conductors, whereas the differential-mode voltage has different polarity on the A and B conductors.

There are a number of synonyms for differential and common modes:

Area of Usage	Differential	Common
Instrumentation	Differential Normal Transverse	Common
IEC	Symmetrical	Asymmetrical
Telephony	Metallic	Longitudinal

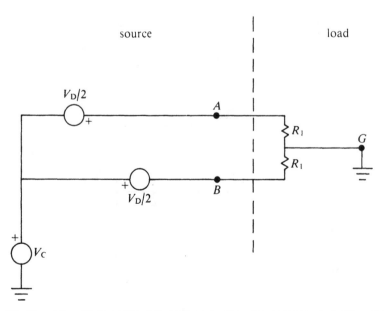

Fig. 4-2. *The same circuit as Fig. 4-1, redrawn to show the common- and differential-mode voltage sources.*

There is another use of the term "common mode" that indicates a voltage between either the *A* or *B* conductor and ground but not necessarily between *both* the *A* and *B* conductors and ground. This usage is not helpful in understanding balanced communication lines but is listed in IEEE Standard 100-1984 as an alternative definition of "common-mode interference."

The IEC term for a voltage between either the *A* or *B* conductor and ground, but *not* both *A* and *B* with respect to ground, is *unsymmetrical* or *nonsymmetrical*. This term is particularly useful in overstress testing in a laboratory where a surge generator is connected between one conductor and ground.

2. Current and Power

The concepts of differential and common mode can also be applied to current. The differential-mode current, I_D, and the common-mode current, I_C, are defined in terms of the currents in the *A* and *B* conductors, I_A and I_B, by Eq. 3. The direction of currents I_A and I_B is shown in Fig. 4-1.

$$I_C = I_A + I_B \tag{3a}$$

$$I_D = (I_A - I_B)/2 \tag{3b}$$

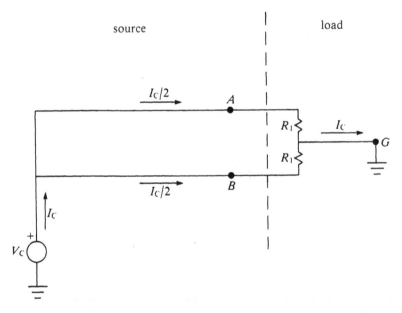

Fig. 4-3. *The same circuit as Figs. 4-1 and 4-2, showing the common-mode current.*

These relations may also be solved for I_A and I_B in terms of I_D and I_C, as Eq. 4.

$$I_A = I_D + (I_C/2) \tag{4a}$$

$$I_B = -I_D + (I_C/2) \tag{4b}$$

Notice that the common-mode current flows in the same direction in *both* the A and B conductors, as shown by Fig. 4-3. The return path for the common-mode current is the grounding conductor. The differential-mode current flows in opposite directions in the A and B conductors, as shown by Fig. 4-4.

The sum of the power, P, that is dissipated in the pair of load resistors R_1 is given by either Eq. 5a or 5b.

$$P = (I_A V_{AG}) + (I_B V_{BG}) \tag{5a}$$

$$P = (I_D V_D) + (I_C V_C) \tag{5b}$$

Either Eq. 5a or 5b can be used; the choice depends only on which is more convenient.

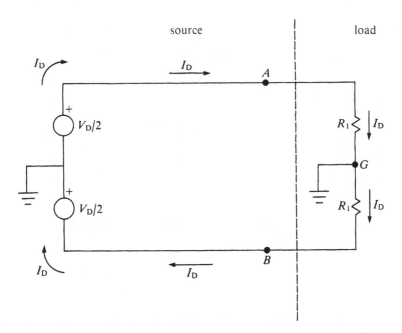

Fig. 4-4. *The same circuit as Figs. 4-1 and 4-2, showing the differential-mode current.*

3. Balance

The *balance* of a pair of conductors is a useful concept when discussing differential and common modes. Figure 4-5 illustrates the Thévenin equivalent circuit of a purely differential mode voltage source and a load. If the two load resistors, R_{L1} and R_{L2}, have the same value, then the load is perfectly balanced. By definition, the two halves of the output resistance of a pure differential-mode voltage source are identical. When a balanced line is cut, as along the dashed line in Fig. 4-5, the same impedance is observed between each of the two wires and ground in each half-cable:

$$R_{L1} = R_{L2} \quad \text{and} \quad R_D/2 = R_D/2$$

If the four resistances are equal,

$$R_{L1} = R_{L2} \quad \text{and} \quad R_D = R_{L1} + R_{L2}$$

then the line is said to be both *balanced* and *matched*. Matching the source and load resistances is required to obtain the maximum power transfer from the source to the load.

Imperfect balance of the load will produce a differential-mode voltage at the load from a common-mode source voltage. In many situations, V_D

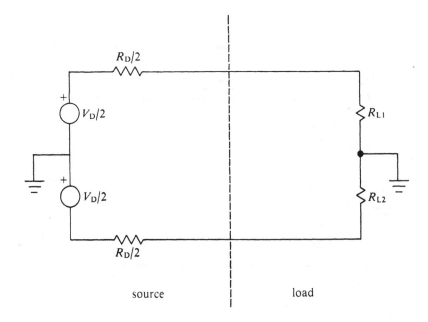

Fig. 4-5. *The concept of balanced impedances in the source and load in a two-wire transmission line.*

represents a desirable signal, and V_C is an "unwanted" voltage or noise that should be rejected. In these situations it is important to use a balanced load to avoid mixing unwanted common-mode voltages with desirable signals. It is common to use a differential amplifier, which has both balanced input resistances and a large value of common-mode rejection ratio (CMRR), as a load to process V_D while rejecting V_C.

Note that a transmission line does *not* need to be balanced to calculate V_C and V_D.

4. Examples of Generation of Common-Mode Voltages

Consider an example of how a common-mode overvoltage is generated on a cable. Figure 4-6 is a schematic drawing of two buildings, one of which contains a voltage source that is connected via a cable to a receiver in the other building. During normal conditions the voltage V_{BG}, at the receiver in Fig. 4-6, is nearly zero; ideally it would be exactly zero. The voltage V_{AG} is usually a few volts larger or smaller than V_{BG}. One building is then struck

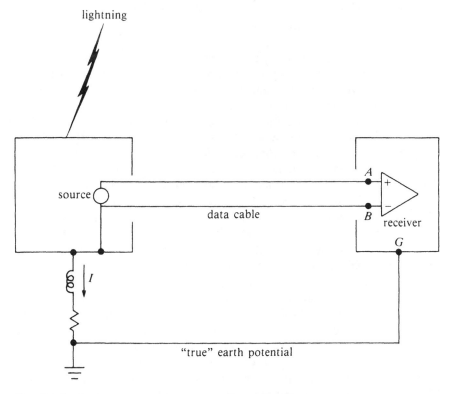

Fig. 4-6. *Production of common-mode overvoltage by lightning strike to one building. A common-mode overvoltage will appear on the data cable in both buildings.*

by cloud-to-ground lightning. The lightning current travels to earth through the inductance and resistance of the building and the soil under and surrounding the building. The voltage dropped across these impedances causes the value of ground potential to differ between the two buildings. The value of V_{BG} may be as large as a few kilovolts if there are no overvoltage protection devices at the input of the amplifier. The value of V_{AG} will be sum of V_{BG} and the signal voltage, V_{AB}, which is essentially the same as the value of V_{BG} during the transient overvoltage. The important thing to understand in this example is that the transient appears as a common-mode voltage, and the desired signal is the differential-mode voltage, V_{AB}. The same situation occurs if the building that contains the receiver is struck by lightning, rather than the building that contains the source.

In Fig. 4-6 the signal comes from a single-ended source that has one terminal connected to local earth ground. This could be an example of computer data transmission on a RS-423 interface or any of a number of other common situations. The desired signal in Fig. 4-6 is actually a mixture of common and differential modes. However, this does not affect the operation of the system, because the output voltage of the receiver is essentially a function only of the differential-mode input voltage, V_{AB}, and it normally ignores the common-mode voltage at its input terminals. It is critical to say "normally ignores," because if the potential difference between one or more input terminal and ground is sufficiently large (e.g., during a transient overvoltage), we expect the differential receiver to be upset or damaged. The example in Fig. 4-6 also demonstrates a second point: a differential receiver can be used with a single-ended source.

Figure 4-7 shows a situation where noise is coupled into a cable and the noise appears as a common-mode voltage. If there is a magnetic field with an appreciable time rate of change, dB/dt, a voltage will be developed in the two loops in Fig. 4-7 indicated by letters $ABCD$ and $EFCD$. The value of the voltage will be proportional to the area of the loop and the value of dB/dt. If the distance AE is small compared with distance AD (and if BF is small compared with BC), then the areas of the two loops are nearly equal. In this situation, the voltage induced between BC and between FC will be nearly equal. In other words, the voltage induced by dB/dt in these loop areas is essentially a pure common-mode voltage.

The concepts of differential and common modes are definitely useful for understanding propagation of voltages and currents on balanced lines. Bodle and Gresh (1961) found that only 10% of transients with peak voltages exceeding 60 V had a differential-mode component greater than 10 V peak to peak. In other words, most transients were essentially common-mode phenomena. However, one might expect such a result, since Bodle and Gresh studied transients on a telephone line, which is balanced. This finding has been misinterpreted to imply that *all* transients are common-mode phenomena and that protection from differential-mode overvoltage is unnecessary.

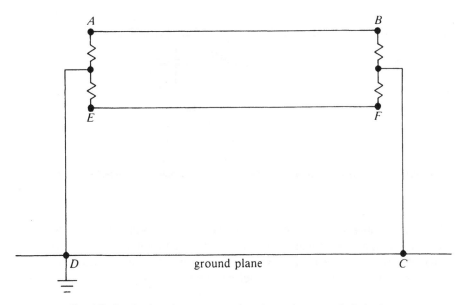

Fig. 4-7. Production of common-mode voltages by magnetic induction.

The concepts of differential and common modes *may* be helpful in understanding the propagation of overvoltages on the single-phase mains. However, the situation is somewhat confused in typical U.S. installations, because the neutral conductor is connected to earth ground at the service panel, which destroys the balance of the line.

C. TRANSMISSION LINE THEORY

The theory for propagation of voltages and currents on transmission lines is well known (Johnson, 1950; Skilling, 1951). We begin by discussing an ideal transmission line that has the following properties:

1. perfect insulators between conductors, no shunt conductance
2. perfect conductors, no series resistance
3. uniform dimensions along its length
4. no ferromagnetic materials (e.g., iron)

After the ideal transmission line is discussed, we will consider the additional effects caused by the nonideal properties.

1. Ideal Lumped Line

A simple way to analyze a transmission line is to consider it as a cascade of lumped elements, as shown in Fig. 4-8. The parameters L and C represent

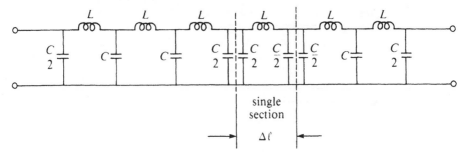

Fig. 4-8. *Representation of transmission line as a cascade of lumped elements.*

the series inductance and shunt capacitance of the pair of conductors that carries the signal. A single section of the lumped element model represents a section of length $\Delta\ell$ in the continuous transmission line. If the continuous line has a total length, L, the model will need to have N sections, where

$$N = L/\Delta\ell$$

It is convenient to let the value of $\Delta\ell$ be a unit length: either 1 m, 1 cm, or 1 ft. This makes the values of L and C in the model have the same numerical value as the inductance per unit length and the capacitance per unit length. The lumped model becomes a better approximation to a continuous line when $\Delta\ell$ is smaller and N is larger.

Propagation of a voltage along the line is analyzed by considering a pulse with a duration that is much longer than the time required to traverse one section of the line, Δt. Then a Taylor's expansion of the voltage waveform as a function of time is made with terms of the order $(\Delta t)^4$ and higher neglected. This restricts the analysis to smooth waveforms with negligible changes during the time $(\Delta t)^4$. It can be shown that the voltage propagates along the line as a wave without distortion so that the output voltage at time t, $V_{out}(t)$, is the input voltage at some earlier time:

$$V_{out}(t) = V_{in}(t - [L/v])$$

where L is the total length of the line and v is the speed of propagation of the wave. The speed is given by

$$v = \frac{\Delta\ell}{\Delta t} = \frac{1}{\sqrt{LC}}$$

When the incident waveform reaches the end of the transmission line, in general, part or all of the incident waveform is reflected. It can be shown that terminating the transmission line with a resistance of the proper value can eliminate the reflection of the wave from the end of the line. Each section of the transmission line stores energy in the magnetic field surrounding the inductor and in the electric field inside the capacitor. Analysis shows that there is a fixed ratio between the voltage and current in the line that is independent of location and time. To eliminate reflections, the line is terminated with a resistance that has the same ratio of voltage and current as a section of the transmission line. This terminating resistance "looks like" another section of transmission line to the incident wave; however, the energy in the wave is converted to heat in the resistance rather than stored for propagation further down the line. The proper terminating resistance has a value R_0 given by

$$R_0 = \sqrt{L/C}$$

The value of R_0 is called the *characteristic impedance of the line* and is one of the most important parameters for applications of a transmission line.

To obtain these results it is necessary to consider smooth waves with relatively small values of slope. The lumped model of the transmission line, shown in Fig. 4-8, is composed of a number of low-pass filters (of the π type) connected in cascade. To find the cutoff frequency, f_c, for the lumped transmission line, one assumes a sinusoidal voltage on the line and obtains

$$f_c = 1/(\pi \, \Delta t) = 1/(\pi\sqrt{LC})$$

For sinusoidal voltages with a frequency greater than f_c, the voltage waveform propagates with appreciable distortion. This calculation applies only to a lumped line. An ideal continuous transmission line does not distort waveforms, regardless of their frequency. To appreciate this, consider approximating a continuous line as a lumped line and take the limit as $\Delta \ell$ goes to zero. The value of f_c for the lumped line approximation then goes to infinity.

2. Continuous Transmission Line

A general transmission line has series and shunt resistances, in addition to the series inductance and shunt capacitance considered in Fig. 4-8. A continuous transmission line (e.g., coaxial cable) has no lumped circuit elements. A schematic diagram of an incremental section of a generalized, continuous transmission line is shown in Fig. 4-9. The length of the transmission line is aligned with the x axis.

To find the voltage and current accurately on a continuous transmission

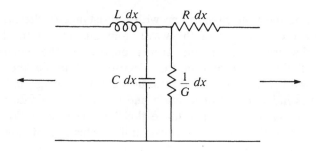

Fig. 4-9. *Incremental section of a generalized, continuous transmission line.*

line requires the solution of partial differential equations (Eq. 6a and b), which are known as the telephone equations.

$$\frac{\partial}{\partial x^2} V(x, t) = LC \frac{\partial}{\partial t^2} V(x, t) + (RC + GL) \frac{\partial}{\partial t} V(x, t) + RG \cdot V(x, t) \quad (6a)$$

$$\frac{\partial}{\partial x^2} I(x, t) = LC \frac{\partial}{\partial x^2} I(x, t) + (RC + GL) \frac{\partial}{\partial t} I(x, t) + RG \cdot I(x, t) \quad (6b)$$

where L = series inductance *per unit length* of transmission line, C = shunt capacitance *per unit length* of transmission line, R = series resistance *per unit length* of transmission line, G = shunt conductance *per unit length* of transmission line, $V(x, t)$ = voltage across transmission line at position x and time t, and $I(x, t)$ = current in transmission line at position x and time t.

The resistance $(1/G)$ in shunt with each capacitor accounts for imperfect insulation between the conductors in the transmission line. With modern plastic insulators, the effect of this shunt resistance is negligible under most conditions. At frequencies above 1 GHz, shunt resistance may be useful to account for heating of plastic insulation, an effect that can be neglected in a book on transient overvoltages.

The solution of Eq. 6a and 6b for a single sinusoidal traveling wave with a frequency ω radians per second, propagating in the positive x direction are given by Eq. 7a and b.

$$V(x, t) = Re[Ae^{-\alpha x} e^{j(\omega t - \beta x)}] \quad (7a)$$

$$I(x, t) = Re[Ae^{-\alpha x} e^{j(\omega t - \beta x)}/Z_0] \quad (7b)$$

The parameters α, β, and Z_0 in Eq. 7a and b are given by Eqs. 8 and 9.

$$\alpha + j\beta = \sqrt{(R + j\omega L)(G + j\omega C)} \quad (8)$$

$$Z_0 = \sqrt{\frac{R + j\omega L}{G + j\omega C}} \tag{9}$$

The parameter α describes the attenuation of the wave as it travels along the line. The attenuation of the voltage waveform in decibels by propagating a distance Δx is given by Eq. 10.

$$20 \log_{10}(e^{\alpha \Delta x}) = 8.7\alpha \, \Delta x \tag{10}$$

The speed at which a particular point on the waveform travels (the phase velocity) is ω/β.

The frequency band used for communication signals on long transmission lines in the early part of the 20th century extended up to a few tens of kilohertz; transmission of electric power uses 50 or 60 Hz. Therefore, the only high-frequency waves on early transmission lines were due to lightning and other transient overvoltages. For this reason, the characteristic impedance of the transmission line, Z_0, was called *surge impedance* (Bewley, 1929). This usage persists today.

We consider briefly the ideal case when R and G are both zero and then consider practical cases.

3. Ideal Case

When R and G are both zero, the speed is $1/\sqrt{LC}$, and the value of the characteristic impedance of the transmission line, Z_0, is a real number, $\sqrt{L/C}$. The value of α is zero, so there is no attenuation as the wave propagates. These are the same results described above for a lumped line.

The value of β depends linearly on frequency, so the phase velocity is independent of frequency. This corresponds to propagation without distortion.

Many practical short transmission lines are nearly ideal at frequencies below about 1 GHz. The speed of propagation in a typical transmission line is of the order of 200 m/μs. Values of Z_0 for common transmission lines range between about 25 and about 500 Ω.

4. Distortion

If $RC = GL$, then α is independent of frequency, and β varies linearly with frequency, so that waveforms propagate without distortion. The ideal transmission line ($R = G = 0$) is a special case of this. However, in general, a waveform is distorted as it propagates down the transmission line. One could, in theory, insert a uniform shunt conductance, G, along the transmission line to avoid distortion. However, this would double the value of α at high frequencies; the increased attenuation makes deliberate insertion of G undesirable.

Transient overvoltages can be represented in the frequency domain (by the Fourier transform) by a wide range of frequencies. There will not be a single value of speed that characterizes the propagation of the overvoltage, because the speed is a function of frequency.

D. EFFECT OF SERIES RESISTANCE

Real cables are made of material with nonzero conductivity that produces a series resistance in the cable. Although the effects of this resistance are usually negligible when normal operations are concerned, overvoltages are often associated with abnormally large currents. The waveshape of a transient overvoltage may be affected by the dc resistance of the cable. At high frequencies the current is concentrated in a very thin layer on the exterior of conductors, which is known as the skin effect. The concentration of current in a thin layer increases the current density, which can be modeled as an increase in the resistance per unit length of the cable. These phenomena act to increase the rise time, decrease the peak voltage, and increase the duration of transient overvoltages as they propagate along the cable. In particular, some of the energy contained in the leading edge of a traveling wave is transferred toward the tail of the wave.

When R is small but nonzero and G is zero, the value of α can be approximated at large values of ω by Eq. 11.

$$\alpha = R/(2Z_0) \tag{11}$$

1. DC Resistance

The resistance per unit length of a circular conductor of radius a is given by

$$R = 1/(\sigma \pi a^2)$$

where σ is the conductivity of the conductor. This relation is valid at dc and very low frequencies, where the current is uniformly distributed over the cross-sectional area of the conductor.

Mains branch circuits in buildings in the United States usually have copper wire that is about 2 mm in diameter, known as American wire gauge 12. This wire has a dc resistance per unit length of 5.2 Ω/km. On a branch that has a length of 30 m and a current of 10 A, a voltage drop of about 1.6 V would be expected on each conductor. This is a small change in voltage compared to the nominal mains voltage of 120 V rms.

Wires for telephone and computer data communications often have a diameter of about 0.5 mm, known as American wire gauge 24. This copper wire has a dc resistance per unit length of about 84 Ω/km.

2. Skin Effect

At high frequencies, the magnetic field associated with the current in the conductor, together with the nonzero conductivity, causes the current to be concentrated in a thin layer on the surface of the conductor, which is the *skin effect*. The distance in which the current per unit area decreases by a factor of $1/e$ is known as the skin depth, δ, given by Eq. 12

$$\delta = \sqrt{1/(\pi\sigma\mu f)} \tag{12}$$

where σ is the conductivity of the conductor, μ is the permeability of the conductor, and f is the frequency. The value of permeability for nonmagnetic materials is $4\pi \cdot 10^{-7}$ H/m.

The resistance per unit length, R, of a circular conductor for which δ is much less than the radius of the conductor, a, is approximately given by Eq. 13:

$$R = \frac{\sqrt{\mu f}}{2a\sqrt{\pi\sigma}} \tag{13}$$

where we have assumed that there is uniform current within a distance δ of the surface of the conductor and zero current further inside the conductor.

The skin depth, δ, is inversely proportional to the square root of frequency, so as the frequency increases, the skin depth decreases. As the skin thickness decreases, so does the cross-sectional area for current flow. Since the conductivity of the metal is independent of frequency, the resistance per unit length increases with frequency. As a consequence, cables become more lossy at high frequencies.

Equations 10, 11, and 13 together can be used to determine the attenuation caused by skin effect. Consider an example typical of the mains: copper wire with a radius of 1 mm (known as American wire gauge 12). The resistance per unit length is 1.3 Ω/m at a frequency of 1 GHz, a factor of about 240 greater than the dc resistance per unit length. When that wire is used in transmission line with a characteristic impedance of 100 Ω, the attenuation at a frequency of 1 GHz for a 100 m length of line is 5.7 dB.

Attenuation of signals (desirable voltages) on transmission lines is undesirable. To reduce the attenuation caused by skin effect, the surfaces of conductors in transmission lines that carry very high frequency signals are often silver plated. At high frequencies, where the thickness of the plating is greater than the skin depth, silver plating produces the minimum attenuation for a given geometry, because silver has the largest conductivity of any known material. However, attenuation of transient overvoltages is generally desirable, so the skin effect may be beneficial in reducing values of dV/dt caused by overvoltages with rise times of no more than a few tens of nanoseconds.

E. REAL TRANSMISSION LINES: NONUNIFORMITY

To obtain a simple analytical solution for current and voltage on a transmission line, it is necessary to assume that the line is symmetrical and uniform—for example, (1) two parallel wires above a perfectly conducting plane, or (2) two coaxial cylinders. This analysis leads to the development of a characteristic impedance, Z_0, that is constant along the line.

Practical problems are not so simple. For example, the mains conductors in industrial buildings are three or more wires stuffed into conduit. The spacing of the wires varies with distance; there are kinks and bends in the wires. These effects greatly change the nature of the problem: the values of inductance and capacitance per unit length are no longer parameters that are independent of position along the line: they are now a function of distance along the transmission line.

One way to cope with this complication is to use a digital computer program to provide a numerical solution for current and voltage as a function of position and time. A simple approach is to use one of the software packages for analog circuit analysis, such as SPICE, and describe the line as a series of lumped elements. One should consider a constant length of a section of the line, $\Delta\ell$, and compute the value of L and C per section by, for example,

$$L_i = \bar{L}(x)\,\Delta\ell$$

where L_i is the value of the lumped inductance in the ith section, and $\bar{L}(x)$ is the average inductance *per unit length* at distance x. Distance x corresponds to the location of the center of the ith section of the lumped line.

An additional complication occurs when the transmission line is contained in a conduit that is made of ferrous material. For common-mode voltages and currents, the effect of the conduit is to introduce an iron core choke in series with the transmission line. This choke is a nonideal inductor. Knowledge of the B versus H curve for the conduit material is necessary to accurately model this situation. However, one does not need an accurate model to predict that the effect of a ferrous conduit will be to attenuate components of traveling waves that have high frequencies. This effect, when combined with nonlinear shunt elements (e.g., spark gaps, varistors, avalanche diodes), may be useful in protecting electronic systems from transient overvoltages with large magnitudes of dV/dt.

F. REFLECTIONS

When a voltage or current traveling wave reaches a discontinuity where the characteristic impedance of the transmission line changes from Z_0 to Z, part of the traveling wave is reflected. If the amplitude of the incident and reflected waves are V_i and V_r, respectively, the voltage reflection coefficient, Γ, is given by

$$\Gamma = \frac{V_r}{V_i} = \frac{Z - Z_0}{Z + Z_0}$$

Common devices shunted across a transmission line to protect circuits from damage by transient overvoltage have values of Z that are much smaller than Z_0 during passage of large surge currents. It is then inescapable that most of the transient voltage will be *reflected*, and *not absorbed*, by the shunt protective device during each passage of the traveling wave down the line.

For example, a spark gap is a common shunt component for diverting very large transient currents (spark gaps are discussed in detail in Chapter 7). When a spark gap operates in the arc region, the voltage, V, across the gap is about 20 V, and the current, I, in the gap can be of the order of 100 A. The effective impedance, V/I, of the spark gap is then of the order of 0.2 Ω, which is almost a short circuit. If this 0.2 Ω termination resistance is placed across a transmission line with a 50 Ω characteristic impedance, 99.2% of the incident voltage (or current) wave will be *reflected*.

If there are conducting spark gaps at both ends of the transmission line, the wave will slosh between the two ends. There are only three dissipation mechanisms: losses in the series resistance of the transmission line, the energy that is converted to heat in the spark gap, and possible radiation of electromagnetic energy by the transmission line.

Nonlinear shunt circuit elements (e.g., spark gaps, varistors, and avalanche diodes) are not the only cause of reflections in transient protection circuits. Low-pass filters are often connected to transmission lines (e.g., filters at point of entry of mains power and communication signals into a building; power filters at each chassis). If these filters have a capacitor shunted across their input terminals, the input impedance of the filter will have a small magnitude at high frequencies. If these filters have an inductor in series with the input terminal, the input impedance of the filter will have a large magnitude at high frequencies. Either way, a low-pass filter can introduce reflections on a transmission line.

The mismatched shunt impedance of a nonlinear device (e.g., spark gap, varistor, diode) or filter will cause reflections of transient voltages. In this view, the use of a nonlinear device did not "solve" the problem of transient overvoltages, it merely dumped the transient back on the line. Although one can hardly advocate casually introducing reflections, one should recognize that reflecting the transient away from a vulnerable load *does protect* the load.

G. TRANSMISSION LINES VERSUS ORDINARY CIRCUIT ANALYSIS

There are two ways to consider the propagation of voltages: transmission line theory and ordinary circuit analysis. Applications of transmission line theory can be messy and complicated, particularly in systems with many branches and discontinuities in impedance. Ordinary circuit analysis is easy to apply but gives the wrong result when applied to nonsteady-state situations.

Consider an example in which the leading edge of a waveform takes 8 μs to go from zero to the peak value and waves travel at 200 m/μs: the leading edge of the waveform spans a distance of 1.6 km. In this example, use of ordinary circuit analysis (Kirchhoff's laws) in lumped circuit elements will be acceptable for a circuit that has a maximum dimension of 10 cm, since the current or voltage wave will be essentially uniform over such a small distance. However, use of ordinary circuit analysis would give erroneous results for a circuit with a dimension of 1 km, because no account is taken of the finite speed of the voltage or current wave. Transmission line analysis *must* be used when the length of any significant feature of the waveform is less than or comparable to the size of the conducting circuit.

Transmission line analysis also *must* be used when there are reflections from discontinuities in impedances. These reflections alter the waveshape, which is discussed below, in the section on generation of ring waves.

In a finite network of transmission lines with impedance discontinuities, excited, for example, by a rectangular voltage wave, there are an infinite number of reflected voltage waves. It can be easily shown that the superposition of the multiple reflections is a geometric series that converges to the result given by ordinary circuit analysis. When the amplitude of the reflected wave at all of the impedance discontinuities in the system has reached a small value, the system is in its steady state. Ordinary circuit analysis is valid only in the steady state.

When the length of the system is longer than the length of the traveling wave, the characteristic impedance of a transmission line can be useful to limit the surge current. A surge with a peak voltage of 1 kV will have a peak current of only 10 A when propagating on a line with a surge impedance, Z_0, of 100 Ω. However, this limiting effect on current does not occur in the long-time regime: surge currents are often of the order of a few kiloamperes. In the long-time regime, Martzloff (1983) has spoken of the *impedance to the surge* formed by the load and series resistance of the conductors, rather than the *surge impedance* of the transmission line.

Martzloff (1983) showed that a 75 m length of two-wire transmission line that was shorted at the load end presented an inductance about 65 μH to a surge generator at the source end of the line. This inductance stretched the normal 8/20 μs current waveform into a 25/70 μs waveform.

In the notation 8/20 μs, the first number, 8 μs, is approximately the time required for the waveform to rise from zero to the peak. The second number, 20 μs, is approximately the duration of the waveform from the initial zero to the half-value point on the trailing edge of the waveform. This notation is explained in Chapter 5.

This example shows that the properties of a moderate length of cable could affect the waveshape of current surges. The length of this cable is only about 1.5% of the length of the rising edge of the current waveform, so transmission line theory is not required. The cable can be considered as a series combination of 0.7 Ω (the dc resistance) and 65 μH, which Martzloff called the *impedance to the surge*. This is markedly different from the 100 Ω value of Z_0, which is called the *surge impedance*.

H. ATTENUATION OF SHORT-DURATION SURGES

Overvoltages with a rise time on the order of a nanosecond can only be observed within a few meters of their source. Inside shielded buildings, far from the point of entry, overvoltages caused by HEMP may have a longer rise time than on overhead lines.

Martzloff and Gauper (1986) studied propagation of surge voltages in three conductors inside steel conduit with a length of 225 m. They showed that a voltage pulse with a full width at half-maximum (FWHM) of 0.1 μs at the generator had an attenuation of amplitude and an increase in rise time after propagation along the line. After traveling 225 m, the peak voltage was only 35% of the value at the source end of the cable. The FWHM had increased from 0.10 μs at the generator to 0.18 μs at the end of the cable.

Martzloff and Gauper (1986) then reported the propagation of a voltage pulse with a FWHM of about 0.9 μs on the 225 m length of wire inside steel conduit. The amplitude of the pulse at the far end decreased to about 80% of the value at the source end, while the FWHM increased to about 1.05 μs. This longer wave is attenuated less than the short wave discussed in the previous paragraph. The reason is simple: the frequency spectrum of the wider pulse contains more energy at low frequencies where the attenuation per unit length, α, in Eq. 7 is less, owing to skin effect. Martzloff and Gauper (1984) concluded that long lines will attenuate pulses with duration of less than 1 μs but would provide "no appreciable attenuation" for pulses with durations longer than 1 μs.

Martzloff and Wilson (1987) extended the previous work by considering pulses with shorter durations. The wires were in steel conduit with a length of 63 m, which was properly terminated with a resistive load. The results are shown in Table 4-1.

TABLE 4-1. Effect of Propagation in 63 m Length of Conduit

Input		Output		
Rise Time	Duration	Amplitude	Rise Time	Duration
0.7 ns	2.0 ns	3%	10 ns	30 ns
2.5 ns	8.0 ns	15%	12 ns	40 ns
12 ns	45 ns	46%	22 ns	90 ns
50 ns	150 ns	60%	60 ns	250 ns

The "amplitude" parameter in Table 4-1 is the ratio of the peak voltage at the end of the line to that at the source end. It is clear that a short-duration voltage wave is squashed as it propagates along the wires: the rise time is increased, the duration is increased, and the amplitude is markedly decreased. It is also clear from these results that waves with a longer duration are not as drastically affected by propagation as shorter-duration waves.

Martzloff and Wilson (1987) compared the propagation of a voltage waveform with a rise time of 5 ns and an FWHM of 50 ns in coaxial cable to wire in steel conduit. The amplitude was reduced to 72% of the value at the source by propagation in a 63 m length of coaxial cable but was reduced to 55% by propagation in wires inside the same length of conduit. The difference was ascribed to the varying orientation of the wires inside the conduit compared to the uniform dimensions of the coaxial cable. The authors did not comment on possible effects of the steel conduit, which might add the effect of a lossy series inductance to the line.

Short-duration overvoltages can be generated by a showering arc (see Chapter 2). The large magnitudes of dV/dt in such events will only affect electronic systems when the source is connected to the vulnerable equipment by conductors with a length of less than a few tens of meters, because nanosecond rise times are stretched by propagation. However, the energy in the voltage traveling wave is not significantly decreased by propagation along the line (Martzloff and Wilson, 1987).

I. GENERATION OF RING WAVES

We now consider an example of how reflections from discontinuities in a network of transmission lines can modify the incident waveform. The particular network in this example, shown in Fig. 4-10, is intended to simulate the mains with seven branches in a small building. For simplicity, all loads are located at the end of a branch. The length of each section of transmission line is specified by the one-way travel time along the line; the travel times are shown in Fig. 4-10. In this example the original overvoltage is a trapezoidal pulse, also shown in Fig. 4-10, that has an origin outside the

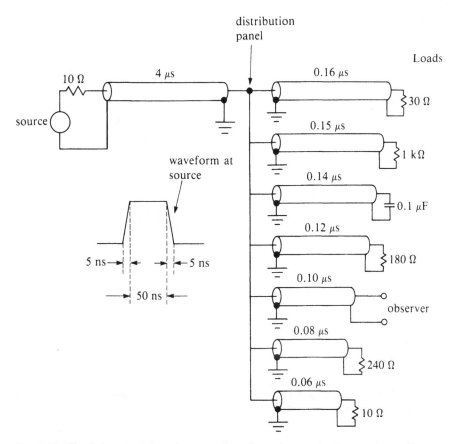

Fig. 4-10. Circuit for simulation of propagation of a voltage pulse in a system with many branches. The value of Z_0 for each cable is 100 Ω.

home. The overvoltage begins at zero time and takes 4 μs to travel to the distribution panel. The cable between the source and distribution panel is inserted so that reflections from the source are not seen during the interval between 4 μs and 8 μs, the duration of this example. The calculations in this simulation were performed by the PSPICE circuit analysis software on a desktop computer. All of the transmission lines are ideal ($R = G = 0$) and have a characteristic impedance of 100 Ω.

The voltage as a function of time at the end of the branch marked "observer" in Fig. 4-10 is shown in Fig. 4-11. Notice that the first event is an attenuated replica of the original overvoltage, a trapezoidal pulse with a duration of 0.05 μs. Thereafter, the voltage at the observer becomes a complicated oscillatory waveform that has little resemblance to the original pulse. The oscillatory waveform is produced by multiple reflections from the mismatched loads at the end of each branch. Spectral analysis of the

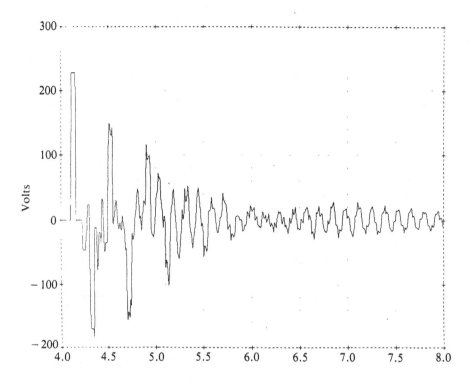

Fig. 4-11. Results of simulation of circuit shown in Fig. 4-10. The voltage is shown for the point in Fig. 4-10 labeled "observer."

oscillatory waveform shows peaks with nearly equal amplitudes near 2.6 and 7.8 MHz, and smaller peaks at 12.5 MHz and higher frequencies.

This example shows how the ring wave that is observed in a network of branch circuits (described in Chapter 3) is generated. It also shows that the waveform that impinges on vulnerable equipment at the end of a line can be dramatically different from the original waveform at the source of the overvoltage. Measurements of overvoltage waveforms made at the source of the overvoltage provide only part of the information needed to characterize the environment. The properties of transmission lines and loads on the lines are also critical information.

J. ARTIFICIAL MAINS NETWORK

Most transmission lines for signals are designed with a particular characteristic impedance, which is matched to the output impedance of the source and the load impedance to prevent reflections. Transmission of electric power is done with different rules because (1) the load impedance is unknown, (2) the wavelength associated with 50 or 60 Hz ac is long compared to the length of most lines so the system is in steady state, and (3) it is desirable to minimize the output resistance of the generator to avoid losses of power there.

It is necessary to characterize the electric power transmission line to understand the propagation of overvoltages on the mains inside buildings. This has already been done by the electromagnetic compatibility community, who needed to make reproducible measurements of the electromagnetic noise conducted out of a chassis and on to the mains.

There have been many measurements of the impedance of the low-voltage mains (Bull, 1968; Nicholson and Malack, 1973; Malack and Engstrom, 1976; Vines et al., 1985). There are several general conclusions that can be drawn from these measurements. Probably the most important is that the impedance of the mains is a function of frequency. At frequencies below about 20 kHz the value of $|Z_0|$ is small—for example, less than 5 Ω. At moderate frequencies, between about 20 kHz and 1 MHz, the impedance of the mains tends to increase as frequency increases. At high frequencies, between about 1 and 20 MHz, the impedance of the mains tends to be approximately independent of frequency; $|Z_0|$ can vary from about 50 Ω to about 300 Ω. The value of Z_0 at high frequencies often contains a substantial reactive part, but the value of the reactive part depends strongly on the loads that are connected to the mains and is therefore not reproducible from site to site. Furthermore, the impedance of the mains will change with time owing to connection and disconnection of loads (Piety, 1987).

In order to make reproducible measurements of conducted emissions of noise from equipment connected to the mains, it is necessary to have a standard mains impedance. Such a circuit is called an "artificial mains network" or a "line impedance stabilization network" (LISN). The official artificial mains network specified in IEC 3-1975 and ANSI Standard C63.4-1981 is shown in Fig. 4-12. In Fig. 4-12 the hot, the grounding, and the neutral conductors are labeled H, G, and N, respectively. The impedance at frequencies of tens of kilohertz is modeled by resistances of 5 and 10 Ω between each line and the grounding conductor. The impedance at somewhat higher frequencies is modeled by inductors in series with each line, together with the resistors. The constant impedance at frequencies of more than 1 MHz is modeled by a 50 Ω resistance shunted between each line and the grounding conductor. One of the 50 Ω resistances is the impedance of the voltmeter that measures the emitted noise, the other 50 Ω resistance is a used to maintain the balance of the circuit. The grounding conductor in the artificial mains network has no series inductance, because it is a high-frequency ground plane in a laboratory.

The artificial mains network is intended to be used to measure conducted noise that is emitted by equipment while it is operating. In order to be operating, power must be supplied by the mains during the measurement of noise. The capacitors, shown in Fig. 4-12, prevent the normal mains

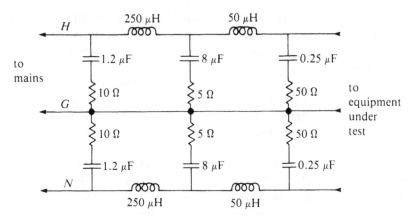

Fig. 4-12. *Artificial mains network.*

waveform, which usually has a frequency of 50 or 60 Hz, from dissipating appreciable power in the shunt resistances. These capacitors cause the impedance of the mains to increase as the frequency decreases below 7 kHz, which is contrary to the actual behavior of the impedance of the mains.

To use this artificial mains network in *calculations* of the impedance of the mains at low frequencies, it is desirable to remove the capacitors. Also one must insert inductors in series with the grounding conductor, because in a cable this conductor has the same geometry as the other two conductors. The circuit in Fig. 4-13 is a simple lumped element approximation of the mains wiring and the output impedance of the generator and transformers in the distribution network.

Three conductor cable that is commonly used in single-phase branches inside buildings has a characteristic impedance of about 100 Ω (Martzloff,

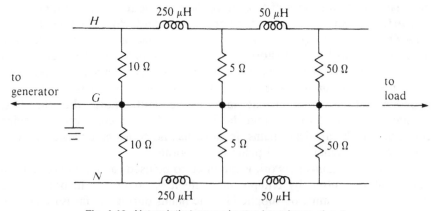

Fig. 4-13. *Network that approximates impedance of mains.*

1983). Vines et al. (1985) measured a line to neutral impedance of 98 Ω for cable in a residence. Standler and Canike (1986) measured a characteristic impedance between the line and neutral of 117 Ω for jacketed two-wire plus ground cable. These measurements agree rather well with the 100 Ω differential-mode impedance at high frequencies of the circuit in Fig. 4-13.

At low frequencies, the differential-mode impedance of the circuit in Fig. 4-13 is 6.25 Ω (the parallel combination of 100, 10, and 20 Ω, because the inductances are short-circuits at low frequencies). This value is too large for use at frequencies below 10 kHz. The electromagnetic compatibility engineers who designed and specified this waveform were not concerned with emissions at frequencies below 10 kHz, so it was not appropriate for them to add additional circuit elements to accurately describe the impedance at low frequencies. A more realistic network will require at least one more section cascaded on the generator side of the line, with a shunt resistance of a 1 Ω or less and an additional series inductance. The author does not wish to propose specific values here.

The common-mode impedance of the circuit in Fig. 4-13 has a value of 25 Ω at high frequencies (this is the parallel combination of the two 50 Ω resistances, because the series inductors are open circuits at high frequencies). At low frequencies, the common-mode impedance is 1.6 Ω (the parallel combination of all six resistors in Fig. 4-13).

5

Standard Overstress Waveforms Used in the Laboratory

A. INTRODUCTION

Overstress testing in a laboratory is a critical and important process in developing and validating survival of equipment in the adverse electrical environment that is described in the previous four chapters. A discussion of overstress testing is given in Chapters 22 to 24. This chapter contains a description of standard overstress test waveforms that are used in laboratory testing. These waveforms represent a simplification of the environments that are described in Chapters 2 to 4 and are used both for testing of components and proof-testing of protective circuits.

Current, not voltage, is the input variable in understanding transient protection. When current or voltage in transient overstresses is plotted, the independent variable is time. When V-I characteristic curves for devices are plotted, current is a reasonable independent variable (although this is not the conventional practice in electron devices, as discussed in Chapters 7 and 8). Martzloff (1982, p. 5) stated:

> Perhaps a long history of testing insulation with voltage impulses has reinforced the erroneous concept that voltage is the given parameter. Thus *overvoltage protection* is really the art of offering low impedance to the *flow of surge currents* rather than attempting to block this flow through a high series impedance.

As long ago as 1933, it was recognized that current, not voltage, was the appropriate variable for testing lightning arresters (McEachron, 1933). Despite this wisdom, most test waveforms for electrical overstresses are specified as a voltage and not a current.

Conventional overstress testing technique specifies the open-circuit voltage from the generator and (sometimes) the output impedance. One might expect to use Thévenin's theorem to reconstruct the actual waveform. But this method is not valid, because the load usually has a nonlinear relation between voltage and current. Most of the common nonlinear protective devices (e.g., spark gaps, avalanche diodes, metal oxide varistors) clamp at a nearly constant voltage, which is approximately independent of current. Therefore, the open-circuit output voltage of the generator is not useful as a single parameter when the generator is connected across one of these nonlinear devices. The short-circuit output current of the generator may be more useful, provided the generator output impedance is much greater than the V/I value of the device under test. Despite this, most descriptions of overstress test waveforms specific only an open-circuit voltage.

B. NOMENCLATURE

Before standard overstress test waveforms can be presented, several waveform parameters need to be defined. Figure 5-1 shows an overstress test waveform and several concepts that are used in the definitions.

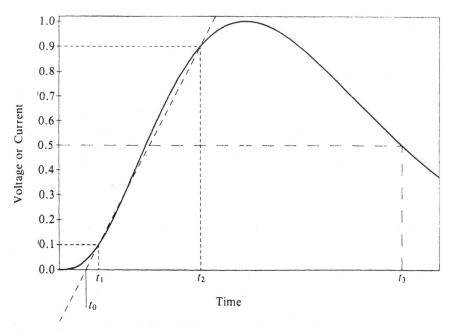

Fig. 5-1. *Graphical illustration of various points used in the definition of waveshape of surge voltage or current. The point t_0 is called the virtual origin.*

1. Virtual Origin

It is useful to have a precise and reproducible definition of the time at which the overstress waveform began. Two points on the initial part of the waveform are chosen, perhaps at 10% and 90% of the peak value (or at 30% and 90% of the peak value: see discussion of "front time" later in this chapter). A straight line is extrapolated through these two points, as shown by the diagonal dashed line in Fig. 5-1. The intersection of this straight line with the voltage or current zero is the *virtual zero time*, t_0.

One might ask why the nomenclature for high-voltage test waveforms is so different from nomenclature for waveforms in electronics. Many overvoltage test waveforms have a gentle turn-on, which does not have an obvious beginning, particularly when viewed on several different voltage or current scales. The definition of virtual zero time is a simple and easily reproducible method of defining the beginning of the waveform. A definite time for the beginning of the waveform is used to define the duration of the waveform, which will be discussed later in this chapter.

When this nomenclature was introduced, many years ago, measurements of waveforms by oscilloscopes were recorded on photographic film. Old-style oscilloscopes had poor trigger circuits (by modern standards) and often had a bright spot at the left edge of the screen while the oscilloscope was waiting to trigger. Fogging of the film by the oscilloscope sweep before and after the overstress sometimes obscured the location of the time the waveform began, which was another reason for the introduction of the virtual zero time.

2. Front Time

The *front time* is defined as 1.25 $(t_2 - t_1)$, where t_2 and t_1 are the 90% and 10% points on the leading edge of the waveform, respectively. This definition with a factor of 1.25 is used for all surge current waveforms and some voltage waveforms. There is an *alternative definition* of front time that is used for some voltage waveforms: 1.67 times the time difference between the 90% and 30% points on the leading edge of the waveform. The front time serves the same purpose as the more familiar parameter, rise time, which is discussed later in this section.

There are two obvious questions: (1) why are factors of 1.25 or 1.67 included in the definition of front time? and (2) why use the 10% point for some waveforms and the 30% point for others? The reasons are historical. Unfortunately, those who established the definitions appear not to have been concerned with mathematical precision or consistent practice.

The original definition of front time was the time difference between the beginning and the peak of the waveform (Sporn, 1933). This is an unsatisfactory definition. The beginning of the waveform is often not well defined, as explained above in the discussion of virtual zero time.

Furthermore, many waveforms have a relatively long duration that gives a broad peak. When the peak is broad, the precise time of the peak is impossible to determine. A specific example is given later in this chapter under the discussion of the $10/1000 \mu s$ waveform.

When this unsatisfactory definition of front time was amended to use two points on the leading edge of the waveform, it was necessary to insert a correction factor of 1.25 or 1.67 to obtain approximate agreement with the previous definition. If the initial portion of the waveform were a linear ramp (it is not), then the front time would be the difference between the virtual zero time and the time of the peak.

Why use the 10% point for some voltage and all current waveforms and the 30% point for other voltage waveforms? Current waveforms tend to be less noisy than voltage waveforms: overstress current waveforms require a conducting loop. The inductance of the loop together with the parasitic capacitance of the device under test forms a low-pass filter that attenuates high-frequency noise. On the other hand, measurement of voltage waveforms is done with essentially open-circuit conditions, which have no filtering. Voltage waveforms that have a large peak open-circuit voltage (i.e., 20 kV or more) tend to have more noise from corona discharge than waveforms that have a relatively small peak open-circuit voltage (i.e., 1 kV or less). Some types of voltage waveform generators are noisier than others; for example, switches that use triggered gas-filled spark gaps tend to produce less noise than mechanical relays with air between the contacts. It appears that the 30% point was used in the definition of "virtual zero" and "front time" if, at the time the waveform specification was first written, the voltage generator circuits in common use for this waveform produced substantial noise. If not, the 10% point was used.

Many engineers with experience in surge testing believe that the intention of the standards (e.g., IEC 60-2, ANSI/IEEE Std 4-1978) is that *all voltage* waveforms have a front time defined by the 30% and 90% points, whereas *all current* waveforms have a front time defined by the 10% and 90% points. This may have been the intention of the writers of the standards, but that is not what the standards say. ANSI/IEEE Std 4-1978 (pp. 42–45) defines front times with the 30% and 90% points for "lightning impulse voltages." The "standard lightning impulse" is the $1.2/50 \mu s$ waveform (IEC 60-2, p. 25; ANSI/IEEE Std 4-1978, p. 47). Nowhere do these documents say that the 30% point is to be used for voltage waveforms that represent other kinds of overstresses. In fact, as described below, many *voltage* test waveforms have a front time (or rise time) defined with the 10% and 90% points.

3. Rise Time

It is conventional in electronic engineering, particularly in signal processing and circuit analysis, to call the time difference between the 10% and 90%

points the *rise time*. Since this is different by a factor of 1.25 from the definition of front time, one must be careful not to interchange the two terms. Several overvoltage test waveforms, which are discussed later in this chapter, are defined with the rise time: the 100 kHz ring wave in ANSI C62.41-1980 and the Electrical Fast Transient in IEC 801-4.

4. Time to Half-Value

The *time to half value* is $t_3 - t_0$, shown in Fig. 5-1, the time between the virtual zero time and the time when the waveform reaches half of the peak value during the decay. The time to half value is the most commonly used measure of the duration of an overstress test waveform.

5. Full Width at Half-Maximum

The time to half-value, defined above, is a measure of the duration of the high-voltage test waveform. Another measure of duration is the *full width at half-maximum* (FWHM). The FWHM is the time difference between the points on the initial and decaying parts of the waveform where the voltage or current is half of the peak value, as shown in Fig. 5-2. It is common in applied mathematics and experimental nuclear physics to specify duration or pulse width in terms of FWHM. The duration of the Electrical Fast

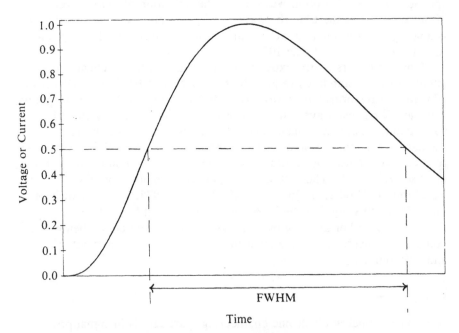

Fig. 5-2. *Graphical illustration of the full width at half-maximum (FWHM).*

Transient in IEC 801-4, which is discussed later in this chapter, is specified in terms of the FWHM. There is no doubt that FWHM is easier to measure than the time to half value, which requires a determination of the virtual zero.

6. α/β Format

Laboratory waveforms for unipolar overstress tests are commonly described in a $\alpha/\beta\,\mu$s format, where α and β are positive numbers. This notation indicates that the waveform has a front time of $\alpha\,\mu$s, followed by an decay that reaches half of the peak value at $\beta\,\mu$s. In older literatue, the notation $\alpha \times \beta$ is found, which means the same thing as α/β.

To have a complete description of a particular test waveform, the peak open-circuit voltage *and* the peak short-circuit current from the generator should also be specified, in addition to the waveshape α/β. However, if only one peak value is given, then the output impedance must be specified.

7. Recommendation: Nomenclature for New Waveforms

As described above, there are several alternative ways for describing the leading edge and duration of the waveform. It is inadvisable to modify the definitions of the nomenclature for the traditional waveforms, because this would change the numerical values in the name of the waveform and increase the confusion. The author believes that if new waveforms are to be specified by two parameters, then rise time and FWHM ought to be used.

Hyltén-Cavallius (1967) proposed that waveforms with a double exponential waveshape ought to be specified by their time constants. He went on to remark about the difference between those who favor the "theoretical aspect" and those with a "practical mind [that] may prefer something which is simpler or more straightforward." His exact words are worth repeating:

> Experience has shown me that all definitions, units of measurements, etc., should be based, as far as possible, on fundamentals or basic physics. When this principle is violated because of practical consideration in a given situation, we will sooner or later run into difficulties, for instance, when we try to extend our definitions beyond the range for which they were originally intended. A *practical* definition will then turn out to be a rather impractical one—this is a well-known fact in standardization work.

An example of this is the original definition of front time as the time between the beginning and the peak of the waveform. This is a practical definition that is not so bad for short-duration waveforms, such as the $1/5\,\mu$s voltage waveform, which was commonly used when front time was originally defined in the 1930s. It is completely unacceptable for long-duration waveforms, such as the $10/1000\,\mu$s waveform that was introduced around 1960.

Although specification of new waveforms with an equation and values of parameters is desirable, it would still be convenient to speak of the waveform by referring to a name that contained the rise time and FWHM (or the front time and time to half-value for older waveforms).

C. STANDARD OVERSTRESS TEST WAVEFORMS

There are many overstress test waveforms that are specified in various standard publications. Some of the more commonly used standard waveforms are discussed below.

Most of these waveforms were initially described to represent a simplification of the environment on the mains; some were initially described for telephone equipment. However, once these waveforms became common, they were used also for additional purposes. For example, waveforms that simulated lightning effects on particular systems were also used to characterize components. Waveforms that were initially specified for telephone systems have been used to test the mains connection of electronic equipment. Rather than capriciously invent new waveforms, appropriate waveforms from the armamentarium should be considered. From this point of view, there is no such thing as a power line waveform or telephone waveform.

The standard documents define test waveforms in terms of front time, time to half the peak value, amplitude, and, when appropriate, oscillation frequency. Since no mathematical relations are given in most of these standard documents, Standler (1988a) provided relations for five commonly used waveforms. Equations are necessary for engineers who wish to perform computer simulations of transient protection circuits. Further, having the time constants (rather than front time and time to half-value) makes it easier to design RLC pulse-forming networks for use in laboratory pulse generators.

Mathematical relations are often used to express an exact and unique view of a phenomenon. The equations that follow in this chapter are neither unique nor exact representations of the corresponding test waveforms. The equations given below are claimed to be a simple mathematical relation that satisfy the requirements given in the appropriate standard(s). However, they are not the only equations that could satisfy the requirements.

It is emphasized that the output of a laboratory waveform generator *cannot* be labeled *un*satisfactory, because it differs from the waveshape given by an equation. The criteria for a satisfactory waveshape are those listed in the appropriate standard: front time, time to half value, oscillation frequency, etc.

Engineers who are familiar with electronics laboratory technique, where tolerances between ±1% and ±5% are common, may be surprised to find that high-voltage waveforms are not nearly as precise. Several of the standards specify tolerances between ±10% and ±30% for the front time and time to half-value. These relatively large tolerances are justified for several reasons. Tolerances of ±10% are common for components used in pulse-forming networks. Parasitic inductances and capacitances in components in both the pulse-forming network and the test fixture may cause additional discrepancy. And it makes little sense to specify great accuracy in the open-circuit voltage waveform or short-circuit current waveform when the waveform can change drastically when a nonlinear device is connected to the output terminals of the surge generator.

1. 8/20 μs Current Waveform

Some of the effects of a direct lightning return stroke to overhead power transmission and distribution lines can be simulated by a current waveform with an 8/20 μs waveshape (or by a voltage waveform with a 1.2/50 μs waveshape, which will be discussed next). However, this waveshape does not simulate the effects of continuing current in cloud-to-ground lightning.

A simple, approximate mathematical expression for the 8/20 μs current waveform that is specified in IEC 60-2, ANSI/IEEE Std 4-1978, and ANSI C62.1-1984 is $I(t)$, as given by Eq. 1.

$$I(t) = AI_p t^3 \exp(-t/\tau) \tag{1}$$

The parameters in Eq. 1 have the following values:

$$\tau = 3.911 \,\mu\text{s}$$

$$A = 0.01243 \,(\mu\text{s})^{-3}$$

In Eq. 1, t is the time in μs ($t \geq 0$), and I_p is the peak value of the current. A plot of Eq. 1 with these values of the parameters is shown in Fig. 5-3.

This approximation gives a front time that is 0.3% too large and a time to half-value that is 0.3% too small. However, this small error is outweighed by having an equation for current that can be integrated analytically, unlike the situation where the parameters for Eq. 1 exactly satisfy the definition of the 8/20 μs waveform (Standler, 1988a). The standards specify a tolerance of ±10% for both the 8 μs front time and the 20 μs time to half-value, so the ±0.3% error in using this approximation is well within acceptable limits.

A simple double exponential relation, which is used for many of the other nonoscillatory test waveforms, does not fit the 8/20 μs waveform.

The total charge, ΔQ, that is transferred by the 8/20 μs waveform in Eq. 1 is given by

$$\Delta Q = 6 \times 10^{-6} A I_p \tau^4$$

When I_p is in amperes, A is in $(\mu\text{s})^{-3}$, and τ is in μs, ΔQ will have units of

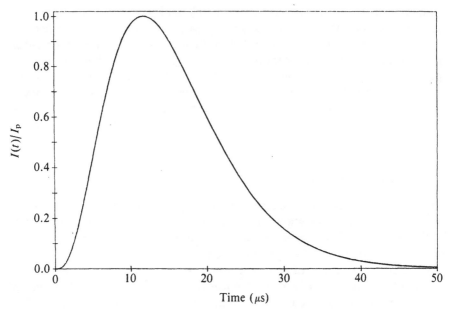

Fig. 5-3. *Plot of short-circuit current in an 8/20 μs waveform.*

coulombs. When numerical values are substituted for A and τ, the relation becomes

$$\Delta Q = 1.746 \times 10^{-5} I_p$$

where I_p is in amperes and ΔQ is in coulombs. This relation is useful for many things, among them avoiding or checking numerical integration routines and calculating the minimum energy storage capacitance for a generator that is to provide the 8/20 μs waveform.

Values of I_p in routine use range from about 50 A to more than 60 kA, depending on the nature of the test to be performed. An 8/20 μs waveform is commonly used to determine the behavior of protection components (e.g., spark gaps or varistors) during currents greater than about 1 kA.

The origin of the 8/20 μs current waveform is obscure. Archival literature between 1920 and 1950 contains many references to production and use of voltage waveforms but few references to current waveforms. American Standards have used a $\overline{10}/20$ μs current waveform since 1936 to test arresters for high-voltage transmission lines (Cornelius, 1953). These arresters commonly contained a silicon carbide varistor in series with a spark gap. The 10 μs front time was defined as the time between the zero and peak of the waveform and is essentially identical to the 8 μs front time specified as 1.25 times the time between the 10% and 90% points

(Armstrong et al., 1960). To distinguish the difference in definition of front time, the notation $\overline{10}/20$ is used in this book instead of 10/20. Some time after 1957, American Standards deleted the $\overline{10}/20\,\mu s$ waveform and added the $8/20\,\mu s$ waveform that was specified by the International Electrotechnical Commission (IEC).

Stoelting (1956) mentions that when the $\overline{10}/20\,\mu s$ current waveform is passed through a silicon carbide arrester, the voltage across the arrester is a "reasonable approximation of the 1.5/40 voltage wave used in testing equipment insulation." Earlier, Cornelius (1953), as corrected by McMorris (1953), showed that the $\overline{10}/20\,\mu s$ current wave produced a 1.5/30 μs voltage waveform across an arrester. This was thought to be a justification for the $\overline{10}/20\,\mu s$ current waveform, since the voltage waveform was well established.

2. 1.2/50 μs Voltage Waveform

The $1.2/50\,\mu s$ waveform is specified in IEC 60-2, ANSI/IEEE Std 4-1978, and ANSI C62.1-1984. The front time for this waveform is defined as 1.67 times the time interval between the 30% and 90% of peak points on the leading edge of the waveform. The standards specify that a tolerance of ±30% applies to the 1.2 μs front time and a tolerance of ±20% applies to the 50 μs time to half-value.

A simple mathematical relation that describes this waveform is $V(t)$, as given by Eq. 2.

$$V(t) = AV_p\{1 - \exp(-t/\tau_1)\}\exp(-t/\tau_2) \tag{2}$$

The parameters in Eq. 2 have the following values:

$$\tau_1 = 0.4074\,\mu s$$

$$\tau_2 = 68.22\,\mu s$$

$$A = 1.037$$

In Eq. 2, t is the time ($t \geq 0$), and V_p is the peak value of $V(t)$. A plot of Eq. 2, with these parameters for a $1.2/50\,\mu s$ waveform, is shown in Fig. 5-4.

It is convenient to write the double exponential relation in the form of Eq. 2, rather than the more conventional form

$$V(t) = k[\exp(-t/a) - \exp(-t/b)]$$

so that there is a time constant for the leading edge of the wave, τ_1, and a time constant for the tail of the wave, τ_2. Equation 2 also has the amplitude of the wave expressed as a product of two terms: the peak voltage, V_p, and a scaling constant, A. The constant A is necessary to make the maximum

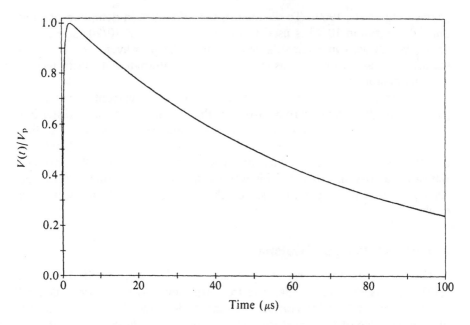

Fig. 5-4. *Plot of the open-circuit voltage in an 1.2/50 μs waveform.*

value of

$$A\{1 - \exp(-t/\tau_1)\} \exp(-t/\tau_2)$$

be unity, so that the peak voltage, V_p, appears directly in Eq. 2.

The 1.2/50 μs waveform is used for testing insulation, particularly of electric motors and transformers. Since this is an open-circuit test (unless the insulation fails), an output impedance is not specified for this voltage generator. Commercial 1.2/50 μs waveform generators have output resistances between about 1 and 500 Ω. If a value of output impedance is required for the 1.2/50 μs waveform and large surge currents are desired, a value of 2 Ω is suggested. This effective value is obtained from the ratio of open-circuit voltage and short-circuit current specified in ANSI C62.41-1980: 6 kV/3 kA, or 2 Ω. The word "effective" is used because the peak open-circuit voltage and peak short-circuit current occur at different times.

The history of the development of the 1.2/50 μs waveform is not clear. A 1.5/40 μs voltage waveform was developed and used in the United States for testing of insulation (Sporn, 1933). The 1.2/50 μs voltage waveform was developed in Europe sometime before 1938 (Allibone and Perry, 1936; Edwards et al., 1951). The two waveforms have approximately the same values of front time and duration: the nominal values for one waveform are near the extreme acceptable values for the other waveform. Sometime after

1957 American Standards deleted the 1.5/40 μs waveform and added the 1.2/50 μs waveform.

3. Combination 8/20 and 1.2/50 μs Waveform

The 8/20 μs waveform is specified under short-circuit conditions, and the 1.2/50 μs waveform is specified under open-circuit conditions. What happens to the waveform when a different kind of load (e.g., a varistor, spark gap, capacitor, etc.) is connected to the generator? The answer depends on the circuit used to generate the waveform, and this is not specified in standards. The 8/20 and 1.2/50 μs waveforms represent different aspects of the same phenomenon: the 8/20 waveform applies to low-impedance circuits and components, and the 1.2/50 waveform applies to high-impedance circuits and components. Richman (1983) described why it is desirable to combine them into a single generator, especially when the nature of the load is not specified.

ANSI C62.41-1980 suggests that a representative level for maximum overstresses on major feeders and short branch circuits inside a building is given by the combination of an 1.2/50 μs waveform with a peak open-circuit of 6 kV and an 8/20 μs waveform with a peak short-circuit of 3 kA.

4. 0.5 μs–100 kHz Ring Waveform

Martzloff and Hahn (1970) observed that transient overvoltages on low-voltage mains often have a rapid rise to a peak value, followed by an oscillation with exponentially decaying amplitude. A typical overvoltage has a 0.5 μs rise time between the 10% and 90% points, followed by an oscillation with a frequency of 100 kHz. The amplitude of the oscillation decays by a factor of 0.6 during each half-cycle. This waveform was written into the IEEE Standard 587 (which became ANSI Std C62.41-1980) that describes transient overvoltages on low-voltage mains, where it was called a *ring wave*.

The peak value of the ring wave is specified by the user during the test. ANSI C62.41 states that the peak value is no larger than 6 kV, owing to flashover of insulation in buildings. A representative peak value for some environments might be considerably smaller than 6 kV.

A simple mathematical expression for the 0.5 μs–100 kHz ring wave is $V(t)$, as given by Eq. 3a

$$V(t) = \begin{cases} BV_p y(1 + \eta y) & 0 \le t \le 2.5 \ \mu s \\ BV_p y & 2.5 \ \mu s < t \end{cases} \tag{3a}$$

where $y(t)$ is given by Eq. 3b.

$$y(t) = A\{1 - \exp(-t/\tau_1)\} \exp(-t/\tau_2) \cos(\omega t) \tag{3b}$$

The parameters in Eq. 3 have the following values:

$$\tau_1 = 0.4791 \ \mu s \qquad \tau_2 = 9.788 \ \mu s$$

$$\omega = 2\pi \ 10^5 \ s^{-1} \qquad A = 1.590$$

$$B = 0.6205 \qquad \eta = 0.523$$

In Eq. 3, t is the time and V_p is the peak value of the voltage, $V(t)$. Although this equation is defined in a piecewise fashion, both $V(t)$ and its derivative are continuous for $t > 0$. Standler (1988a) explained why such a complicated relation was necessary to represent this ring wave. A plot of Eq. 3 with these values of the parameters is shown in Fig. 5-5.

The ring wave can also be used as a current waveform by specifying a peak current instead of a peak voltage. ANSI C62.41 suggests that a representative peak current is about 500 A for Category B (major feeders and short branch circuits inside buildings). This 500 A peak current corresponds to an open-circuit peak voltage of 6 kV, which implies that the generator that simulates Category B environments has an effective output impedance of 12 Ω.

On Category A (long branch circuits: more than 20 m from the point of entry and more than 10 m from major feeders and distribution panels), a

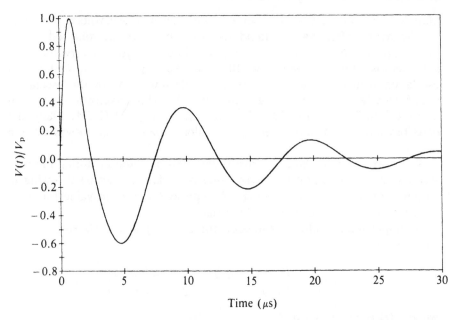

Time (μs)

Fig. 5-5. Plot of the open-circuit voltage in the 0.5 μs–100 kHz ring wave defined in ANSI C62.41-1980.

representative peak short-circuit current is 200 A. This 200 A peak current corresponds to an open-circuit peak voltage of 6 kV, which implies that the generator that simulates Category A environments has an effective output impedance of 30 Ω.

5. 1.25 MHz Ring Waveform (SWC)

ANSI C37.90a-1974 defines a ring wave that is representative of the transient overvoltage generated when high-voltage switchgear operates in power switchyards, which affects the control circuit relays. This ring wave is part of the tests prescribed by this standard for "surge withstand capability" (SWC). The standard shows a circuit diagram that produces this test waveform and describes the important parameters as follows:

oscillation frequency 1.25 ± 0.25 MHz

peak voltage 2.75 ± 0.25 kV

In addition, it is stated that the amplitude of oscillation decays "to 50% of crest value in not less than 6 μs," but it does not give a preferred value. The information in the standard implies an infinite value of dV/dt at $t = 0$, which is not realistic. Because specific values of time to half-value and rise time are not specified in the standard, there can be no official values. A time to half-value of about 10 μs and a rise time of 0.005 μs, though arbitrary, appear reasonable. Equation 4 describes these features.

$$V(t) = AV_p\{1 - \exp(-t/\tau_1)\} \exp(-t/\tau_2) \cos(\omega t) \tag{4}$$

The parameters in Eq. 4 have the following values:

$\tau_1 = 2.35$ ns $\tau_2 > 8.7\ \mu$s (15 μs suggested)

$A = 1.0096$ $\omega = 2.5\pi\ 10^6$ rad/s

In Eq. 4, t is the time, and V_p is the value of the voltage at the first peak. The first peak of $V(t)$ has a rise time of 5 ns when Eq. 4 is used with these values of parameters. A plot of $V(t)$, given by Eq. 4 with $\tau_2 = 15\ \mu$s, is shown in Fig. 5-6.

The event $V(t)$ is repeated at least once every cycle of the normal mains waveform for a duration of at least 2 seconds.

The output impedance of the voltage generator is specified to be 150 Ω, which is one of the largest output impedances of the commonly used overstress test waveforms.

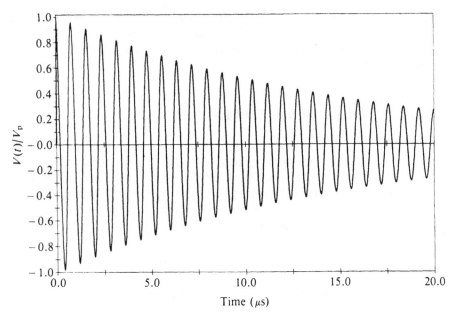

Fig. 5-6. Plot of the open-circuit voltage in the ring wave defined in ANSI C37.90.1.

6. Electrical Fast Transient (EFT)

IEC 801-4 defined an Electrical Fast Transient (EFT) test waveform. For testing equipment connected to the mains, the peak voltage, V_p, of this waveform is selected between 0.5 and 4 kV when delivered to a 50 Ω load. The peak value is decreased by 50% for tests of conductors that carry data or control signals. The capacitors that couple the EFT waveform to the equipment under test form a capacitive divider with the capacitance of the equipment under test. Therefore, the peak voltage across the equipment may be much less than the peak voltage across the output terminals of the EFT generator.

A simple mathematical expression for this waveform is $V(t)$, as given by Eq. 5:

$$V(t) = AV_p\{1 - \exp(-t/\tau_1)\} \exp(-t/\tau_2) \tag{5}$$

The parameters in Eq. 5 have the following values:

$$\tau_1 = 3.5 \text{ ns}$$

$$\tau_2 = 55.6 \text{ ns}$$

$$A = 1.270$$

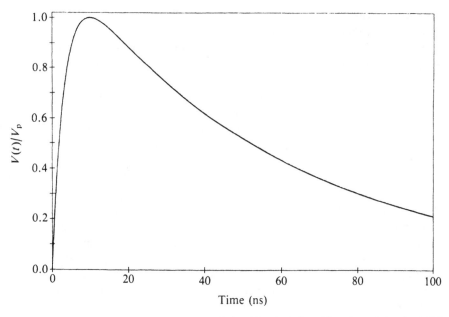

Fig. 5-7. *Plot of the open-circuit voltage of the Electrical Fast Transient defined in IEC 801-4.*

In Eq. 5, t is the time, and V_p is the peak value of $V(t)$. The standard specifies tolerances of $\pm 30\%$ for both the 5 ns rise time and the 50 ns duration. A plot of Eq. 5, with the parameters for the EFT, is shown in Fig. 5-7.

Each of the pulses described by $V(t)$ is repeated a number of times during a 15 ms interval, which is called a *burst*. The burst is repeated every 300 ms for the duration of the test.

The output impedance of the generator is specified to be $50\,\Omega \pm 20\%$. The choice of a nominal output impedance of 50 Ω was made so that the voltage generator is properly terminated in a common measurement system where the characteristic impedance of cables and the input impedances of voltmeters are all 50 Ω. This eliminates reflections from the source.

7. 10/1000 μs Waveform

Bodle and Gresh (1961) measured transient overvoltages that appeared on teltephone lines during thunderstorms. The 10/1000 μs waveform was designed by Bodle and Gresh to have a duration that was greater than 95% of the measured surges and a front time that was less than that of 99.5% of their measured surges. In other words, the 10/1000 μs waveform represents a reasonable "worst case": it has a lesser front time and greater duration than most overvoltages on telephone lines. Bodle and Gresh state quite clearly that 10/1000 μs is not a typical waveform on telephone lines.

The 10/1000 μs waveform has a slower front time by about two orders of magnitude than cloud-to-ground lightning. Transients that propagate on telephone lines have been distorted by the transmission line. High-frequency components of waveforms are attenuated by the skin effect and the loading coils in combination with the shunt capacitance of the line.

Although it is clear from the designation "10/1000 μs" that the front time has a duration of 10 μs, it is not clear how this word is properly defined for this waveform. Three different definitions have been used. Bodle and Gresh appear to have used the term "front time" to indicate the time between the initial zero and the peak value. As a second definition, Bennison et al. (1973) used the standard definition of rise time in electronic engineering, the time interval between the 10% and 90% points of the waveform. The third definition, which is present Bell Communications practice (Bellcore TR-EOP-000 001) and used in the August 1987 draft of what is to become ANSI C62.36, is to use 1.25 times the interval between the 10% and 90% points on the leading edge of the waveform as the 10 μs front time.

A simple expression for this voltage waveform, with the third definition of front time, is $V(t)$, as given by Eq. 6.

$$V(t) = AV_p\{1 - \exp(-t/\tau_1)\} \exp(-t/\tau_2) \qquad (6)$$

The parameters in Eq. 6 have the following values:

$$\tau_1 = 3.827 \ \mu s$$

$$\tau_2 = 1404 \ \mu s$$

$$A = 1.019$$

In Eq. 6, t is the time ($t \geq 0$) and V_p is the peak value of $V(t)$. A plot of Eq. 6, with these values of parameters, is shown in Fig. 5-8.

Actually, the question about the proper interpretation of the 10 μs front time of this waveform has little practical significance. This waveform is used as a long-duration unipolar stress. On the scale shown in Fig. 5-8, details of the leading edge of the waveform are barely discernible.

For waveforms with a duration that is much greater than the front time, the exact time at which the peak occurs is impossible to determine from an oscilloscope photograph or plot because the voltage appears nearly constant in the vicinity of the peak. For example, the 10/1000 μs waveform has a width of about 44 μs at 98% of the peak value, about four times larger than the front time. Therefore, the time of the peak cannot be used to specify the front time of the waveforms with a long duration, such as the 10/1000 μs waveform.

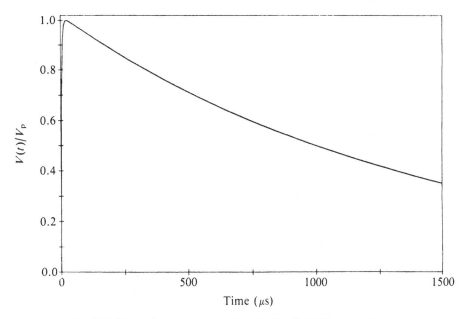

Fig. 5-8. *Plot of the open-circuit voltage in the 10/1000 μs waveform.*

There is no single specification for the output impedance of the 10/1000 μs voltage waveform. Various overvoltage generators for use in a laboratory that are in commercial production have output resistances between 10 and 600 Ω. The IEEE Standard for Gas Tube Protectors, C62.31-1984, calls for a 10/1000 μs waveform with a peak current as large as 500 A. If such a current is to be obtained from a generator with a peak open-circuit voltage of 1.5 kV, the effective output resistance of the generator is 3 Ω.

As expected from its origins to simulate reasonable "worst-case" overvoltages in the telephone environment, the 10/1000 μs waveform is commonly used to test equipment that is to be connected to the telephone network. The 10/1000 μs waveform is also commonly used to test avalanche diodes that are intended to be exposed to transient overvoltages.

8. 10/700 μs Waveform

The 10/700 μs waveform is specified by The International Telegraph and Telephone Consultative Committee (CCITT) Recommendation K17 for use in testing overvoltage protection of repeater amplifiers. The waveform is defined by the schematic of a surge generator shown in Fig. 5-9. The nominal values of the components in this circuit produce a front time of about 9.2 μs, which is determined by 1.67 times the difference between the 30% and 90% points on the leading edge of the waveform, and a duration

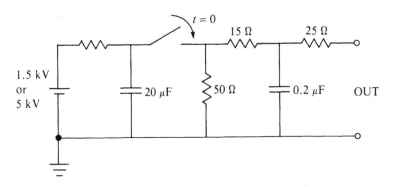

Fig. 5-9. *Surge generator for 10/700 μs waveform.*

of 722 μs. There are no tolerances given in CCITT K17, but the 8% error in front time and the 3% error in duration seem reasonable.

A simple expression for the 10/700 μs voltage waveform is $V(t)$, as given by Eq. 7.

$$V(t) = AV_p\{1 - \exp(-t/\tau_1)\} \exp(-t/\tau_2) \tag{7}$$

The parameters in Eq. 7 have the following values:

$$\tau_1 = 4.919 \ \mu s$$
$$\tau_2 = 827.6 \ \mu s$$
$$A = 1.163$$

In Eq. 7, t is the time ($t \geq 0$) and V_p is the peak value of $V(t)$.

CCITT K17 specifies a 40 Ω output impedance for this surge generator (which is the sum of the 15 Ω and 25 Ω resistors), so that a peak open-circuit voltage of 1.5 kV corresponds to a peak short-circuit current of 37.5 A.

9. FCC Part 68

Equipment that is sold in the United States must comply with Federal regulations. The ability of telephone terminal equipment to withstand transient overvoltages is specified in Title 47 of the Code of Federal Regulations, Part 68, Section 302 (cited as 47 CFR 68.302).

The 10/560 μs metallic voltage surge has a *maximum* front time of 10 μs and a *minimum* time to half-value of 560 μs. The word "metallic" indicates that the surge is to be applied between two nongrounded conductors, which is called *differential mode* in electronics. The peak voltage is 800 V, and the peak current must be at least 100 A.

The $\overline{10}/160\,\mu s$ longitudinal voltage surge has a *maximum* front time of $10\,\mu s$ and a *minimum* time to half-value of $160\,\mu s$. The word "longitudinal" indicates that the surge is to be applied between earth ground and all of the nongrounded conductors, which is called *common mode* in electronics. The peak voltage is $1.5\,kV$, and the peak current must be at least $200\,A$.

The longitudinal voltage surge specified by the FCC for the mains connection of telephone equipment has a *maximum* front time of $2\,\mu s$ and a *minimum* time to half-value of $10\,\mu s$. The peak voltage is $2.5\,kV$, and the peak current must be at least $1\,kA$.

The front time is used in a nonstandard way in the Code of Federal Regulations. The actual expression is "rise time to crest." The term is not defined or explained further in the Code, but the confusion over the proper interpretation of the $10\,\mu s$ front time of the $10/1000\,\mu s$ waveform comes to mind.

10. HEMP Test Waveform

It is difficult to specify a voltage or current waveform that is a reasonable simulation of EMP from nuclear weapons. There is a standard waveform for the electric and magnetic fields in free space during HEMP, which was given in Chapter 2. However, the current in a long cable will have a different waveshape from that of the electromagnetic field illuminating the cable.

As noted in Chapter 2, there is no standard waveform for electromagnetic fields during EMP from surface bursts of nuclear weapons. Therefore the appropriate EMP test waveform is system-specific.

There are two overstress test waveforms that are commonly used to simulate NEMP. One is a damped oscillatory waveform with a frequency of $1\,MHz$, a $10\,ns$ rise time, a time to half-value of $600 \pm 200\,ns$, and a peak value of $12\,kV$. The nominal output impedance of the generator is $12\,\Omega$, so the peak short-circuit current is $1\,kA$.

NATO has specified a $5/200\,ns$ unipolar test waveform to simulate NEMP.

11. Constant *dV/dt* Waveform

For tests of the time required for a spark gap to conduct, it is conventional to use a voltage waveform that has a constant rate of rise, dV/dt. Before the spark gap conducts, it behaves like an ideal open circuit, so there is no loading of the generator. When the spark gap conducts, the magnitude of the voltage across the gap decreases suddenly, and it approximates a short circuit, as described in Chapter 7. The voltage across the gap must still be increasingly linearly with time when the gap conducts. One is not interested in the behavior of the voltage and current after the gap conducts, so only the rate of rise and the maximum voltage at the end of the linear rise need to be specified. If the maximum voltage is too small, the spark gap will not conduct while the waveform is increasingly linearly with time, and the measurement will be difficult to interpret.

Common values of dV/dt are 0.1, 0.5, 1, 2, 5, 10, and 100 kV/μs. For very fast spark gaps, it is conventional to use a dV/dt value of 1000 kV/μs, which is equivalent to 1 kV/ns. Additional information on testing spark gaps is contained in ANSI/IEEE C62.31.

12. Constant *dI/dt* Waveform

For tests of semiconductors with currents of about 10 A or less, it is simple to use a controlled current source whose current waveform is known, and measure the voltage across the device under test. In this way, devices with very different V-I characteristic curves can be compared. Standler (1984) suggested a linear ramp with a slope of about 1 A/μs. This small slope compared to most surges makes the effect of inductance in the package negligible. To prevent device destruction due to heating, a low duty cycle should be used (e.g., ramp for 10 μs, followed by 25 ms of zero current). For a device with a constant 100 V across its terminals, the steady-state power input from such a test waveform would be only about 0.2 W.

D. ENERGY TRANSFER

The selection of an appropriate test waveform is discussed in Chapter 23. One of the most important considerations in selecting a test waveform is the energy transferred to the device under test. The energy transferred to a load can be calculated from the equations for the overstress test waveforms presented earlier in this chapter. Calculations are presented below for two different types of loads: (1) a resistive load, and (2) a device that can be modeled as a constant voltage across its two terminals when it is conducting. The second type of load is a crude approximation for the behavior of several types of devices used to protect circuits from overvoltages.

The circuit for calculating the energy transferred to a resistive load is shown in Fig. 5-10. The energy, W, transferred to the load, R_L, is given by

$$W = \frac{1}{(R_{out} + R_L)} \frac{R_L}{(R_{out} + R_L)} \int_0^\infty V^2(t)\, dt$$

If the value of $\int V^2(t)\, dt$ is known, it is a simple matter to calculate the energy transferred to any load resistance, R_L. The value of the integral for several of the overvoltage test waveforms discussed above is given in Table 5-1.

The energy, W, transferred to a resistive load by a current waveform,

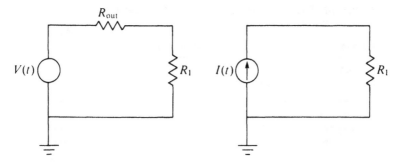

Fig. 5-10. Schematic diagram for calculation of the energy transfer to a resistive load.

using the circuit shown in Fig. 5-10, is given by

$$W = R_L \int I^2(t)\, dt$$

Again, it is a simple matter to compute the value of W when the value of the integral is known. The value of $\int I^2(t)\, dt$ is given for the 8/20 μs waveform in Table 5-1.

It is a simple matter to calculate the values of the integral for other peak voltages or currents than those given in Table 5-1. The value of the integral is proportional to V_p^2 or I_p^2, as shown by the examples for several waveforms in the table.

For convenient comparisons of the relative energies in these waveforms, Table 5-2 gives the total energy transferred to a resistive load. A 1 Ω resistive load was selected for use in Table 5-2, because the voltage across a 1 Ω resistance is reasonable for all of the transient currents. The maximum

TABLE 5-1. Values of Integrals Required for Calculation of Energy Transferred to a Resistive Load: General Case

Waveform	Peak	$\int V^2\, dt$ (V^2 s)	$\int I^2\, dt$ (A^2 s)
8/20 μs	3 kA		110
8/20 μs	500 A		3.0
100 kHz ring	6 kV	90	
100 kHz ring	1 kV	2.5	
1.25 MHz ring*	2.75 kV	29	
EFT	4 kV	0.6	
10/1000 μs	1 kV	720	
10/1000 μs	0.6 kV	260	

* SWC ring wave with $\tau_2 = 15\ \mu$s.

TABLE 5-2. Energy Transferred to a Resistive Load: Specific Case

Waveform	Peak	$R_{out}(\Omega)$	W in 1 Ω (joules)	max(W) in R_L (joules)
8/20 μs	3 kA	—	110	—
8/20 μs	500 A	—	3.0	—
100 kHz ring	6 kV	12	0.53	1.9
100 kHz ring	6 kV	30	0.093	0.75
100 kHz ring	1 kV	12	0.015	0.052
100 kHz ring	1 kV	30	0.0026	0.021
1.25 MHz ring*	2.75 kV	150	0.0013	0.048
EFT	4 kV	50	0.0002	0.003
10/1000 μs	1.0 kV	10	6.0	36
10/1000 μs	0.6 kV	10	2.2	13

*SWC ring wave with $\tau_2 = 15\ \mu$s.

value of energy transferred to a resistive load occurs when the value of the load resistance, R_L, and the output resistance, R_{out}, are equal. For this special case,

$$\max(W) = \int V^2(t)\, dt \Big/ (4R_{out})$$

Notice that the energy deposited in a 1 Ω resistance ranges over almost 6 orders of magnitude. Because of its long duration, the 10/1000 μs waveform can transfer larger amounts of energy than waveforms of shorter duration and the same peak voltage. Similarly, the very short duration of the EFT waveform provides a very small energy transfer. The peak currents of the 8/20 μs waveforms are larger than the peak short-circuit currents of other waveforms listed in Table 5-2, so the energy deposited by the 8/20 μs waveform with a 3 kA peak current in a 1 Ω resistance is larger than that deposited by other waveforms in this table.

It is also of interest to calculate the energy that is transferred to a load that has a constant magnitude of voltage across it, when the load is conducting. This type of load is a crude approximation for the behavior of several types of components that are used to protect circuits from overvoltages (e.g., varistors, avalanche diodes). The circuit used to do this calculation is shown in Fig. 5-11. The parameter V_L is the constant *magnitude* of voltage across the conducting device, which is always positive. In Fig. 5-11a, the device is not conducting, because the magnitude of the source voltage is less than V_L, and no power is dissipated in the device. In Fig. 5-11b, the device is conducting. The total energy, W, transferred to the

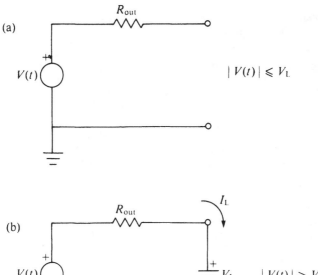

Fig. 5-11. *Circuit for calculation of energy in a device with a constant magnitude of voltage.*

device is given by

$$W = \int P(t)\, dt$$

where $P(t)$ is the power in the device as a function of time. The power is given by

$$P(t) = \begin{cases} 0 & |V(t)| < V_{\mathrm{L}} \\ V_{\mathrm{L}}[V(t) - V_{\mathrm{L}}]/R_{\mathrm{out}} & V(t) > V_{\mathrm{L}} > 0 \\ V_{\mathrm{L}}[-V(t) - V_{\mathrm{L}}]/R_{\mathrm{out}} & V(t) < -V_{\mathrm{L}} < 0 \end{cases}$$

The third line in the equation for $P(t)$ is used when the overvoltage, $V(t)$, is oscillatory or when either V_{p} or I_{p} is negative. The value of V_{L} remains positive even in this third case.

When the overstress waveform is a current, the energy, W, transferred to

TABLE 5-3. Energy Transferred to a Load with Constant Voltage: Specific Case

Waveform	Peak	R_{out} (Ω)	W in 500 V load (joules)
8/20 μs	3 kA	—	26
8/20 μs	500 A	—	4.4
100 kHz ring	6 kV	12	0.98
100 kHz ring	6 kV	30	0.39
100 kHz ring	1 kV	12	0.025
100 kHz ring	1 kV	30	0.0010
1.25 MHz ring*	2.75 kV	150	0.036
EFT	4 kV	50	0.0017
10/1000 μs	1.0 kV	10	11
10/1000 μs	0.6 kV	10	0.70

*SWC ring wave with $\tau_2 = 15\ \mu$s.

a load with a constant voltage, V_L, is given by

$$W = \int I(t)V_L\ dt = V_L\,\Delta Q$$

It is assumed that the current source can always supply a voltage V_L.

The results of these calculations for energy transferred to a load with constant voltage are given in Table 5-3. The value of V_L was chosen to be 500 V, which is typical of some overvoltage protection devices for use on the mains with a nominal voltage of 120 V rms.

E. FREQUENCY SPECTRA

It has been traditional among engineers engaged in research on transient overvoltages to consider the phenomena only in the time domain. However, electrical engineers in general do most of their analysis in the frequency domain. This section explains the reasons for this difference in choice of domains and discusses circumstances for which frequency domain analysis of transient overvoltages would be appropriate.

The goal of mathematical analysis is to present a description of a physical situation in the simplest terms. Expressions can be converted between the time and frequency domains by using the Fourier transforms. There is no question that an accurate representation can be made in either domain. The choice of the independent variable, time or frequency, depends only on which provides the simpler or more convenient analysis.

For unique, nonrepetitive phenomena, the time domain often provides a simpler description than the frequency domain. For example, a single, narrow rectangular pulse that is so simple in the time domain is represented in the frequency domain by a wide spectrum of sinusoidal waves that interfere destructively, except during the duration of the pulse.

For periodic phenomena, there is no doubt that the frequency domain provides a simpler description than the time domain. For example, the response of an electrical circuit or system that is driven by a continuous sinusoidal voltage or current source can be neatly presented in a Bode plot. Solutions for electrical circuit problems are often easier in the frequency domain, where Laplace or Fourier transforms can be used.

Transient overvoltages tend to resemble a narrow, rectangular pulse more than a continuous sinusoid, so one would expect time domain analysis of transient overvoltages to be simpler than analysis in the frequency domain. Further, calculations of important variables such as charge and energy transfer are simpler in the time domain than in the frequency domain.

There is another reason to prefer time domain analysis for transient overvoltages. Measurements in high-voltage laboratories are nearly always done in the time domain, by using either oscilloscopes or digitizing voltmeters. This topic is discussed further in Chapter 23, on laboratory technique.

There is one situation in which frequency domain analysis of transient overvoltages is useful. Filters (usually low-pass) are often used as part of protective measures from transient overvoltages. Sometimes the filters are included deliberately; sometimes they occur by virtue of parasitic inductance and capacitance. These topics are considered further in Chapters 13, 17, and 19. For this reason, it is useful to have frequency spectra for standard overstress test waveforms. Even if the reader has no interest in calculations in the frequency domain, a look at these spectra will give a general view of the frequency band for which a filter should have appreciable attenuation.

The analytical expressions for the Fourier transforms of the relations given above for standard overstress test waveforms are simple to obtain, with the exception of the 100 kHz ring wave. The transform, $F(\omega)$, for all of the double exponential relations is given by Eq. 8. The values of the parameters A, τ_1, and τ_2 in Eq. 8 are found earlier in this chapter in the presentation of the analytical relation for each overstress waveform.

$$F(\omega) = AV_p \left[\frac{1}{\dfrac{1}{\tau_2} + j\omega} - \frac{1}{\dfrac{1}{\tau_1} + \dfrac{1}{\tau_2} + j\omega} \right] \tag{8}$$

The damped cosine function that is used for the 1.25 MHz ring wave, Eq. 4, has the Fourier transform given by Eq. 9. In Eq. 9, the parameter ω_0 denotes the angular frequency of oscillation of the time domain waveform;

ω is the independent variable in the frequency domain. Equation 9 can also be used as an approximate Fourier transform of the 100 kHz ring wave when $A = 1.59$, $\tau_1 = 0.53\ \mu s$, $\tau_2 = 9.78\ \mu s$, and $\omega_0 = 2\pi 10^5$ rad/s. This approximation and its error was discussed in Standler (1988a).

$$F(\omega) = A\omega_0 V_p \left[\frac{1}{\left[\dfrac{1}{\tau_2} + j\omega\right]^2 + \omega_0^2} - \frac{1}{\left[\dfrac{1}{\tau_2} + \dfrac{1}{\tau_1} + j\omega\right]^2 + \omega_0^2} \right] \qquad (9)$$

The Fourier transform for the 8/20 μs waveform is given by Eq. 10. The values of the parameters A and τ are given with the discussion of Eq. 1.

$$F(\omega) = \frac{6AI_p}{\left[\dfrac{1}{\tau} + j\omega\right]^4} \qquad (10)$$

Plots of the magnitude of the Fourier transform for several of the standard overstress test waveforms are shown in Fig. 5-12. With all of the previous discussion of front times of the order of a microsecond, one might

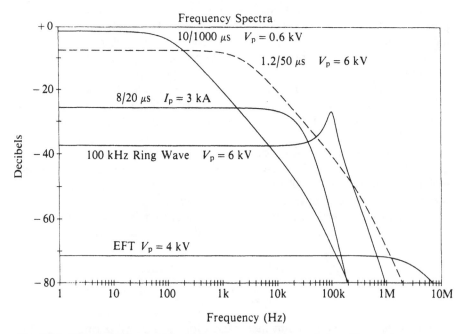

Fig. 5-12. *Plot of Fourier transform of various nominal waveforms. The peak voltage or current is labeled with the name of each waveform. The zero dB reference is 1 V or 1 A. The spectrum of the 1.2/50 μs waveform is shown with a dashed line for clarity.*

get the impression that the spectra of overstresses have a maximum in the vicinity of 1 MHz. From this erroneous viewpoint, it seems surprising how much energy exists at relatively low frequencies for the standard overstress test waveforms. To understand why these spectra appear as they do, consider the idealized pulse, the Dirac delta function, $\delta(t)$, that is defined by

$$\delta(t - t_0) = \begin{cases} 0 & t \neq t_0 \\ \infty & t = t_0 \end{cases}$$

The frequency spectrum of $\delta(t)$ is a nonzero constant, which indicates that the spectral density is independent of frequency. The behavior of the spectrum of the electrical fast transient (EFT) resembles that of the delta function below 1 MHz. The 5 ns rise time of the EFT is much slower than the discontinuity presented by the delta function, and so the spectral density of the EFT decreases as frequency increases above about 3 MHz. The other overstress test waveforms shown in Fig. 5-12 have longer rise times and durations than the EFT, so their spectra have less resemblance to that of the delta function.

The approximate spectrum for the 100 kHz ring wave has a peak at 100 kHz, as expected. The peak is not narrow, because of the exponentially decaying amplitude of the ring wave. In fact, this waveform is not periodic, because the amplitude of the oscillation changes with time.

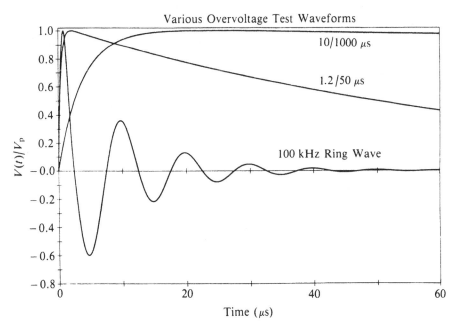

Fig. 5-13. *Three different overvoltage test waveforms.*

F. CONCLUSION

A plot of three different overvoltage test waveforms is shown in Fig. 5-13. The EFT waveform is not shown in Fig. 5-13, because, on this scale, it would be a vertical line at $t = 0$. The 8/20 μs waveform is also not shown in Fig. 5-13, because it is a current waveform, not a voltage waveform.

The discussion of which overstress test waveform to select for a given application is in Chapter 22.

6

Overview of Surge Protection

This chapter contains a general overview of circuits for protection against damage by electrical overstresses. This overview is helpful before characteristics of specific devices and details of applications are discussed in the following chapters.

A. BLOCKING OR DIVERTING

Protection from surge currents can be attained by either (1) blocking or limiting them with a large series impedance or (2) diverting them with a small shunt impedance. The two methods can be, and often are, used together to provide attenuation of overvoltages by voltage division.

Example of components for limiting currents include inductors and resistors. Optical isolators and isolation transformers block common-mode currents but not differential-mode currents. Examples of commonly used components to divert surge currents include spark gaps, varistors, and avalanche diodes. These components will be discussed further in Chapters 7 through 14. This chapter contains a general overview of types of surge protective devices.

It is desirable to use elements that have a nonlinear V-I relation for protection circuits. In that way one can have devices that present a small series impedance or a large shunt impedance during *normal* system operation, which is desirable to minimize the effect of the protective circuit on normal system operation. However, during surges the nonlinear elements have a large series impedance or a small shunt impedance, which is desirable to effectively block or divert the surge. The use of highly nonlinear components enables effective surge protective circuits to be designed that have minimal effect on normal system operation.

There are many nonlinear shunt elements, which are discussed in Chapters 7 through 11. It is conceivable that a nonlinear series component could be developed that would present a small impedance during normal signal voltages but that would become a large impedance during an overstress. There are no known nonlinear series elements, except for PTC resistors which are discussed in Chapter 12.

B. GENERAL FORM

Circuits for protection against electrical overstresses can be divided into three broad classes which describe how the circuit functions:

1. voltage discrimination
2. frequency discrimination
3. state discrimination

1. Voltage Discrimination

The most common kind of circuits for protection against electrical over-stresses discriminate between normal signals and overstresses on the basis of voltage levels. A nonlinear shunt component is used that is essentially nonconductive at normal signal voltages and is highly conductive at larger voltages that are encountered during transient overstresses.

2. Frequency Discrimination

In this discussion the term "signal frequency" refers to the band of frequencies that is present during the normal operation of the circuit. The term "transient frequencies" refers to the range of frequencies in the overvoltage. In other words, the "signal frequency" is the "good information," and the "transient frequencies" contain the undesirable waveform that causes damage or upset.

When the range of signal frequencies is appreciably different from the overstress frequencies, it is convenient to use frequency discrimination in the circuit that protects against overstresses. Economical filters usually do not provide enough attenuation of the overstresses, as discussed in Chapter 13, so voltage discrimination circuits will also be required. Frequency discrimination is commonly used to protect circuits and systems from transient overstresses that propagate on power buses, which normally operate at frequencies between dc and 400 Hz. Frequency discrimination is also useful to protect radio frequency (rf) circuits. The rf signal may have a narrow bandwidth; furthermore, the frequency spectrum of many transient overstresses do not contain much energy at frequencies above a few tens of megahertz.

3. State Discrimination

The third class of circuits for protection against overstresses is state discrimination. There are two states: (1) the normal operation of the system, and (2) the state when overstresses are present or anticipated. During the overstress state, the protected system is instructed to ignore information that is present on the input data lines. This helps to prevent temporary malfunctions (upsets) that arise when the overstress corrupts the input data or interrupts power to the system. State discrimination techniques are discussed further in Chapter 21, on upset.

C. GENERAL OVERVOLTAGE SUPPRESSION CIRCUIT

The general form of a suppression circuit is shown in Fig. 6-1. The series impedance, Z_1, is usually a resistor in circuits with small signal currents. However, if the signal frequency is less than that of the transient, Z_1 may contain an inductor. If the frequency of the signal is greater than that of the transient, Z_1 may be a capacitor. In power circuits, Z_1 can be the parasitic resistance and inductance of the line as well as that of a fuse or circuit breaker.

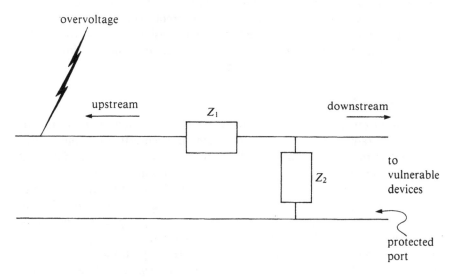

Fig. 6-1. *General overvoltage suppression circuit.*

The shunt impedance, Z_2, is usually a nonlinear component that normally has a large impedance but that during an overstress has a small impedance (e.g., avalanche diode, spark gap, or varistor). However, Z_2 may be a capacitor or inductor that forms a low-pass or high-pass filter with Z_1, provided that the signal frequency is appreciably different from the transient frequencies.

1. Voltage and Current Division

The circuit of Fig. 6-1 acts like a voltage divider. When properly designed, this circuit will provide negligible attenuation for the signal but will have appreciable attenuation of the transient. If nonlinear shunt devices are used for Z_2, the voltage across the protected port should be clamped to a value that will not cause damage to the vulnerable devices downstream.

The circuit of Fig. 6-1 also acts like a current divider: transient current that would otherwise pass through the vulnerable devices is shunted away by Z_2. When the shunt element is nonconducting, there is no current division. During the overstress the shunt element should have a much smaller resistance than elements downstream, so the shunt element protects by diverting current away from the devices downstream.

The actions of voltage division and current division are complementary. The two mechanisms work simultaneously to protect vulnerable devices downstream from damage by transient overstresses. If one thinks of the transient overstress as a voltage, then voltage division may be more appealing. If one thinks of the transient overstress as a surge current, then current division may be more appealing.

Overvoltage protection devices for use on ac supply mains are commonly called an *arrester* in the United States. This is a misleading name, because the device does *not* arrest or block the surge current. The German word for overvoltage protection device is *ableiter,* which translates as "diverter." The British also use the word "diverter." This is an accurate description of the current division in the protective device.

2. Parasitic Elements

Parasitic inductance is often overlooked by circuit designers who are not familiar with radio frequency circuit design practices. In fact, the way a component is mounted on a printed circuit board can be more important than the properties of the component itself when a transient with a rise time of less than a microsecond is encountered. When the transient has frequencies greater than that of the signal, inductance *must be minimized* in the *shunt* path that includes components such as Z_2. Specific suggestions about how to accomplish this are given in Chapter 15.

However, inductance in *series* with the line (e.g., parasitic inductance of Z_1) is a desirable feature, provided that the signal frequencies are sufficiently small that the voltage drop across the inductance is negligible during normal operation.

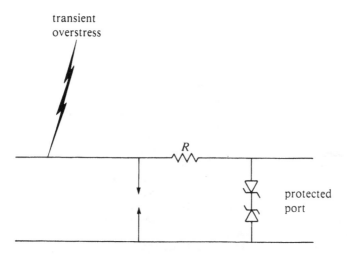

Fig. 6-2. *Hybrid protection circuit.*

The parasitic capacitance of some nonlinear shunt devices (e.g., varistor or avalanche diode), which can be of the order of 1000 pF, automatically forms a low-pass filter when these devices are used for Z_2. In this way, protection circuits that discriminate on the basis of voltage levels may also be low-pass filters.

3. Hybrid Circuits

It is difficult to design a single-stage circuit, as shown in Fig. 6-1, that can protect integrated circuits from transient overstresses with peak currents greater than about 50 A. When such large transients are anticipated, one usually forms a hybrid (two-stage) circuit, such as shown in Fig. 6-2. The first stage is designed to remove most of any large transient and protect the second stage. The purpose of the second stage is to protect the load. The two-stage circuit shown in Fig. 6-2 appears to have been first described by Bodle and Hays (1957) and has become the standard circuit for protection of analog and digital data lines. This circuit is discussed in detail in Chapter 17.

D. OVERVIEW OF NONLINEAR COMPONENTS

The next eight chapters describe the properties of various components that are useful in circuits that protect against overstresses. Table 6-1 presents a very brief overview of the major families of components. These components can be organized in three groups: clamps, crowbars, and isolators. A *clamp* has an approximately constant voltage across it during conduction of surge current. A *crowbar* is a device that changes state from an insulator to a

TABLE 6-1. Components

Family	Advantages and Disadvantages
	Clamps
Metal oxide varistors (discussed in Chapter 8)	Fast response (<0.5 ns)
	Large energy absorption
	Can safely conduct large currents (1 kA for 20 μs)
	Inexpensive
	Large parasitic capacitance (1 to 10 nF)
Avalanche diodes (discussed in Chapter 9)	Fast response (<0.1 ns)
	Selection of precisely determined clamping voltages (between about 6.8 and 200 V)
	Small maximum allowable current (≤100 A for 100 μs)
	Large parasitic capacitance (1 to 3 nF)
Switching and rectifier silicon diodes (discussed in Chapter 10)	Small clamping voltage (0.7 to 2 V)
	Inexpensive
	Small parasitic capacitance
	Crowbars
Spark gaps (discussed in Chapter 7)	Can safely conduct large currents (5 kA for 50 μs)
	Low voltage in arc mode
	Very small parasitic capacitance (<2 pF)
	Requires large voltage (≥100 V) to conduct
	Can be slow to conduct
	Possible "follow current" (sustained short circuit)
SCRs and triacs (discussed in Chapter 11)	Small voltage (0.7 to 2 V) across conducting switch
	Can tolerate sustained large currents
	Slow to turn on or turn-off (2 μs)
	Possible sustained conduction
	Isolators
Optoisolators (discussed in Chapter 14)	Large isolation voltages (5 kV)
	Good common-mode rejection
	Easy to use to receive data
	Difficult to use to transmit data
	Fast devices (<1 μs switching time) are expensive
Isolation transformers (discussed in Chapters 14; 20)	Large isolation voltages (5 kV)
	Good common-mode rejection
	No attenuation of differential-mode overvoltages
Common-mode filters (discussed in Chapters 13, 19, 20)	*Can* be effective in attenuating short-duration common-mode overvoltages
	Many problems when used alone as SPD

nearly perfect conductor during the overstress. It has the effect of dropping a metal crowbar across the circuit, which is why this type of surge protective devices is called a crowbar. An *isolator* offers a large series impedance to common-mode voltages.

In order to divert surge currents, clamps, and crowbars are connected in shunt with the equipment to be protected. Isolators must be inserted between the source of the overstress and the equipment to be protected.

Part 2

Protective Devices

7

Gas Tubes

A. INTRODUCTION

The oldest transient overvoltage protection circuit in common usage is the carbon block protector in telephone installations. The two carbon blocks were the electrodes of a spark gap. The spacing of blocks was adjusted to give a gap that would conduct at about 600 V. The spark occurred in the air between the two blocks; the gap was not sealed. The gap spacing was between 80 and 150 μm (0.003 to 0.006 inches). This was a very inexpensive protection circuit. However, it had several disadvantages. Following exposure to transients with large energy content, the gap spacing widened owing to erosion of the carbon electrodes. After the gap widened, the gap would no longer conduct at relatively small peak voltages. Thus there was no longer any protection against these transients. Another disadvantage was that after conduction, carbon dust would accumulate between the electrodes and cause erratic conduction during normal operation of the telephone. This erratic conduction was a source of noise.

These disadvantages were overcome by developing a sealed spark gap that used metal electrodes and a mixture of noble gases (neon, argon, etc.). The sealed spark gap was called a *gas tube* to distinguish it from the older carbon block protectors, which were called either a *spark gap* or an *air gap*. The use of carbon block protectors is now obsolete except for minimal-cost applications in protecting telephone company exchanges. Except in the telephone protection community, the word "spark gap" is often used as a generic term to refer to gas tubes.

In many circuits, spark gaps are the principal component for shunting large transients away from vulnerable circuits. Typical miniature spark gaps in a ceramic case can conduct transient current pulses of 5 to 20 kA for 10 μs without appreciable damage to the spark gap. Spark gaps have the smallest shunt capacitance of all known nonlinear transient protection devices at present: a typical value is between 0.5 and 2 pF. This low capacitance makes spark gaps one of the few nonlinear devices that can be used to protect circuits in which the signal frequency is greater than 50 MHz.

The operation and construction of spark gaps has been described by Kawiecki (1971a, 1974), Cohen et al. (1972), Bazarian (1980), and Standler (1988c).

B. *V-I* CHARACTERISTIC

The spark gap is inherently a bipolar device; it makes no difference which electrode has positive charge. Therefore, only one quadrant of the *V-I* curve is needed to understand the behavior of the spark gap. However, some gas tubes (particularly low-voltage neon indicator lamps or dc voltage regulator tubes) have small differences in behavior depending on the polarity of the voltage across the tube. These differences are owing to a coating of barium or strontium on the electrode that is intended to be the cathode. Such details will not be discussed further, because they are irrelevant to transient protection.

The *V-I* characteristic for a representative low-voltage spark gap is shown in Fig. 7-1. Notice that, unlike conventional characteristic curves of electronic devices, the vertical axis is the logarithm of current. A logarithmic axis is necessary to display behavior for currents that span six orders of magnitude. Figure 7-1 shows voltage as the independent variable and current as the dependent variable. This convention for presenting data on electronic devices goes back to the days of vacuum tubes. As discussed in Chapter 5, there is no doubt that current is a more convenient independent variable with devices such as spark gaps that have an approximately constant voltage over a wide range of currents. The author's choice for the format of Fig. 7-1 reflects the usual convention for electronic devices, which is probably more familiar to most readers than is plotting current as the independent variable.

The characteristic curve (Fig. 7-1) is rather complicated. Each segment of the curve is associated with a particular physical process, each of which is discussed below.

We shall follow the characteristic curve in Fig. 7-1 for a slowly increasing potential difference across the gap, starting at point *A*. At point *A* the gap switches from an insulating to a conducting state. The potential at *A* is called the *dc firing voltage of the gap*.

Between *A* and *B* the incremental resistance, dV/dI, is negative; thus this portion of the curve is called the *negative resistance region*. If the gap is

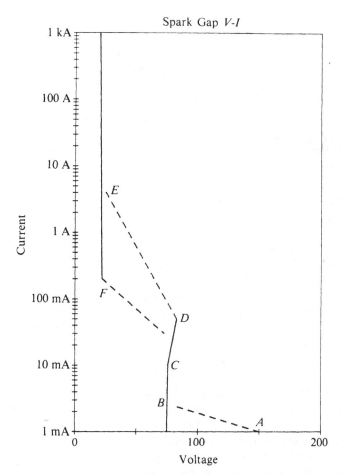

Fig. 7-1. *V-I relationship for a spark gap with a dc firing voltage of 150 V. The significance of each lettered point is discussed in the text.*

operated in this region (e.g., in series with a 1 MΩ resistor and a 200 V dc voltage source), the discharge will flicker. This can produce intermittent noise pulses.

The portion of the curve between points *B* and *C* is known as the *normal glow region*. Voltage regulator tubes are operated in this region, since the voltage across the tube is approximately independent of the current. The light that is emitted from the tube comes from a *cathode glow*, a thin region of the excited gas atoms that cover part of the cathode surface. The area of the cathode glow is approximately proportional to the current.

When the glow covers nearly the entire cathode surface and the current increases, the discharge enters the *abnormal glow regime*. This is the portion of the characteristic curve between *C* and *D*, where *dV/dI* is positive.

Electron-ion pairs are accelerated in the intense electric fields that exist between the electrodes when the spark gap is operated in either the normal or abnormal glow region. When the charged particles attain sufficient energy, they can form additional ion pairs by collision with either neutral atoms or ions. This process is called an *avalanche*. The avalanche is limited by the formation of space charge layers and by recombination of electron-ion pairs. The light from the discharge in either the normal or abnormal glow region contains spectral lines that are characteristic of the atoms of gas and ions. For neon, the color is orange-red; for argon, it is blue-violet.

For typical spark gaps (with dc firing voltages between 90 and 300 V), the maximum current in the glow region is usually between 0.2 and 1.5 A. When the current is increased to a sufficiently large value, the discharge abruptly enters the *arc* regime, at point *E*. The value of this glow to arc transition current is not precisely reproducible, even for the same gap.

In the arc regime the light comes from excited metal atoms that have been removed from the cathode by positive ion bombardment. For a tube with nickel electrodes, the light from the arc appears blue. Positive ions are accelerated in the electric field between the electrodes and collide with the cathode in both the abnormal glow and arc regimes. The cathode material is blasted away by this bombardment, a process that is known as *sputtering*. This is the source of the black deposits that are seen on the walls of old neon lamps and spark gaps with glass cases. At greater currents the cathode material is removed more rapidly. The electrodes are eroded and consumed by this process. One result is that the gap spacing (and hence the dc firing voltage) increases after appreciable sputtering has occurred, and the tube may no longer perform satisfactorily. Therefore, all spark gaps are consumable components. However, by making the electrodes suitably massive and taking care to deal with degradation of insulation by sputtering, one can make spark gaps that can conduct hundreds of large transients (e.g., 8/20 μs waveform with a peak current of 10 kA).

The potential across the spark gap is essentially constant at about 20 V (independent of current) when the tube is in the arc regime. For tubes with different gases and at different pressures, the arc voltage can be between 10 and 30 V. In overvoltage protection applications it is highly desirable to operate the tube in the arc regime, because the voltage is clamped at a relatively small value.

When the current is reduced to the *arc extinguishing current*, shown as point *F* in Fig. 7-1, the arc stops and is replaced by a glow discharge. The value of the extinguishing current is usually between 0.1 and 0.5 A. Notice that there is hysteresis in Fig. 7-1: the current to initiate an arc is greater than the minimum current that will sustain the arc. Therefore, each of the two paths between the arc and glow regions, as shown in Fig. 7-1, can be traversed in only one direction.

When a tube is operated in the arc regime, the gas and the metal electrodes becomes very hot. This heating can cause failure of the seal between the insulating case (glass or ceramic) and the electrodes. Alternatively, the large internal pressure of the hot gas can shatter the case. A ceramic case can probably withstand higher temperature and pressure than a glass case. Many spark gaps use Kovar™ (an iron-nickel alloy) sealed to an alumina ceramic case. The thermal expansion coefficient of these two materials is nearly the same, which reduces mechanical stress on the seal during operation. The electrode metal is attached to the Kovar seal.

If atmospheric air at near sea-level pressure leaks into the gas tube, the dc firing voltage of the tube may increase by as much as a factor of 20. In many situations this will destroy the tube's protective function. If the tube shatters or explodes, of course, the protective function is lost. Some tubes have a metal with a relatively low melting point inside the gap, which will cause the gas tube to fail as a short circuit if it is operated in the arc regime for a prolonged period (e.g., $I \ \Delta t = 20$ to $50 \, \text{A s}$). Such gas tubes are identified by the manufacturer as being "fail-safe." A better label would be "fail-short," which is discussed in Chapter 16.

C. FOLLOW CURRENT

If the normal voltage or current source can maintain the discharge, then the gas will not return to the nonconducting state after the transient is completed. This condition, which is known as *power follow* or *follow current*, can ruin the spark gap and damage the wires and the source of follow current. Follow current can maintain conduction in either the arc or glow regime. The source of the follow current must be able to supply at least 60 V to maintain operation in the glow regime. The source of the follow current must be able to supply at least 20 V across the spark gap *and* a current greater than the arc-extinguishing current (shown as point F in Fig. 7-1) in order to maintain operation in the arc regime.

1. Arc Regime

When spark gaps are connected across the ac supply mains, one should not blindly rely on the periodic zero crossings of the sinusoidal voltage to extinguish follow current in the gap. It is true that the magnitude of the voltage on the 120 V rms 60 Hz mains is less than 20 V for 0.63 ms during each zero crossing. However, thermionic emission of electrons from hot electrodes may sustain the arc during these brief zero crossings.

When spark gaps are connected across a dc power circuit that has a potential greater than about 15 V and a maximum current greater than about 50 mA, follow current in the arc regime is possible.

2. Glow Regime

Most treatments of follow current concentrate on follow current in the arc regime and ignore the possibility of follow current in the glow regime.

Follow current is also possible in the glow regime. The minimum current necessary for operation in the glow regime is of the order of 10 µA; the exact value depends on gas pressure and composition, gap spacing, and other details. The voltage across the gap during operation in the glow regime is usually between about 50 and 100 V. Operation of gaps with large values of follow current in the glow regime can dissipate appreciable power and cause thermal damage to the spark gap. However, follow current of the order of a milliampere in the glow regime that persists for many months or years can also damage the spark gap by sputtering. Sputtering can bury active materials on the surface of the electrodes beneath a layer of electrode material. The active materials (e.g., alkali halides such as KCl, oxides of alkaline earths such as MgO) are necessary for fast response of the spark gap.

There are two sets of conditions that assure that spark gap will not continue to operate in the glow regime after a transient overvoltage—one for dc sources, the other for ac sources. For dc sources, the magnitude of the normal system voltage should be less than 50 V. For ac sources, the dc firing voltage of the spark gap should be greater than the peak system voltage. The glow will extinguish during a zero crossing of the ac source, and the gap will not conduct during the next half-cycle.

3. Prevention of Follow Current

When a spark gap is connected across a power supply (or other circuit) that can furnish follow current, one must prevent follow current. There are three simple ways to do this, which are shown in Fig. 7-2a.

A power resistor in series with a spark gap, as shown in Fig. 7-2a, is often suggested to prevent follow current. Although this technique can prevent follow current, it also greatly increases the clamping voltage owing to the voltage drop across the resistor. The IR drop can be quite substantial; for example, consider $R = 1\ \Omega$ and $I = 5\ \text{kA}$. In many situations the value of R needs to be much greater than $1\ \Omega$, a condition that is even less favorable to protecting a circuit from transient overvoltages. A series resistor is unsuitable for preventing follow current in the glow regime. Including a resistor in series with a spark gap is definitely *not* a recommended technique.

A good way to interrupt follow current is to use a varistor in series with a spark gap, as shown in Fig. 7-2b. This method is discussed in the next chapter. Use of a series varistor might be adequate to prevent dc follow current in the glow regime.

A fuse or circuit breaker may be included between the spark gap and the source of the follow current, as shown in Fig. 7-2c. The fuse should be specified to interrupt an arc follow current. The disadvantage of this method is that it causes upset of devices downstream from the spark gap by disconnecting power. A fuse is unsuitable for interrupting small levels of

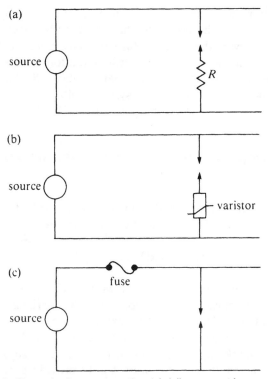

Fig. 7-2. *Three simple ways to extinguish follow current in a spark gap.*

follow current in the glow regime. The fuse could be combined with the varistor as an additional precaution.

Another method of interrupting follow current in a spark gap is to put a normally closed relay upstream from the spark gap and relay coil, as shown in Fig. 7-3 (De Sousa, 1967). Large follow current through the relay coil will cause the relay contacts to open, which extinguishes the follow current. There will then be no current in the relay coil, so the contacts will return to

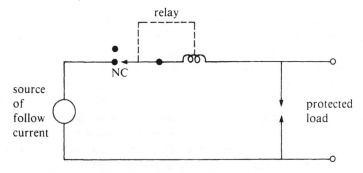

Fig. 7-3. *Automatically reclosing follow current interrupter.*

the closed position, and the system resumes normal operation. That this circuit briefly interrupts the power to the load may be viewed as a serious disadvantage. However, this disadvantage is really owing to the arc in the spark gap and not the method of extinguishing the follow current. When the spark gap is operating in the arc mode, there is about 20 V across the load, essentially independent of the usual system voltage. Therefore, the arc discharge has already interrupted the power to the load.

D. SPECIFICATION OF SPARK GAP PARAMETERS FOR RELIABLE OPERATION

A spark gap should be selected that will not conduct during normal operation of the system. This is done by requiring that

$$1.2 \times \max(V_p) \le \min(V_f)$$

where $\max(V_p)$ is the peak magnitude of voltage during normal operation, and $\min(V_f)$ is the minimum dc firing voltage of the spark gap. The effects of tolerances should be included in both $\max(V_p)$ and $\min(V_f)$.

There are several suggestions for increasing the reliability of circuits that are protected by spark gaps. Popp (1968) put two spark gaps in parallel for redundancy: if one shattered, the other would provide protection. Palmer (1977, p. 28) and Sherwood (1977, p. 35) both advised that the spark gap with the smallest tolerable dc firing voltage is *not* necessarily the best choice. They both choose 230 V gaps for protection of overhead cable television coaxial lines instead of 90 V gaps. In their situation, mains voltages could be injected into the coaxial line. In effect, these engineers have defined the mains voltages as "normal" on the television cable, so that the spark gap will not conduct during a power cross.

E. TRANSIENT OPERATION

We now consider the transient operation of a spark gap. For rapidly changing waveforms (e.g., dV/dt of $1\,\mathrm{kV}/\mu\mathrm{s}$ or more), the actual firing voltage is observed to be several times greater than the dc firing voltage (Trybus et al., 1979). The dc firing voltage alone is sufficient to produce an electric field between the electrodes that is necessary to get ions and electrons to energies that will produce ionization by collision (a glow or arc discharge). So why does the tube require greater voltages to fire? The time delay comes from two effects: (1) some electron-ion pairs must be created in the gas by random processes, and (2) a finite time is required for these initial charged particles to form an avalanche process. These effects are called the *statistical* and *formative* times, respectively.

The initial creation of electron-ion pairs is made more frequent by introducing a beta-emitting radioactive isotope into the gas tube. Common choices are tritium gas (^{3}H), radioactive krypton gas (^{85}Kr), or radioactive nickel electrodes (^{63}Ni). The initial rate of decay is usually between 0.01 and 5.0 μCi; about 0.1 μCi is typical (1 curie [Ci] $= 3.7 \times 10^{10}$ disintegrations per second). Although one beta particle (electron) is emitted, on the average, every 0.3 ms from a 0.1 μCi source, the secondary electron-ion pairs produced by the collision of the beta particle with atoms of the gas will persist for a much longer time. Thus the beta-emitting isotope provides, indirectly, a source of ions that is continually present. This eliminates the "statistical" time delay.

In recent years, several companies have developed spark gaps that will conduct in less than 1 ns when a voltage is applied with a rate of change of at least 1 kV/ns. Standler (1988c) discussed theoretical limits to the speed of response of spark gaps and various practical ways to make spark gaps conduct faster.

As mentioned in Chapter 5, the response time for spark gaps is commonly measured with a voltage source that has a constant rate of rise, dV/dt. For rates of rise larger than about 0.5 kV/μs, the firing voltage of the spark gap is appreciably greater than the dc firing voltage. The response time, Δt, is given by

$$\Delta t = V_f / (dV/dt)$$

where V_f is the measured firing voltage of the spark gap with a constant rate of rise, dV/dt. A plot of V_f as a function of Δt is shown in Fig. 7-4 for a modern, but not state-of-the-art, spark gap. Lines of constant dV/dt are also shown in Fig. 7-4. The value of Δt decreases as the value of dV/dt increases, up to about 100 kV/μs. As values of dV/dt increase above 100 kV/μs, Δt approaches an asymptotic limit.

The curve in Fig. 7-4 for the gap with a dc firing voltage of 150 V has the following relation for V_f in volts and Δt in seconds:

$$\Delta t = \frac{6.2 \times 10^8}{(V_f - 150)^{5.67}}$$

and

$$4 \times 10^{-8}\,\text{s} \leq \Delta t \leq 10^{-3}\,\text{s}$$

Singletary and Hasdal (1971) found minimum values of Δt between about 1.8 and 3.0 ns at rates of rise of about 1 MV/μs for a dozen different spark gaps with dc firing voltages between 90 and 800 V. There was little correlation between dc firing voltage and the response time. A few models

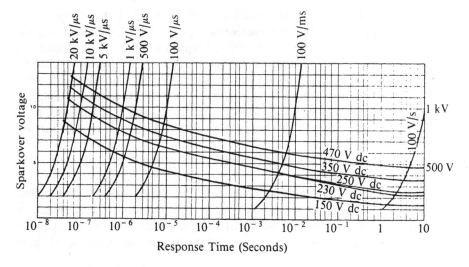

Fig. 7-4. *The firing voltage as a function of response time for spark gaps with a dc firing voltage of 150, 230, 250, 350, and 470 V. This plot is reproduced courtesy of Joslyn Electronic Systems.*

of state-of-the-art spark gaps are claimed by their manufacturer to have a response time of less than 1.0 ns at a rate of rise of 1 MV/μs.

The ratio of V_f to the dc firing voltage is called the "impulse ratio." Spark gaps with values of the impulse ratio less than 3 would be greatly desirable for protecting vulnerable equipment against overvoltages with short rise times.

Spark gaps conduct more quickly when $|dV/dt|$ is increased. Older literature cites dynamic or impulse breakdown voltages for a rate of rise between about 1 and 10 kV/μs. Literature from manufacturers of state-of-the-art "fast" spark gaps gives the maximum breakdown voltage at a rate of rise on the order of 1 MV/μs, a factor of 10^2 or 10^3 greater than that used to specify the old, "slow" spark gaps. Although there is no doubt that the new "fast" gaps conduct more quickly, one should recognize that the old "slow" gaps would be faster responding if they were tested with the new steeper waveform. It is suggested that complete specifications for a spark gap should, among other parameters, specify the maximum breakdown voltage, V_f, at three different rates of rise: 1 kV/μs, 10 kV/μs, and 1 MV/μs. This would allow users to compare performance of devices from various manufacturers.

F. SPARK GAPS

1. Problems with Spark Gaps

Spark gaps are the principal component for diverting very large transient currents: they are rugged, are relatively inexpensive, and have a small shunt capacitance, so they do not limit the bandwidth of high-frequency circuits as much as other nonlinear shunt components. However, there are three major problems that must be considered when applying spark gaps in an effective manner:

1. They *can* be slow to conduct, and all spark gaps require *at least* 60 to 100 V across the gap before the gas will conduct.
2. In *some* situations they are difficult to turn off after the transient has ended (power follow current).
3. The large magnitude of dI/dt when the spark switches from the insulating to conducting state can cause problems.

That the user must consider these problems does not imply that spark gaps are bad or undesirable, but rather that spark gaps must be used carefully.

When transient overvoltages with relatively small magnitudes of dV/dt appear across the spark gap, the gap conducts several microseconds, or more, after the overvoltage has exceeded the dc firing voltage of the gap. When overvoltages with large magnitudes of dV/dt apppear across the spark gap, conduction may occur a few nanoseconds after the overvoltage has exceeded the dc firing voltage of the gap. While the delay time, Δt, of a few nanoseconds seems small, the large magnitude of dV/dt makes the magnitude of $(dV/dt)\,\Delta t$ be typically several times the dc firing voltage. Therefore, there will always be an appreciable remnant that propagates downstream from a spark gap. The remnant may have a peak voltage slightly greater than the dc firing voltage of the gap and a duration of microseconds, or a peak voltage at least several times greater than the dc firing voltage and a duration of a few nanoseconds, depending on the value of dV/dt of the overvoltage. Additional protective devices should be connected downstream, as shown in Chapters 6 and 17, to further attenuate the remnant from the spark gap.

The mechanism for power follow current in a spark gap was discussed earlier in this chapter. Because of the possibility of power follow current, spark gaps alone are not suitable for use on the ac supply mains, dc supply buses greater than about 20 V, or to protect radio transmitters. If spark gaps are to be used in these situations, a current-limiting device must be inserted in series between the spark gap and source of the follow current, as discussed in Chapters 8 and 19.

Another problem with spark gaps is the large magnitude of dI/dt when the gap changes from the insulating to the conducting state. The change in current, ΔI, may be of the order of 10 kA, and Δt may be of the order of 10 ns, which gives dI/dt of the order of 10^{11} A/s. Such large values of dI/dt are not encountered by the typical electronic circuit designer, and it is therefore not surprising that the large values of dI/dt cause unexpected problems. In particular, this large value of dI/dt in an unshielded system may radiate appreciable electromagnetic energy to adjacent cables and systems. This effect has been noted by Lord (1963) and Martzloff (1982). Techniques for coping with these problems are discussed in Chapter 15 (on parasitic inductance and inadvertent transformer effect) and in Chapter 13 (on filters and lossy line).

When the spark gap conducts, it essentially short-circuits the transmission line, and most of the incident transient is reflected rather than absorbed in the spark gap. This is a general problem with all shunt protective devices and is not unique to spark gaps. The problem may be more severe with spark gaps than with varistors and avalanche diodes, because spark gaps can have a smaller impedance during a surge. The use of a nonlinear shunt device, such as a spark gap, can protect the node(s) downstream from where the spark gap is located. However, reflections of the transient may result in continued radiation of energy into other systems and general electromagnetic compatibility problems. Such problems might be solved with lossy lines, which is discussed in Chapter 13.

Spark gaps may be compared to a potent medicine that has adverse side effects. When used appropriately, such medicine can be life-saving; used inappropriately, it only increases suffering.

2. Three-Electrode Spark Gaps

Spark gaps are also available with three-electrodes, as shown in Fig. 7-5. In the usual application, one connects the middle electrode to local ground (earth) and the other two electrodes to a pair of signal conductors. The three-electrode spark gap is particularly useful for the protection of a balanced transmission line (e.g., telephone signals, RS-422 computer data

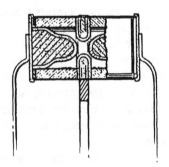

Fig. 7-5. *An exposed view of a three-electrode spark gap (Kawiecki, 1971a).*

communications). The middle electrode of these spark gaps has a small hole that allows plasma (highly conducting gas) from one chamber to reach the other chamber within 0.1 μs. The three-electrode spark gap provides both differential- and common-mode protection in a single component. Similar protection would require three separate two-electrode spark gaps, but there is no simple way to get separate spark gaps to conduct essentially simultaneously. In fact if two independent gaps are used, one may not conduct at all. The three-electrode spark gaps provides superior protection for a balanced line *and* lower component and assembly costs.

Prior to the introduction of three-electrode spark gaps, it was common practice to connect one two-electrode gap between each line and ground. Two two-electrode gaps were used to protect a balanced pair of lines. Bodle and Gresh (1961) found that large differential-mode transients could be produced when one two-electrode gap, but not the other, fired. Most of the original transient overvoltages were essentially common-mode. Use of the wrong type of transient protection components (i.e., two-electrode gaps instead of three-electrode gaps) introduced a new threat, differential-mode overvoltages, to the equipment.

3. Coaxial Spark Gap

Since antennae are usually situated in elevated, exposed locations, they are commonly struck by lightning. Moreover, they can collect appreciable electromagnetic radiation from nearby lightning or EMP from nuclear weapons. It is therefore desirable to protect the electronic circuits that are connected to an antenna.

The most common protective device for antenna lines is a spark gap in a coaxial mounting. The spark gap has a very small shunt capacitance and large resistance in the nonconducting state, which gives the spark gap a small insertion loss for frequencies up to about 500 MHz. In addition to these desirable passive properties, the spark gap excels at conducting large peak currents that would be encountered in a direct lightning strike to the antenna.

In the case of a transmitting antenna, the protective circuit must be able to withstand the relatively large voltages that are encountered in routine operation. For example, a transmitter power of 50 W requires a peak voltage of about 90 V (assuming sinusoidal waveform and a perfect match between the 75 Ω transmission line and a dipole antenna). Larger transmitter power requires larger peak voltages across the line. The relatively large firing voltage of a spark gap, compared to other components used for transient protection applications, makes it possible to protect transmitting antennae with spark gaps.

The breakdown voltage of a spark gap can be much less than its dc value when a steady-state radio frequency voltage is applied across the gap.

To conveniently install the spark gap and minimize insertion losses, spark gaps are mounted in fixtures with coaxial connectors (types N, BNC, and UHF are particularly common). The coaxial spark gap was patented by Cushman (1960).

G. NEON LAMPS

Neon lamps have been used for many years as indicator lamps in electronic circuits. The GaAs solid-state light-emitting diode (LED) has made neon lamps obsolete for most electronic applications, although neon lamps continue to be commonly used for pilot lights that are connected to the mains. Neon lamps were also used as electronic circuit components, but this application is also obsolete owing to the superior performance of semiconductor devices such as DIACs and various integrated circuits. Neon lamps have been used for transient suppression by engineers who were familiar with them. The following advantages were claimed for neon lamps, compared to spark gaps:

1. lower firing voltage (e.g., 60 V dc for a neon lamp vs. 150 V dc, or more, for a spark gap)
2. transparent case allows condition of electrodes to be inspected
3. lamps are available with a bayonet base that allows use of a socket, for convenient replacement
4. lamps are less expensive than spark gaps

There is no doubt that neon lamps do have a dc firing voltage that is smaller than that offered by nearly all spark gaps. However, C.G. Clare and Siemens each make one spark gap (models CG-75 and B1-C75, respectively) that has a dc firing voltage of 75 ± 15 V, and many manufacturers make models that have a dc firing voltage of 90 ± 20 V. During rapidly rising surges, the maximum voltage across the neon lamp *or* spark gap will be at least several times the dc firing voltage. The value of the breakdown voltage during the surge is more important than the value of the dc firing voltage.

To obtain a small dc breakdown voltage (e.g., between 60 and 100 V), the Penning effect must be used, and this further increases the time required to initiate a gas discharge in low-voltage gas tubes, such as neon lamps. To obtain the Penning effect, two different kinds of gases are mixed together (e.g., 0.1% argon in 99.9% neon). Photons released from metastable atoms of the more common species can produce photoionization of the less common species. However, the release of photons from the excited metastable atoms is a slow process that can take tens of milliseconds. Therefore, spark gaps that use the Penning effect will *not* be fast responding. It has been shown that at values of dV/dt of 5 MV/μs, a spark gap with a dc breakdown voltage of 230 V has a smaller value of V_f than a gap with a dc breakdown voltage of 90 V (Singer et al., 1987). It appears that spark gaps with subnanosecond response time at 1 MV/μs rate of rise will have a nominal dc breakdown voltage of at least 150 to 250 V. Therefore, the small dc breakdown voltage of neon lamps is not an advantage during overvoltages with large magnitudes of dV/dt.

Favoring neon indicator lamps because they have a transparent case is probably not a good idea. The thin glass cases on lamps can shatter during surge currents. Replacement of the lamp without the glass bulb is then difficult. The ceramic case of typical spark gaps is much more robust. If a spark gap with a transparent case is desired, one manufacturer (Siemens) makes a variety of models of "button gaps" that have transparent glass cases. However, conditioning the electrodes during the final stages of manufacturing sputters metal on the inside of the glass case, which makes it difficult to see the electrodes even in a new spark gap.

The use of neon lamps with bayonet bases in transient protection circuits for "ease of replacement" is *not* a good idea. Moreover, the inductance in the wires to the lamp socket is definitely worth avoiding, as discussed in Chapter 15. If the use of sockets is desired, spark gaps for protection of telephone systems are commonly available in models that fit into sockets.

The neon lamps are considerably less expensive than spark gaps, a feature that merits considerable attention. Although neon lamps are generally inferior to spark gaps for overvoltage protection, the limited protection from a neon lamp may be preferable to no protection in an application where a spark gap is too expensive.

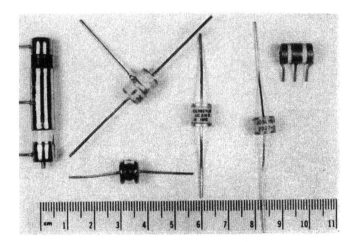

Fig. 7-6. Several two- and three-electrode spark gaps. A scale with minor divisions of 1 mm is included for reference.

If neon lamps are to be used for transient protection, the model 5AH (formerly NE-83) appears to be desirable. It was designed for use in pulse circuits and with currents that are greater than those of other neon lamps. However, it is still more fragile than spark gaps.

H. PACKAGING OPTIONS

Most modern spark gaps are enclosed in small ceramic tubes whose ends are metal electrodes, as shown in Fig. 7-6. There are two ways of mounting these spark gaps: wire leads and friction fit in special sockets. Use of wire leads is common when the spark gap is to be mounted on a printed circuit card or installed across the terminals of some device. The socket mounting is common in protecting many conductors at the point of entry of a building, such as in telephone cable applications.

A list of manufacturers of spark gaps is given in Appendix B.

8

Varistors

"Varistor" is the generic name for a voltage-variable resistor. These devices obey the relations $V = I \times R$, where R is a function of V or I, and for which R decreases as the magnitude of V (or I) increases. Strictly speaking, all nonlinear devices may be called varistors. However, the conventional use of the word "varistor" is restricted to devices that dissipate energy in a solid, bulk material (not a semiconductor junction). A varistor is inherently a bipolar device: its characteristics are not dependent on polarity. Modern commercially available varistors are fabricated from various mixtures of metal oxides (of which zinc oxide is the principal ingredient). An older type of varistor is fabricated from silicon carbide. This chapter will concentrate on the metal oxide varistor; some remarks about the silicon carbide varistor are located at the end of this chapter.

A. *V-I* CURVE

A typical *V-I* curve for a metal oxide varistor is shown in Fig. 8-1. Notice that the varistor is a symmetrical, bipolar device: it clamps either positive or negative voltages.

Figure 8-1 shows voltage as the independent variable and current as the dependent variable. This convention for presenting data on electronic devices goes back to the days of vacuum tubes. As discussed in Chapter 5, there is no doubt that current is a more convenient independent variable with devices such as varistors that have an approximately constant voltage over a wide range of currents. The author's choice for the format of Fig. 8-1 reflects the usual convention for electronic devices, which is probably more familiar to most readers than is plotting current as the independent variable.

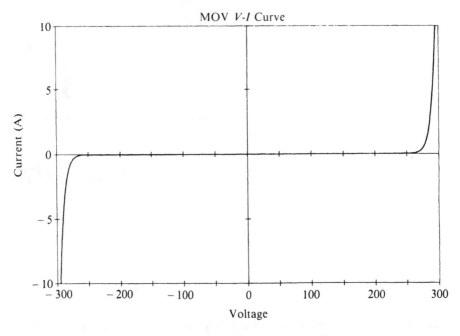

Fig. 8-1. *V-I relationship for a metal oxide varistor (linear scale).*

Varistors are not precision components. The voltage, V_N, at which the current in the varistor is 1 mA dc, is used as a nominal conduction voltage. V_N is specified with a tolerance of $\pm 10\%$ for models with $V_N \geq 33$ V. For smaller values of V_N, tolerances can be as large as $\pm 30\%$.

1. Varistor Device Model

When the *V-I* characteristic curve is plotted on a log–log scale (Fig. 8-2), three characteristic regions of operation become apparent (Harnden et al., 1972). At very small currents, less than 0.1 mA, the varistor behaves like a simple resistor, called R_{leak}. At very large currents, more than 100 A, the varistor response is dominated by the bulk resistance of the device, R_{bulk}. In between, the varistor obeys Eq. 1.

$$I = kV^{\alpha} \tag{1}$$

2. Circuit Model

The circuit diagram that corresponds to this model of a varistor is shown in Fig. 8-3. The inductance shown in Fig. 8-3 is due to packaging and the length of the leads in a particular application. The voltage V, in Eq. 1 is across the ideal varistor and does not include the voltage drop across R_{bulk}.

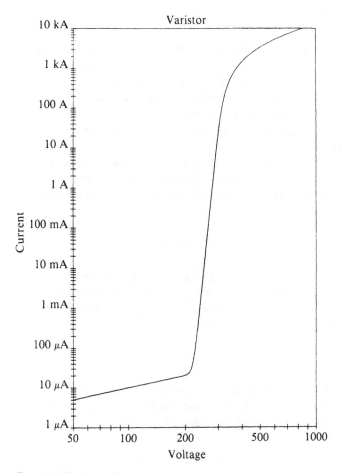

Fig. 8-2. V-I relationship for a metal oxide varistor (log–log scale).

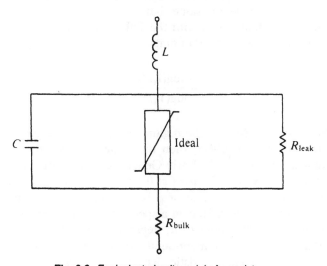

Fig. 8-3. Equivalent circuit model of a varistor.

The current in Eq. 1 is through the ideal varistor and does not include the current in R_{leak}.

The value of α characterizes the nonlinear V-I characteristic. An ordinary resistor would have $\alpha = 1$. As a general rule, the greater the value of α, the "better" the varistor. Modern metal oxide varistors have values of α between about 25 and 60. Metal oxide varistors with larger values of V_N tend to have greater values of α.

Equation 1 with the parameter k often causes problems when implemented in a computer program. The value of k is often less than 10^{-100}, which causes "underflow." This difficulty is eliminated by using the form of the varistor relation,

$$I = (V/D)^\alpha$$

with the parameter D equivalent to $k^{-1/\alpha}$. If α is 50, then the value of V/D is between about 0.87 and 1.2 for I between 1 mA and 10 kA.

Alternatively, one can avoid underflow and overflow during computation by using logarithms to evaluate Eq. 1:

$$\log|I| = \log(k) + [\alpha \log|V|]$$

After the computation with the logarithms is completed, the dependent variable, which is either I or V, can be found by the appropriate inverse operation. For example, if base e logarithms were used,

$$|I| = \exp(\log|I|)$$

After the calculation is completed, the sign of the dependent variable can be assigned: both I and V have the same sign.

The value of α is determined with the following relation, where (V_1, I_1) and (V_2, I_2) are two measured data points.

$$\alpha = \frac{\log(I_1/I_2)}{\log(V_1/V_2)}$$

If either (V_1, I_1) or (V_2, I_2) is in the region where leakage resistance or bulk resistance has a significant effect, then the value of α calculated from the above relation will be too small. Early General Electric varistor data sheets (c. 1975) specified the value of α for I at 0.1 and 1 mA. This appears to be a good way to ensure that the value of α is determined consistently. This is important if the value of α appears in a specification, since errors of ± 1 can easily be made in determinations of α, depending on technique.

In addition to the effects mentioned above, there is a temperature dependence in the V-I characteristic. The varistor voltage at a constant current ($I > 1$ mA) changes by about -0.04% per kelvin increase in temperature (Smith and McCormick, 1982, p. 50). This is of little consequence in transient protection applications, but if the overstress is sustained, this property can produce thermal runaway. Thermal runaway and its implications for specifying varistors are discussed later in this chapter.

Characteristic V-I curves for varistors must be determined from measurements made with brief pulse currents, such as an $8/20\,\mu s$ waveshape, in order to avoid effects of heating the varistor. Further, the interval between the consecutive surges in the laboratory must be sufficiently long to allow the varistor to return to room temperature before the next surge is applied.

It is emphasized that varistors must *not* be used for steady-state voltage regulation. A varistor with a diameter of 20 mm commonly has a maximum steady-state power rating of only 1.0 W. Silicon avalanche diodes are much better suited for steady-state voltage regulation. Diodes with a diameter of less than 3.5 mm can have a maximum steady-state power rating of 5 W. As discussed in Chapter 9, avalanche diodes also tend to have much larger values of α than varistors, which makes dV/dI smaller for the diodes. This means that the voltage across an avalanche diode is more nearly independent of the current in the device than for a varistor.

Varistors that are designed for large surge currents (more than 1 kA peak, $8/20\,\mu s$ waveshape) are fabricated as a disk with a diameter of at least 14 mm. This large cross-sectional area makes the current per area small. However, this also makes the capacitance of the varistor rather large. Varistors have the largest capacitance of the common nonlinear transient protection devices; common values of parasitic capacitance range between 0.2 and 10 nF. Varistors with smaller values of V_N or larger diameters have larger values of parasitic capacitance. This capacitance is not always a bad feature; for example, the shunt capacitance of varistors can be used to construct nonlinear low-pass filters.

The value of either V_N or k and the value of α completely specify an *ideal* varistor. The values of C, R_{leak}, and R_{bulk} complete the model in Fig. 8-3. Most manufacturers, however, do not specify values for R_{leak} and R_{bulk}. The effect of R_{leak} is usually negligible. A maximum value of R_{bulk} may be deduced from a plot of the maximum clamping voltage at large surge currents, which is often found on manufacturers' data sheets.

3. Response Time

The response time of the varistor is less than 0.5 ns (Levinson and Philipp, 1977; Philipp and Levinson, 1981). The literature contains remarks about the "slow" response of MOVs, with typical response times cited as about 50 ns. The apparent slow response time is due to parasitic inductance in the package and leads when the varistor is not connected with minimal lead length (Fisher, 1978). One should be careful to distinguish the response time of the device from the response time owing to parasitic inductance and capacitance of the varistor when combined with the output impedance of the surge generator.

4. Fabrication

Metal oxide (i.e., ZnO) varistors were invented in Japan by workers at Matsushita Electric Corp. and licensed to other companies. The fabrication of ZnO varistors has been described by Matsuoka et al. (1970). Manufacturers of metal oxide varistors for low-voltage (less than 1 kV rms) service are listed in Appendix B.

B. FAILURE MODES

Varistors degrade gradually when subjected to surge currents. The "end of life" is commonly specified when V_N has changed by $\pm 10\%$. However, the varistor is still functional after the "end of life." In nearly all cases, the value of V_N *decreases* with exposure to surge currents. This degradation manifests itself as an increase in idle current at the maximum normal operating voltage in the system. Excessive idle current during normal, steady-state operation will cause heating in the varistor. Because the varistor has a negative temperature coefficient, the current will increase as the varistor becomes hotter. Thermal runaway may occur, with consequent failure of the varistor (Vicaud, 1986).

When subjected to excessive continuous currents, the varistor usually fails with a very low resistance. However, when "blasted" with excessive surge current, the varistor may rupture and fail as an open circuit. When connected across an ac or dc power source with a large short-circuit current, the fault current in the low resistance of a failed varistor can also rupture the varistor.

C. APPLICATIONS

There is general agreement that varistors are at present the best of the available nonlinear devices for protection of electronic systems from transient overvoltages that propagate on the mains. Varistors are also useful for clamping the voltage across inductive loads when the current in the inductor is switched off. Other applications are, of course, also possible.

1. Selection of V_N

How should one specify a varistor that is to be used alone for suppression of overvoltages? The choice of the value of V_N for a varistor is a compromise between two conflicting concerns. Lesser values of V_N provided smaller clamping voltages during a surge, which is desirable. However, greater values of V_N provide a longer lifetime. There are several issues that are relevant to selection for the minimum reasonable value of V_N.

1. The varistor should not dissipate appreciable power during normal system operation. The effects of tolerance on both V_N and the system voltage must be included in this determination.
2. Immediately after a surge, when the varistor is still hot, the varistor must not conduct appreciable current during the normal system voltage.
3. After a varistor is exposed to surge current, the value of V_N is reduced. Appreciable degradation can occur after exposure to millions of small surges or after exposure to one very large surge. This reduction in V_N with age must be considered.
4. If the varistor may be operated in a hot environment, the negative temperature coefficient of V_N must be considered. The manufacturer specifies the value of V_N at a temperature of 25°C.

In applications where exposure to large surge currents are anticipated or where reliability is a principal concern, there are two general ways to obtain reliable performance from a varistor.

1. Specify the minimum initial value of V_N to be appreciably greater than the peak voltage expected during normal operation of the system. This gives a larger margin for degradation before the varistor will be destroyed.
2. Select a model of varistor with a suitably larger diameter, or the largest diameter that will fit in the available space. For a given current, the current density is less in varistors with larger diameters. This allows the varistor to have less degradation after large surge currents.

A simple method of determining the smallest acceptable value of V_N for a particular application is to use the relationship

$$V_p(1 + \beta) < V_N(1 - \varepsilon)$$

where V_p is the peak value of the *nominal* system voltage, β is a parameter to account for excursions in system voltage, as well as a safety factor, and ε is the tolerance on the *nominal* value of V_N. Typical values of β are between 0.10 and 1.0. Larger values of β give a longer lifetime for the varistor.

In products where ambient temperatures may be significant greater than 25°C, one *must* also consider the effect of temperature on the varistor V-I relation.

Specific examples and recommendations for use of varistors on the mains are given in Chapter 19.

2. Parallel Connection of Two Varistors

Another possibility for increasing reliability during and after exposure to large surge currents is to connect two varistors in parallel. This may be less expensive than purchasing one large-diameter varistor.

As an example, consider two varistors in parallel that have the following properties.

$$V_1: \quad V_N = 230 \text{ V} \quad \alpha = 48 \quad R_{bulk} = 0.04 \text{ } \Omega$$
$$V_2: \quad V_N = 250 \text{ V} \quad \alpha = 42 \quad R_{bulk} = 0.04 \text{ } \Omega$$

These two varistors both meet specifications for $V_N = 240$ V ± 5%, which is closer matching than the manufactuerer's specification of ±12%. A log-log plot of the V-I curves for these two varistors is shown in Fig. 8-4. The two solid lines in Fig. 8-4 are the individual V-I curves for the two varistors; the dashed line is the V-I curve for the parallel combination of these varistors.

When the voltage across these two varistors is 300 V, the current in V_1 will be about 345 A, while the current in V_2 will be about 2.1 A. In this example, 0.6% of the total current flows in V_2 when 300 V is across the pair of varistors. The second varistor is ineffective in this combination at small voltages.

However, at larger voltages across the varistors, the bulk resistance begins to denominate the V-I characteristic curve and force the current in V_2 to approximate that in V_1. For example, when the voltage across the pair of varistors is 500 V, the currents in V_1 and V_2 are about 3.7 and 2.9 kA, respectively. About 43% of the total current passes through V_2 when there is 500 V across these two varistors. The second varistor is definitely effective in this combination when more than 500 V is across the varistors. There are two benefits to connecting V_1 and V_2 in parallel for service at large surge currents: lower clamping voltage and longer lifetime compared to the use of V_1 alone.

If the parallel combination of these two varistors is used in an environment where surge currents greater than about 1 kA are common, then placing the two varistors in parallel may be a reasonable design. However, one should compare both the cost and performance of (1) two varistors with a particular diameter that are connected in parallel, and (2) a single varistor with a larger diameter.

If the parallel combination of these two varistors is used in an environment where surge currents are usually less than 1 kA, the second varistor has little effect. The first varistor has an initial value of V_N that is smaller than that of the second varistor. Since the first varistor carries most of the surge current, it will be degraded more than the second varistor.

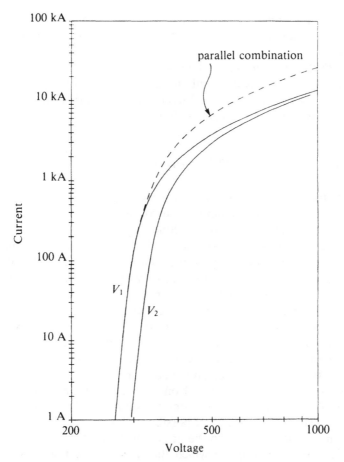

Fig. 8-4. *V-I relationship for two metal oxide varistors, V₁ and V₂. The dashed line shows the V-I relationship for the parallel combination of these two varistors (log–log scale).*

Therefore, with increasing age, the first varistor's value of V_N will decrease faster than the second varistor's value of V_N. In environments where the surge currents are relatively small, V_1 would reach its end of life at a specific time whether V_2 is present or not. Putting V_2 in parallel is not harmful, but the money spent on V_2 might be better used elsewhere.

The manufacturer specifies that varistors of the type used in this example have a rating of either 1000 surges with a 8/20 μs waveshape and a peak current of 400 A or 10 surges with a peak current of 2 kA. Is it reasonable to place two of these varistors in parallel? Let us assume that all surges have an 8/20 μs waveshape. If they are used in an environment where 10 surges with a peak current of 2 kA will be experienced *before* 1000 surges with a peak current of 0.4 kA occurs, it is a good design to place these varistors in parallel. As mentioned in Chapter 3, there are no statistics on surge currents in varistors, so there are no data available for environmental calculations.

The ineffectiveness of placing two nearly identical varistors in parallel in environments with relatively small surge currents is not a problem that is unique to varistors. The same effect occurs with other highly nonlinear devices such as zener and avalanche diodes.

D. USE OF VARISTORS WITH SPARK GAPS

Two ways to combine varistors with a spark gap are shown in Fig. 8-5. These combinations are sometimes advocated to avoid the undesirable features of spark gaps and varistors alone.

Spark gaps can be slow to conduct, which produces a large remnant that travels downstream from the spark gap. Varistors can be degraded by large surge currents. The shunt combination of a varistor and a spark gap seeks to avoid these undesirable features. The varistor clamps the overvoltage before the spark gap conducts. After the gap conducts, it shunts current away from the varistor, which avoids degradation of the varistor. When the value of V_N of the varistor is too small, the spark gap will not conduct. The value of V_N should certainly be greater than the dc breakdown voltage of the spark gap. One might specify the varistor and spark gap so that the gap will conduct before the peak current is attained on an $8/20\,\mu s$ waveshape with a peak current of 50 A. Notice that the shunt combination of a varistor and spark gap does not solve the problem of possible follow current in the gap, so this circuit alone cannot be used on the ac supply mains.

There are two situations in which the series connection of a spark gap and a varistor may be useful.

Varistors have large parasitic capacitance that can have unacceptably large currents when connected across ac supply mains in some applications.

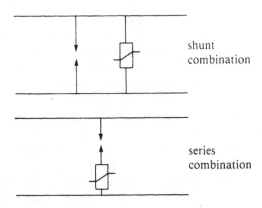

shunt combination

series combination

Fig. 8-5. Use of a varistor with a spark gap.

For example, UL 544 specifies that the maximum line-to-ground current is to be less than 0.1 mA rms for patient care equipment used in medical or dental offices. A varistor with a parasitic capacitance of 2 nF has an rms current of 0.09 mA when installed between line and ground on mains with a nominal voltage of 120 V rms and a frequency of 60 Hz. Such a varistor has an inadequate safety margin to meet the specifications in this application. Connecting a spark gap in series with the varistor can decrease the shunt capacitance of the combination to a few pF, which is a trivial value.

The spark gap can act as a switch to prevent steady-state power dissipation in the varistor. This would allow a varistor with a relatively small value of V_N to be connected to the mains without concern about thermal runaway of the varistor. Provided the spark gap is fast responding, one might use this circuit to achieve smaller clamping voltages during surges than is otherwise possible with a varistor.

Specifications for the spark gap and varistor are difficult to determine. Criteria for preventing follow current in the spark gap were given in Chapter 7. For applications on ac supply mains, the spark gap should have a dc breakdown voltage that is greater than the peak system voltage in order to extinguish follow current in the glow regime during a zero crossing of the mains voltage. The varistor should be specified to avoid follow current in the arc regime. When the spark gap is conducting in the arc regime, there is about 20 V across the gap. The remainder of the system voltage appears across the varistor. The 20 V across the gap can be neglected, since it is much smaller than the peak system voltage; this approximation gives a conservative design. When the entire system voltage is across the varistor, the current in the varistor must be less than the arc-extinguishing current for the spark gap. A conservative value for arc-extinguishing current is on the order of 50 mA.

E. VARISTOR PACKAGES

Common metal oxide varistors for low-voltage service are fabricated as a disk with a diameter of either 3, 5, 7, 10, 14, 20, 32, 40, or 60 mm. A photograph of some common varistor packages is shown in Fig. 8-6. Varistors with a diameter between about 5 and 20 mm are commonly used in a radial lead package, which looks like a ceramic disk capacitor. Varistors with a diameter of less than 5 mm are commonly used in an axial lead package, which looks like a semiconductor diode in an epoxy case. These small packages are suitable for mounting on printed circuit boards, but they can be also connected across terminals of equipment. Varistors with a diameter of more than 20 mm are available in a large package with lugs for connection of electrical leads. Such large packacges are normally used only for applications involving suppression of transient overvoltages on the mains.

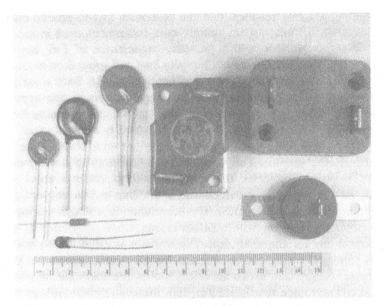

Fig. 8-6. *Metal oxide varistor packages for varistors with diameters between 3 and 40 mm. A scale with minor divisions of 1 mm is included for reference.*

Recently, varistors have become available in various packages for surface-mount applications. This package has the advantage that a careful designer can minimize the effects of inductance in series with the varistor, since some surface-mount varistors are in a leadless package. Metal oxide varistors are also commercially available in hollow coaxial cylinders for use surrounding pins in electrical connectors.

A list of manufacturers of metal oxide varistors for low-voltage applications is given in Appendix B.

F. SILICON CARBIDE VARISTORS

Silicon carbide varistors were introduced in a paper by McEachron (1930) and are commonly known in the United States by the General Electric tradename Thyrite. Silicon carbide varistors have a value of α of about 4, so they are much less nonlinear than metal oxide varistors. This property makes silicon carbide varistors unsuitable for tight clamping of voltage during electrical overstresses. Silicon carbide varistors would be completely obsolete and only of historical interest except for two features: they are inexpensive and quite robust. These features are still particularly attractive in applications where overhead transmission lines must be protected from direct lightning strikes.

The silicon carbide varistor for use on mains below 600 V rms has an approximate characteristic curve

$$I = 4.5 \times 10^{-11} V^4$$

where I is in amperes and V is in volts. Because the value of α is so small for silicon carbide varistors, if their clamping voltage during a 20 kA current is set to be 20 times the peak of the normal system operating voltage, then the silicon carbide varistors will conduct appreciable current (about 0.1 A) during normal system operation. The steady-state conduction current causes heating of the varistor, which will destroy it. In addition, the steady-state conduction current in the varistor is a loss in the power distribution system: there may be hundreds of these varistors on a long distribution line.

To prevent steady-state conduction in the varistor, a spark gap is connected in series with a silicon carbide varistor to form an arrester. The dc breakdown voltage of the spark gap is usually selected to be several times the amplitude of the normal operating voltage. The series varistor extinguishes power follow current in the gap after an overstress, as explained earlier in this chapter. The silicon carbide arrester has proved to be an effective way to prevent damage to insulation of overhead transmission lines and transformers. The use of a series spark gap may produce a relatively large remnant that propagates downstream from the arrester, which is an undesirable property. This makes the silicon carbide arrester unsuitable as the sole device for protecting electronic systems.

9

Avalanche and Zener Diodes

Silicon diodes that are intended for use in the reverse-breakdown region are familiar components to analog electronic circuit designers. The device physics, details of the characteristics, and conventional circuit applications of these diodes are discussed by Todd (1970).

A. AVALANCHE VERSUS ZENER DIODES

Avalanche diodes at present have the smallest value of dV/dI at small surge currents (i.e., less than 50 A) of available nonlinear devices. In other words, they offer the tightest voltage clamping of available devices. This makes avalanche diodes the preferred device for the final stage of multistage protection circuits where the device to be protected is relatively fragile (e.g., integrated circuits).

Although all of the diodes intended for use in the reverse breakdown region are commonly called *zener diodes,* only devices with a reverse-breakdown voltage of less than about 5 V usually use the Zener mechanism. Diodes with a reverse breakdown voltage of more than 8 V use the avalanche mechanism. For diodes with a reverse breakdown value between about 5 and 8 V, both mechanisms operate in the same device. The mechanism of a particular diode in the laboratory can be quickly identified with the following rule. In zener breakdown the magnitude of the conduction voltage *decreases* as the temperature increases; for avalanche breakdown, the magnitude of the conduction voltage *increases* as temperature increases. Avalanche diodes are widely used as voltage regulators, because in the reverse-breakdown region, the magnitude of dV/dI is very small.

1. V-I Curve

A V-I curve for a typical avalanche diode is shown in Fig. 9-1. There are two regions of operation: forward and reverse biased. In the forward-biased direction, the voltage across the diode is positive, and the diode responds like an ordinary silicon rectifier diode, which is discussed in Chapter 10. In the reverse-biased direction, the voltage across the diode is negative, and the diode conducts with an approximately constant voltage. Avalanche diodes are intended to be operated in the reverse breakdown region, where V and I are both negative, as shown in Fig. 9-1, so it is customary to ignore the negative sign on V and I in the reverse breakdown region.

The relationship between voltage and current for zener and avalanche diodes in the reverse breakdown region can be fit to the same mathematical expression that was used for varistors:

$$I = (V/D)^\alpha$$

The value of α for some particular zener and avalanche diodes has been given by Standler (1984). Zener diodes with a nominal breakdown voltage

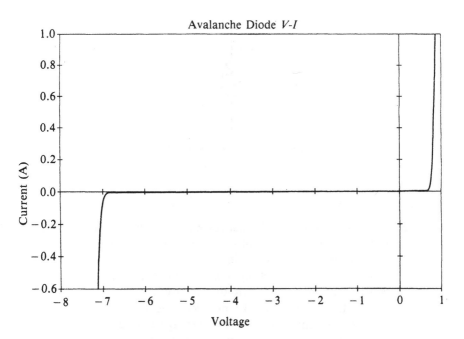

Fig. 9-1. V-I relationship for an avalanche diode.

between 3.3 and 3.9 V have values of α between 7 and 9. Avalanche diodes with a nominal breakdown voltage between 5.6 and 200 V have values of α between 50 and 700.

Zener diodes have much smaller values of α than avalanche diodes. This issue is important in protecting integrated circuits with small operating voltages, such as high-speed CMOS. Avalanche diodes have too large a breakdown voltage to be suitable for protection of all low-voltage integrated circuits. Zener diodes have the appropriate range of breakdown voltages but are not highly nonlinear.

2. Bipolar Circuits

To clamp overvoltages of either polarity or to hold off normal system voltages of either polarity, two avalanche diodes are connected back to back in series, as shown in Fig. 9-2. It is common for the two diodes in Fig. 9-2 to have the same breakdown voltage (within ±10% tolerances), but this is not required to use the circuit of Fig. 9-2. Because this configuration is so common in transient protection applications, many manufacturers provide two identical avalanche diodes connected back to back inside one package. Such devices are called *bipolar* avalanche diodes. This packaging concept has the advantages of reducing parasitic inductance in series with the diodes and minimizing both the assembly time and mass of the circuit.

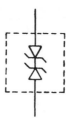

Fig. 9-2. *Bipolar avalanche diode.*

3. Response Time

Avalanche diodes are particularly fast-responding. Claims have been made that transient suppressor diodes have a response time of less than 1 ps. These claims are impossible according to the laws of physics, because a signal traveling at the speed of light *in vacuo*, the fastest speed in the universe, would require about 30 ps to traverse the length of the plastic package that contains the diode. In practice, the response time is determined by the parasitic inductance of the package and leads. This parasitic inductance can produce appreciable overshoot for 50 ps after the beginning of a rectangular pulse (Clark and Winters, 1973). Such a brief response time is obtainable only with a leadless device, as explained in Chapter 15.

4. Diode Capacitance and How to Reduce It

Diodes that can tolerate large pulse currents also have a large capacitance, as a consequence of the large cross-sectional area that it necessary to reduce the current density. Transient suppression diodes commonly have capacitance values between 500 pF and 10 nF. The capacitance value is a strong function of voltage across the diode: the capacitance decreases as the diode is more strongly reverse-biased. Diodes with smaller values of breakdown voltage have larger capacitance given constant cross-sectional area and bias voltage. The relatively large values of capacitance make these diodes alone unsuitable for use in circuits with high-frequency signals.

However, these diodes can be placed in series with forward-biased switching diodes, which have a capacitance of the order of 50 pF, to reduce the total shunt capacitance, as shown in Fig. 9-3. If the voltage between points A and B in Fig. 9-3a is greater than about 0.6 V during normal system operation, the switching diode will conduct,[*] and the system will supply charge to the parasitic capacitance of the avalanche diode. Once the parasitic capacitance of the avalanche diode is fully charged, the switching diode will be nonconducting, and the protection circuit draws no current from the system. The same effect occurs in the circuit shown in Fig. 9-3b, except that the voltage between points A and B must be greater than about 1.2 V in order to charge the parasitic capacitance of the avalanche diode.

The use of switching diodes in series with avalanche diodes prevents current flow when the avalanche diode is forward-biased. Therefore, the bidirectional clamping circuit of Fig. 9-2 will not work when switching diodes are included. To clamp overvoltages of either polarity, use one of the circuits shown in Fig. 9-3 (Popp, 1968; Clark, 1975).

The circuit of Fig. 9-3b is less expensive than the circuit of Fig. 9-3a, since the former uses only one avalanche diode, which is much more expensive than several rectifiers or switching diodes. However, when the circuit of Fig. 9-3b is fabricated from discrete components, the additional parasitic inductance of the extra switching diode in the conducting path may be a serious disadvantage.

The circuits in Fig. 9-3 are suggested for use between two nongrounded conductors (differential mode) or for use between one conductor and ground. To provide common-mode protection for a balanced line, use the circuit in Fig. 9-4 (Abramson et al., 1984).

Forward-biased diodes with large pulse current ratings are relatively slow to conduct, a phenomenon that will be discussed in Chapter 10. Although

---

[*] See Chapter 10 for discussion of V-I characteristics of a switching diode.

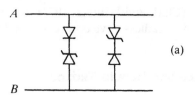

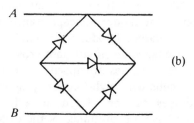

Fig. 9-3. *Two techniques for reducing the effective shunt capacitance of an avalanche diode while preserving bipolar clamping.*

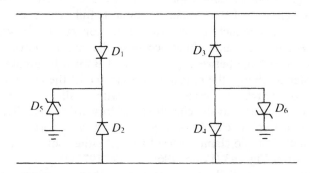

Fig. 9-4. *Another technique for reducing the effective shunt capacitance of an avalanche diode while preserving bipolar clamping.*

the circuits of Fig. 9-3 do decrease the shunt capacitance, they introduce another problem—that of slower response (Clark and Winters, 1973, p. 83).

5. Clamping Voltage

Zener and avalanche diodes are characterized by a *breakdown voltage*, V_z, at a current I_z. The value of I_z is usually 1 mA for transient suppressor diodes. For ordinary diodes, I_z is specified so that the V_zI_z is usually between about 20% and 25% of its maximum steady-state power rating. The value of

V_z usually has tolerances of $\pm5\%$ or $\pm10\%$ and is a useful parameter for voltage regulation applications.

During the passage of large surge currents, which may be 10 to 200 A, the voltage across the diode, V_c, may be substantially greater than the value of V_z. The difference between V_c and V_z is mostly due to bulk resistance in the semiconductor material, to the effects of heating on the breakdown voltage, and to the fact that the parameter α is not infinite.

6. Power Ratings

Steady-state power dissipation ratings of common zener and avalanche diodes are between 0.4 and 5 W. During brief transients (durations up to a few tens of microseconds), most diodes can have an instantaneous power level that is at least a hundred times greater than their steady-state rating.

A *peak (pulse) power rating* is commonly given in data sheets for avalanche diodes that have been characterized for transient overvoltage protection. Values for common transient suppression diodes are either 600 or 1500 W, depending on the size of the internal heat sink. The peak pulse power is defined as the product of the *maximum* peak pulse current, I_{pp}, and the *maximum* clamping voltage, V_c, at current I_{pp} during a $10/1000\,\mu s$ waveform. The actual clamping voltage is likely to be somewhat less than V_c owing to component tolerances and margins in the specifications. Further, the maximum voltage across the diode often occurs after the peak current (for long-duration waveforms such as the $10/1000\,\mu s$) because of the heating of the avalanche diode. Therefore, the "peak pulse power" is a data sheet parameter and not the actual maximum power rating of the diode.

Use of the peak power may be misleading when waveforms are considered that are different from the $10/1000\,\mu s$ waveform. The avalanche diode can survive larger peak currents for waveforms of shorter duration than the $10/1000\,\mu s$ waveform. The maximum current for an avalanche diode during an $8/20\,\mu s$ waveshape is typically 6 to 10 times greater than the rated peak current for a $10/1000\,\mu s$ waveshape. A maximum energy absorption rating for a nonrepetitive, short-duration overstress might be more useful for general applications. The peak power rating is an established tradition, perhaps because most electrical engineers are more familiar with power than energy.

7. Noise

A particular disadvantage of avalanche diodes is that, when operated at relatively small reverse currents (e.g., $1\,\mu A$ to $1\,mA$), many avalanche diodes produce significant noise with a bandwidth from less than 1 Hz to over 0.5 MHz. In fact, special avalanche diodes are used as a source of white noise. Care must be taken to ensure that the normal operating voltages of the system do not push the avalanche diode into a weakly conducting state.

Diodes that are characterized for transient suppression applications have a *reverse standoff voltage*, V_R, and *reverse leakage current*, I_R, listed on the manufacturer's data sheet. The value V_R is the maximum reverse-bias voltage that can be applied across the avalanche diode with a magnitude of current that is less than I_R. A typical value of I_R is 5 μA, although for diodes with a small breakdown voltage, I_R may be as large as 1 mA. In most overvoltage protective applications, the value of the voltage during normal system operation should be less than V_R.

8. Failure Modes

Diodes usually fail as a short circuit. However, sustained large currents can rupture the epoxy package that is used for commercial-grade diodes. If a short-circuit failure mode is essential, one should use devices in a hermetically sealed metal package (e.g., DO-13).

In some tests with large steady-state currents, avalanche diodes were observed to fail with a behavior similar to that of second breakdown (Standler, 1984). In one such test, 300 mA was passed through a 1N5364 diode. The power dissipation was 11 W, substantially greater than the 5 W maximum steady-state rating. After about 22 seconds at 300 mA, the diode failed. Thereafter, the former 33 V diode had a characteristic of a constant voltage of 15 V at currents greater than about 10 mA and of a 2 kΩ resistor at smaller reverse currents.

9. Comparison with Other Surge Protective Devices

Silicon diodes do not exhibit degradation of electrical parameters after prolonged service (unlike gas tubes or varistors). This fact is often stated as a reason to use silicon diodes rather than varistors. However, one should also consider the fact that the maximum tolerable surge current (for durations of less than about 1 ms) for a zener or avalanche diode in reverse breakdown is rather small when compared to tolerable surge currents for varistors and spark gaps.

The truth of the matter is there is no uniquely superior device for all applications. The experienced circuit designer needs to be familiar with a number of different types of devices and choose the appropriate one for each application.

B. MODELS OF AVALANCHE DIODES

Several types of avalanche diodes are available. The common, ordinary types are designed for steady-state voltage regulation and clamping at small values of current. These ordinary avalanche diodes may or may not be suitable for overvoltage protection applications, because they are neither designed nor characterized for behaviour during large surge currents. In contrast, special transient suppressor diodes are designed and characterized for surge currents. This section closes with a brief mention of diodes that may be suitable for applications with repetitive overvoltages that require a diode with a large steady-state power rating. There is also a discussion of low-voltage avalanche diodes for applications that require breakdown voltages of less than 6.8 V.

1. Special Transient Suppressor Diodes

Special transient suppressor avalanche diodes are available from several different manufacturers. These differ from "ordinary" avalanche diodes of the same maximum steady-state power rating in three ways:

1. Transient suppressor diodes have a larger cross-sectional area.
2. Transient suppressor diodes have larger internal heat sinks, which are often made of special materials.
3. The manufacturer's data sheet for transient suppressor diodes lists parameters that are relevant to overvoltage protection applications.

The larger cross-sectional area of transient suppressor diodes gives a smaller current density during surges. The smaller current density increases the likelihood that the diode will not be damaged by the surge and also decreases the contribution to the clamping voltage caused by the bulk resistance of the diode. One manufacturer's 1N5333 series of voltage regulator diodes, which have a 5 W steady-state power rating, has an area of 1.7 mm^2, whereas transient suppressor diodes of the same steady-state power rating may have an area that is three times larger.

Several different diodes were examined with X-ray fluorescence to determine the heat sink material (Standler, 1984). Some of the special diodes for overvoltage protection had solid silver internal heat sinks, whereas generic equivalents had nickel-plated copper or large aluminum heat sinks. Silver and copper have the greatest and second-greatest thermal conductivity of any known material and would be the best choices for rapidly removing heat from the junction. However, the coefficient of thermal expansion of both silver and copper is poorly matched to that of silicon. Molybdenum or tungsten have a thermal expansion coefficient that is near that of silicon and may also be an appropriate choice for an internal heat sink.

Special avalanche diodes that are designed and specified for transient suppression are available from many different companies. These devices are available by specifying either JEDEC registered part numbers or proprietary part numbers. The following list gives three commonly used generic JEDEC series of avalanche transient suppression diodes.

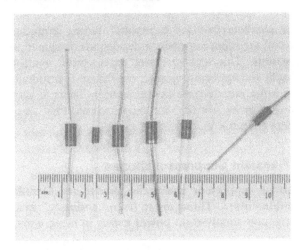

Fig. 9-5. Several transient suppression avalanche diode packages. The diagonal diode is a 1N5364 "ordinary" avalanche diode with a steady-state power rating of 5 W. A scale with minor divisions of 1 mm is included for reference.

Diode Series	Voltage	Package Type
1N5629–1N5665	6.8 V–200 V	DO-13 hermetic
1N6036–1N6072	7.5–220 V bipolar	DO-13 hermetic
1N6267–1N6303	6.8–200 V	Epoxy plastic

Special transient suppression diodes have already been characterized by the manufacturer for performance during an overvoltage with a 10/1000 μs waveshape. The performance of other types of avalanche diodes during transient overvoltages is usually unspecified. Regardless of the type of diode (or other surge-protective device) used in protection circuits, extensive testing should be performed to verify proper performance.

Figure 9-5 shows some transient suppression diodes. A list of manufacturers of avalanche diodes for suppression of transient overvoltages is given in Appendix B.

2. Models with Large Steady-State Power Ratings

Most avalanche diodes are in axial lead packages. Heat conduction along the leads is the principal mechanism for removal of thermal energy from the function following a transient overvoltage. In applications where repetitive transients with large surge currents are expected, it may be useful to consider avalanche diodes in metal packages that can be mounted on a large external heat sink. This arrangement would allow more time-averaged power to be dissipated in the diode junctions. Particularly attractive models include the 1N2804-46 series in a TO-3 package. These devices can dissipate a steady-state power of 50 W when the case temperature is less than 75°C, whereas most axial lead packages are limited to a steady-state power dissipation of 5 W or less.

3. Low-Voltage Avalanche Diodes

Special avalanche diodes with breakdown voltages between 3.3 and 6.8 V are available (models 1N5518-26 and 1N6082-87) and are widely used as voltage reference circuits in battery-operated equipment where small diode currents *and* a small value of dV/dI are required. These low-voltage avalanche devices have a steady-state power rating of only 0.4 W. However, these devices may be attractive for protecting high-speed CMOS and other low-voltage integrated circuits, provided the magnitude of surge currents in these small diodes can be held to acceptable levels by careful circuit design.

C. SELENIUM DIODES

Prior to about 1970, avalanche diodes fabricated from selenium were commonly used for transient overvoltage protection. These selenium diodes were commonly known in the United States by the General Electric trade name Thyrector and by the International Rectifier trade name Klip-Sels. Selenium diodes had a value of α of about 8. One of the interesting features of these diodes was their "self-healing" or restoration after an overstress that exceeded their maximum ratings. Neither silicon diodes nor metal oxide varistors have this feature.

Herbig and Winters (1951) described how to use two selenium diodes in series, in opposite directions, to suppress transient overvoltages generated by interrupting current in an inductive load. Selenium diodes are now obsolete; better performance can be obtained with silicon avalanche diodes or metal oxide varistors.

10

Semiconductor Diodes and Rectifiers

A. FORWARD-BIASED DIODES

Forward-biased semiconductor diodes are nonlinear devices that are useful for protection against transient overvoltages. Switching diodes are intended for applications in electronic circuits where the average current is less than 0.5 A. Rectifier diodes are intended for applications in power supplies with steady-state currents of at least 0.5 A. As mentioned in Chapter 9, zener and avalanche diodes can also conduct when forward biased.

1. *V-I* Curve

A *V-I* characteristic curve for a typical silicon diode under steady-state conditions is shown in Fig. 10-1. The relation between current, i_D, in a semiconductor diode and potential difference, v_D, across a semiconductor diode is usually given in the form of Eq. 1.

$$i_D = I_S\{\exp[v_D/(mV_T)] - 1\} \tag{1}$$

I_S, V_T, and m are three parameters that characterize the diode's *V-I* relationship. With a knowledge of semiconductor physics one can derive Eq. 1 and the relation

$$V_T = kT/e$$

where k is Boltzmann's constant (about 1.38×10^{-23} joule per kelvin), e is the elementary charge (about 1.60×10^{-19} coulombs), and T is the tempera-

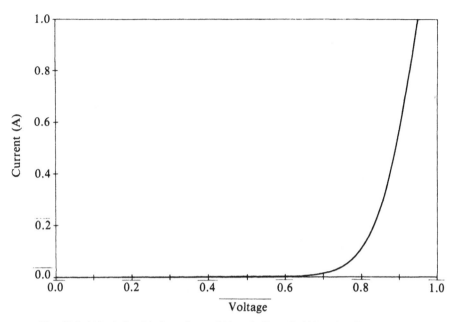

Fig. 10-1. *V-I relationship for a forward-biased silicon P–N junction (linear scale).*

ture of the diode (in kelvin). The value of kT/e is about 25 mV at room temperature, about 300 kelvin. An empirical constant, m, is multiplied times V_T to give better agreement between the current predicted from Eq. 1 and experimental data. Typical values of I_s and m at room temperature are:

	I_S	m
Silicon switching and rectifier diodes	5 nA	1.9
Silicon zener and avalanche diodes (forward-biased)	0.1 pA	1.1

A more realistic model of a diode would include a bulk resistance, R, of the semiconductor material. The resistance of the bulk semiconductor material produces a voltage drop across the diode's terminals that is in addition to the voltage v_D given in Eq. 1. As the current becomes large, appreciable power ($P = v_D i_D$) is dissipated inside the diode. This increases the temperature of the diode, and thus the values of both I_s and V_T will change. Remarkably, the behavior at large currents can be modeled by using Eq. 1 with I_s and V_T *treated as constants* plus an additional voltage drop due to a resistance R, as shown in Eq. 2.

$$v_D = mV_T \ln(i_D/I_S) + i_D R \qquad (2)$$

Typical values of R are 0.05 Ω for a rectifier with a maximum current of 1 A

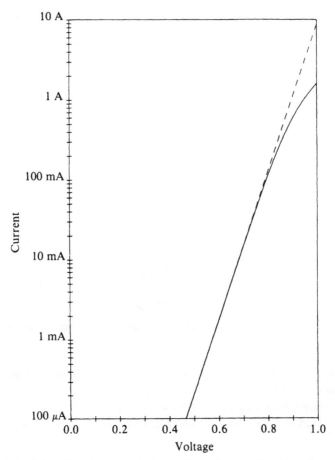

Fig. 10-2. *V-I relationship for a forward-biased silicon P–N junction. The dashed line is the extrapolation of the behavior at smaller currents (logarithmic current scale, linear voltage scale).*

A plot of $(v_D, \log_{10}(i_D))$ for a realistic model diode is shown in Fig. 10-2. The solid line in Fig. 10-2 is a plot of Eq. 2, and the dashed line is a plot of Eq. 1. Notice that the solid curve "rolls over" somewhere in the vicinity of 0.8 to 0.9 V. The different slope at greater values of v_D is due to two different causes: temperature and bulk resistance.

2. Switching Time

Diodes do not switch instantly. When a semiconductor diode is suddenly switched from zero current to a strongly forward-biased condition (e.g., 500 mA constant current), the voltage across the diode reaches a peak value that can be more than 10 times the steady-state value. The time required for this transient voltage to decay is known as the *forward-recovery time*. This phenomenon has been discussed by Armstrong (1957) and Cooper (1962).

An "ordinary" 1N4007 industry-standard rectifier was connected across the output terminals of a Hewlett-Packard 214B pulse generator so that the diode was forward-biased (Standler, 1984); the voltage across the rectifier was recorded on an oscilloscope. The open-circuit voltage (without the rectifier) was a 60 V rectangular pulse that was on for 2 μs and off for 500 μs. Peak current in the diode was 1.4 A. A plot of the voltage versus time is shown in Fig. 10-3. In this situation the voltage across the forward-biased diode was greater than 2 V for 0.4 μs. (There are two curves shown in Fig. 10-3: one for 0 to 0.20 μs; the other for times greater than 0.15 μs. These curves overlap between 0.15 and 0.20 μs.)

This experiment shows that forward-biased silicon diodes do not always have a 0.6 or 0.7 V drop across then, as is commonly assumed. The forward-recovery time is particularly long in high-voltage rectifiers (the diode used in this experiment has a reverse-breakdown voltage of at least 1 kV). Therefore, when diodes with a fast response are needed in transient protection circuits, the diodes should have a small reverse-breakdown voltage specification (e.g., 75 to 200 V).

When a conducting diode is suddenly reverse-biased, the current does

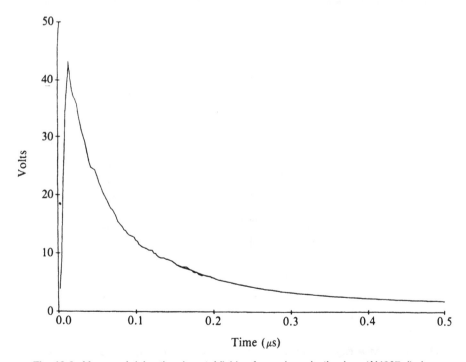

Fig. 10-3. *Measured delay time in establishing forward conduction in a 1N4007 diode.*

not drop to zero instantly. For a brief period (which can be as long as tens of microseconds, depending on the diode construction), the diode will conduct current in the reverse direction. The duration of this brief period during which the diode fails to rectify is known as the *reverse-recovery time*. In most cases, the reverse-recovery time is greater than the forward-recovery time, so the reverse-recovery time can be used as a measure of the diode's speed.

B. USE OF DIODES AS CLAMPS

One of the advantages of forward-biased diodes is that they clamp at a very small voltage, about 0.6 to 2 V. At present this is the smallest clamping voltage of any presently available semiconductor. Diode clamps are particularly useful to prevent the base-emitter junction of a transistor from being reverse-biased.

Two diodes connected in antiparallel form a bipolar clamping circuit, as shown in Fig. 10-4a. Such a configuration is available in a single package from several companies. Such devices are often called *silicon varistors*, but the use of the word "varistor" to describe a semiconductor junction should be deprecated.

When a larger clamping voltage is desired, but the large capacitance of

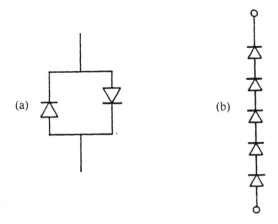

Fig. 10-4. Connecting rectifiers for (a) bipolar response and (b) greater conduction voltages.

zener or avalanche diodes is unacceptable, one can form series strings of two or more forward-biased diodes, as shown in Fig. 10-4b. For bipolar protection, two series strings are connected in antiparallel. Such a configuration is available in a single package from several companies.

C. TYPES OF DIODES

Four classes of silicon diodes are discussed in this section:

1. core switching diodes (e.g., 1N4447)
2. ordinary rectifiers (e.g., 1N4004)
3. fast-recovery rectifiers
4. Schottky barrier diodes

The 1N914 and 1N916 switching diodes were developed for use in computer core memories. The 1N4447 is the same semiconductor as the 1N916A, but the 1N4447 has an improved internal heat sink, so it can tolerate about twice the power or current of the 1N916A. The 1N4447 diode features:

1. a small capacitance (less than 2 pF at 0 V across the diode)
2. fast response (reverse recovery time < 4 ns for $I_F = I_R = 10$ mA)
3. quite inexpensive (about \$0.15 each in quantities of 100)
4. relatively low leakage (less than 25 nA at 20 V reverse bias)

The JEDEC ratings specify a maximum dc current of 0.4 A, a maximum current of 1 A for 1 s, and a maximum pulse current of 4 A for 1 μs. It is unusual to find 1 μs pulse current ratings in JEDEC specifications. The reason is that in the original application for these diodes, which was switching magnetic core memories, the current was applied in brief pulses instead of a continuous current.

The 1N4448 is an improved version of the 1N914B. The 1N4448 is similar to the 1N4447, but the 1N4448 has less bulk resistance and slightly greater capacitance (less than 4 pF at 0 V across the diode). The 1N4148 is an industry-standard switching diode and may be somewhat easier to obtain than the 1N4447 or 1N4448 types.

1. Rectifiers

A rectifier is a diode that is designed to conduct steady-state currents of the order of 1 A or more. Ordinary rectifiers are inexpensive and have large surge current ratings. For example, a 1N4004 rectifier that is rated to carry

1 A steady state can tolerate a peak current of 30 A for one pulse of half-wave rectified 60 Hz sinusoidal waveform but can be slow (e.g., reverse recovery times of a few tens of microseconds). Because of this slowness, it is not clear that ordinary rectifiers are useful in overvoltage protection circuits.

Ordinary rectifiers are often an object to be protected. Nearly every electronic circuit that is powered from the ac supply mains has a dc power supply. Overvoltages that propagate on the ac supply mains often damage rectifier diodes in the dc power supply. The design and hardening of dc power supplies are discussed in Chapter 18. Here we are concerned with the vulnerability of rectifier diodes. Martzloff (1964) and Chowdhuri (1965; 1973) found that the ability of a rectifier to survive an overvoltage has *no correlation* with the peak reverse voltage rating specified by the manufacturer. When the rectifiers were conducting, an overvoltage with an open-circuit waveshape of $0.15/6.5$ μs was superimposed so that the rectifier was reverse-biased. Several of the rectifiers failed when pulsed with an overvoltage that was *less* than the peak reverse rating of the rectifier (Chowdhuri, 1973). Increasing the peak reverse voltage rating of rectifiers does *not* increase their ability to survive transient overvoltages.

Fast-recovery rectifiers overcome some of the disadvantages of ordinary rectifiers but are more expensive and have a larger capacitance and larger reverse leakage current than ordinary rectifiers of the same steady-state current and reverse breakdown ratings. State-of-the-art fast-recovery rectifiers in 1984 have reverse recovery times of the order of 30 ns. Other fast-recovery rectifiers have reverse recovery times of the order of 200 ns.

2. Schottky Barrier Diodes

Schottky barrier diodes and rectifiers are among the fastest diodes available. However, they have a rather small reverse breakdown voltage rating (20 to 80 V is a common range) and a rather large capacitance (200 pF at 0 V is typical for a diode that is rated to conduct 1 A steady state). Schottky diodes begin to conduct at about 0.3 V forward bias, compared to about 0.6 V for regular silicon diodes. Schottky diodes also have a much larger leakage current than regular silicon diodes when reverse-biased.

D. LOW-LEAKAGE DIODES

1. JFET Gate to Channel Diode

Electronic instrumentation often requires a very large input impedance (e.g., 10^9 to 10^{12} Ω). An operational amplifier with a junction field effect transistor (JFET) or metal oxide semiconductor field effect transistor (MOSFET) input stage is commonly used. Transient protection of the input

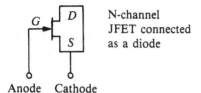

N-channel
JFET connected
as a diode

Anode   Cathode

Fig. 10-5. Connection of N-channel JFET as a low-leakage diode.

terminals is constrained by the specification that leakage currents during normal operation must be less than 1 pA. Very few devices can meet this requirement.

Spark gaps can meet this leakage current requirement, but the firing and glow voltage of a spark gap will kill a semiconductor operational amplifier. Furthermore, spark gaps can be slow to conduct. Thus, spark gaps alone are inadequate protection.

Among semiconductor devices, it has been customary to use the gate to channel diode of a JFET as a low-leakage diode, as shown in Fig. 10-5. These diodes are used in the forward-bias mode during transient protection. Two JFETs connected in antiparallel as a bipolar diode are available in a single package from several manufacturers. For unipolar operation, one can connect a JFET that has a gate to channel leakage current that meets the following specification: $I_G < 1$ pA when $V_{GS} = 0$ at a temperature of 25°C.

A popular model that meets this specification is the 2N4117A (available in a TO-72 metal can package). The same device is available in a less expensive TO-92 plastic package as part number PN4117A. The maximum steady-state gate current for the 2N4117A in a TO-72 metal can package is 50 mA. This is a relatively large gate current, compared to most other JFETs. The V-I characteristic curve for the 2N4117 gate to channel diode is described by Eq. 2 with the following typical values of the parameters:

$$I_S = 0.0019 \text{ pA} \qquad V_T = 28 \text{ mV} \qquad R_{\text{bulk}} = 20 \text{ }\Omega$$

2. GaAs Diodes

In an important article, Damljanovic and Arandjelovic (1981) suggested that GaAsP diodes be used for protection of input terminals of amplifiers with large input impedance. They found that at 0.44 V forward bias, a red GaAsP light-emitting diode (LED) had over seven orders of magnitude less current than a base-emitter junction of a Si NPN transistor. A green LED was even better. This difference is due to both lesser values of reverse saturation current and greater threshold voltage for conduction in GaAs compared with Si.

To confirm and extend this work, Standler (1984) selected four parts to test:

1. Optron OP290, an infrared LED that is designed for operation with large current pulses (maximum 5 A for 250 μs on, 20 ms off) and a maximum steady-state current of 125 mA.
2. General Instrument MV5053, a red LED that has a relatively large steady-state current rating (100 mA max) and is quite inexpensive (about $0.25 each in small quantities in 1983).
3. General Instrument MV5094, a bipolar red LED. This device has a maximum steady-state current rating of 70 mA and is relatively expensive (about $1.25 each in small quantities in 1983). This device may be preferable to the MV5053 only when the available space is small, since just one MV5094 is required for bipolar protection.
4. General Instrument MV5253, a green LED that is specified for 35 mA maximum steady-state current. Green LEDs with greater current ratings are not commercially available.

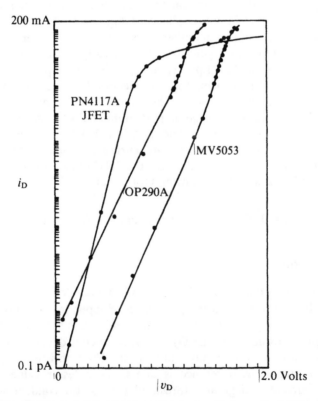

Fig. 10-6. *Plot of measured voltage and current for two GaAs semiconductor diodes and a silicon JFET connected as a diode (logarithmic current scale, linear voltage scale).*

All three of the visible LEDs listed above are rated for a 1 A repetitive pulse current of 1 μs on, 333 μs off.

The relationship between current and voltage in these LEDs and in a JFET gate to channel diode are described by Eq. 2 with the following typical values of parameters at room temperature:

Model and Type	I_S	mV_T	V at $I = 10$ mA
PN4117A JFET	1.9×10^{-15} A	28.3 mV	1.03 V
OP290 infrared	3.4×10^{-13} A	51.5 mV	1.24 V
MV5053 red	1.2×10^{-17} A	47.4 mV	1.63 V
MV5253 green	1.1×10^{-19} A	48.8 mV	1.90 V

Figure 10-6 shows the V-I characteristic curve for the LEDs and a PN4117 JFET used as a diode. The relatively large bulk resistance of the JFET makes it inappropriate for use where the current could exceed about 100 mA. The red and green LEDs have a much smaller current at a given voltage than silicon diodes and are suitable for protecting electrometer inputs, as suggested by Damljanovic and Arandjelovic (1981).

E. PIN DIODES

A PIN diode has a layer of intrinsic (I) semiconductor material between the positive (P) and negative (N) regions. Such diodes are usually employed for switching very high frequency signals ($f > 200$ MHz) or for photosensitive applications. PIN diodes are particularly interesting because they have extremely small capacitance (0.2 to 2.0 pF) and therefore might be useful to protect input terminals of RF receivers. Most other semiconductor devices are unsuitable for such applications owing to an unacceptably large capacitance.

Singletary and Hasdal (1971, p. 31) suggested using a PIN diode in series with a zener diode to reduce the effective capacitance. This valuable suggestion appears to have been ignored for a decade. Lesinski (1983) patented a circuit that used a IN5719 PIN diode in series with a biased avalanche device to provide transient protection for a UHF signal conductor. Apparently PIN diodes have not been characterized with large pulse currents that would be encountered during transient suppression.

11

Thyristors

A. DESCRIPTION OF DEVICES

Semiconductor thyristors are silicon PNPN structures that are useful for switching very large currents. The two most common types of thyristors are the silicon controlled rectifier (SCR) and the triac. The SCR can only conduct current in one direction. The triac can conduct current in either direction and is useful in ac applications. The SCR and triac each have two terminals that act like a switch, and a third terminal (which is called a *gate*) that is used to turn on the device. The switch is normally open. The SCR is turned on by injecting positive charge into the gate and out of the cathode terminal. The triac can be turned on by injecting either polarity of charge into the gate; however, triggering can be accomplished at lesser values of gate current when the gate and main terminal Nr 2 (MT2) have the same polarity.

When the SCR or triac is on, the voltage across the conducting diode is between about 0.7 and 2.0 V, depending on the value of the current, with greater voltages for greater currents. Once the SCR or triac has been turned on, it continues to conduct until the magnitude of the current in the device is decreased to less than the holding current, I_H. Values of I_H at a temperature of 25°C range between about 5 and 50 mA for devices that are rated to conduct a steady-state current between 5 and 15 A rms.

Both the SCR and triac are readily available in models that can conduct rms currents on the order of 5 to 50 A and are widely used as motor speed controllers and light dimmers. The thyristor is designed to routinely tolerate abnormally large currents that occur during starting of motors and the switching on an incandescent lamp.

B. BASIC CROWBAR CIRCUITS

1. SCR and Triac Circuits

Thyristors belong to the family of "crowbar" devices. They are particularly useful in applications where the overstress could have a relatively long duration. The basic SCR transient protection circuit with an avalanche diode and resistor in series with the gate terminal, shown in Figs. 11-1 and 11-2, was patented by Gutzwiller (1965). De Souza (1967) added additional features, which will be discussed in Chapter 18 on protection of dc power supplies. Voorhoeve (1975) used a triac and bipolar avalanche diodes, as shown in Fig. 11-3, to reduce the number of components necessary to protect a circuit from overvoltages of either polarity.

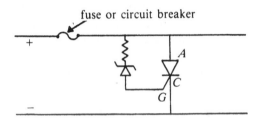

Fig. 11-1. Basic SCR crowbar circuit.

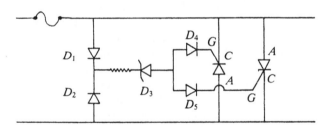

Fig. 11-2. Bipolar SCR crowbar circuit (Gutzwiller, 1965).

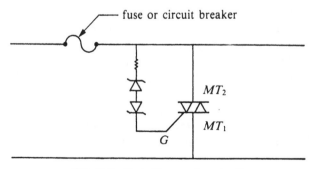

Fig. 11-3. Basic triac crowbar circuit.

A fuse or circuit breaker is required upstream from the thyristor, as shown in Figs. 11-1 through 11-3, whenever the short-circuit current from the normal supply may damage the conducting thyristor. The fuse is supposed to interrupt the current before the thyristor is damaged by prolonged fault current.

All of these circuits have the disadvantage that as the thyristor begins to conduct, the magnitude of the gate current decreases. A large value of dV/dt across any of these circuits will cause the thyristor to conduct. Both of these effects increase the thyristor's vulnerability to damage, as explained below.

Chowdhuri (1974) suggested using a varistor in series with a SCR to turn off the SCR after the transient. This suggestion raises the question "Why not use a varistor alone?" The conducting SCR would have less than 2 V across it, and the varistor might have 300 V across it, depending on the varistor model and the value of the surge current. The varistor will absorb nearly all of the energy and essentially determine the clamping voltage. If a switch is desirable in series with the varistor, a spark gap would probably be less expensive and more robust during kiloampere surge currents. One concludes that there is little advantage in Chowdhuri's suggestion.

2. Failure Mechanisms

An SCR or triac will conduct spontaneously, without deliberately injecting charge into the gate, when the rate of change of voltage, dV/dt, across the main terminals of the thyristor (anode-cathode for the SCR) is sufficiently great. The triggering is caused by currents in the parasitic capacitance of the thyristor. Typical values of dV/dt that will cause triggering are on the order of 5 to 50 V/μs. Such conduction may damage the thyristor, if the magnitudes of the anode current and dI/dt exceed specified ratings for the condition of a marginal gate trigger current. This failure mechanism of thyristors can be avoided by using a trigger circuit that can provide a large gate current or possibly by placing an RC snubber network across the thyristor to reduce the value of dV/dt.

Another failure mechanism for thyristors is caused by a large initial current between anode and cathode in a SCR (or between the main terminals in a triac) before the thyristor is completely turned on. When the thyristor begins to conduct, some regions begin to conduct before other regions. If dI/dt is too great, the higher conductivity regions will develop into "hot spots" with resulting thermal failure of the device. The time required for the thyristor to be completely turned on is a function of the gate current; however, values between 1 and 10 μs are common.

SCRs have been developed that can tolerate values of dI/dt of at least 200 A/μs. One model, the 2N4204, is specified to withstand a value of dI/dt as large as 5000 A/μs.

One can decrease the turn-on time and increase the maximum tolerable

dI/dt by pulsing the gate of the SCR or triac with a very large gate current, 5 to 20 times the minimum gate current that is required to trigger the device when the main terminals carry the maximum steady-state current. Such gate triggering circuits are available in integrated circuit form and are discussed in Chapter 18.

3. Comparison with Other Surge Protective Devices

Because the voltage across a conducting thyristor is less than that for a spark gap, varistor, or avalanche diode, the thyristor operates at a lower power dissipation and temperature than these other devices for a given current. This places less stress across the thyristor, which is desirable for long-duration overstresses.

The thyristor requires several microseconds to become fully conducting. Before the thyristor has become fully conducting, it is vulnerable to damage by large surge currents. Circuits to avoid these problems tend to be complicated and expensive, with the exception of the integrated avalanche diode and SCR discussed below.

A possible disadvantage of SCRs and triacs is that once they are turned on, they can be difficult to turn off unless the current changes polarity regularly (e.g., 50 to 400 Hz mains application) or naturally decays to zero (e.g., capacitor discharge application). In this respect they have the same problem as the power follow of a spark gap that is operating in the arc mode. However, the holding current of a SCR or triac is usually smaller than the arc-extinguishing current for a spark gap.

C. INTEGRATED AVALANCHE DIODE AND SCR

A novel overvoltage protection component was introduced in 1983 that incorporates an avalanche diode and SCR in a single integrated structure. The avalanche diode provides fast response and clamps the voltage across the device until the SCR conducts. The avalanche diode is connected internally in series with the gate of the SCR, so current in the avalanche diode triggers conduction in the SCR. This arrangement gives a large gate current in SCR, which avoids failure mechanisms in the SCR caused by a large initial value of dI/dt. Once the SCR has turned on, there is only a small current in the avalanche diode. In this way the SCR protects the avalanche diode from overstresses that may have a long duration.

The integrated avalanche diode and SCR device is marketed for protection of telephone equipment, but it may have many other applications. Various models are available with breakdown voltages between 60 and 250 V, and a few models are available with breakdown voltages of 18 or 30 V. The minimum value of the holding current, I_H, is usually between 100 and 250 mA, depending on the particular model.

The integrated avalanche diode and SCR structure is inherently unipolar. However, two of these devices can be connected in antiparallel to provide bipolar response. There are several models that have the two devices inside the same package for convenience where bipolar response is required.

Because the integrated avalanche diode and SCR device has been designed and characterized by the manufacturer for use to suppress overvoltages, it is more convenient to design protection circuits with this component than with ordinary thyristors.

A list of companies that manufacture integrated avalanche diode and SCR devices or other thyristors that are useful for suppression of overvoltages is given in Appendix B.

12

Impedances and Current Limiters

This chapter contains information on a variety of useful components in transient protection circuits. Some of these components (e.g., resistors, inductors, capacitors, and fuses) are familiar items to electronics engineers and technicians. Specific details that are important in transient protection applications, which are often rather specialized knowledge, will be described here. Other components, such as positive temperature coefficient resistors and common-mode chokes, are specialized components that will be described more completely.

A. RESISTORS

Resistors are often included in hybrid transient protection circuits—for example, between a spark gap and an avalanche diode. Although this is not a precision application for a resistor, the resistor must remain undamaged by exposure to overstresses. There are several ways that the resistor can fail to function as intended.

1. A spark discharge could form in the air surrounding the resistor (a phenomenon called *flashover*). If the spark goes to ground, the electronic devices downstream are not likely to be damaged. If the spark were to shunt the resistor, catastrophic damage to electronic devices downstream would be possible.
2. Dielectric breakdown could occur inside the resistor package. The resulting arc could shunt part or all of the resistive material, with a

resulting increase in surge current. The increased current may damage devices that are downstream from the resistor.

3. Overheating produced by very large power dissipation during a surge could vaporize the resistance material and cause the resistor package to explode. If the resistor is in series with the normal signal path, such an event could cause the system to fail, just as if a wire was broken.

4. Overheating produced by large transient power dissipation could permanently damage the resistance material (e.g., cause a permanent change in the resistance value of the order of 10% or more), without producing a short or open circuit. The resistor would probably continue to be functional for transient protection applications after a few such changes.

Consider a 22 Ω carbon composition resistor with a steady-state maximum power rating of 0.5 W that is connected between a spark gap and a 15 V avalanche diode. If a rectangular pulse impressed 500 V across the spark gap, the resistor would have an instantaneous power dissipation of about 11 kW before the spark gap conducts. Most engineers who are unfamiliar with transient design would look at the 11 kW power dissipation and conclude without hesitation that the resistor would be destroyed. If the spark gap were to remain nonconductive for 1 μs (not an unrealistic assumption), about 11 mJ of energy would be deposited in the resistor. The energy deposited in the resistor by the pulse is small, and the resistor would almost certainly survive. Once the spark gap goes into the arc mode (20 V across the spark gap), the power dissipation in the resistor is only about 1.1 W, a modest overload.

Several empirical studies have been done on the ability of resistors to tolerate large pulse currents (Lennox, 1967; Domingos and Wunsch, 1975). The conclusion of this research is that carbon composition and wire-wound resistors are able to survive large transient power dissipations. Metal film and carbon film resistors are not as suitable owing to concentration of current in a thin layer of material and owing to internal dielectric breakdown between adjacent turns of resistive material in a helix pattern.

A type RN65 100 Ω metal film resistor, which has a steady-state power rating of 0.25 W, can survive a pulse with a 4 mJ energy content, whereas a carbon composition resistor with the same ratings that is manufactured by Allen Bradley can survive 50 mJ (Lennox, 1967). Allen Bradley carbon composition resistors with a steady-state power rating of 1 W and a resistance between 51 Ω and 20 kΩ can survive pulses of at least 200 mJ. Anecdotal evidence suggests that carbon composition resistors from other manufacturers are less robust.

In addition to the maximum power rating, which is familiar to all electronic engineers and technicians, resistors also have a maximum working voltage rating. Table 12-1 gives the maximum steady-state working voltage rating between the two terminals of the resistor for some common

TABLE 12-1. Maximum Steady-State Working Voltage Ratings for Resistors

MIL style	Description	Max Working Voltage
RC07	0.25 W carbon composition	250 V
RC20	0.5 W carbon composition	350 V
RC32	1.0 W carbon composition	500 V
RC42	2.0 W carbon composition	750 V
RW67V	6.5 W wire wound	400 V

devices. For resistors with a small value of resistance, the maximum power dissipation, P_m, may establish a lower maximum steady-state voltage,

$$\sqrt{RP_m}$$

than that given in Table 12-1.

Transient operation of the resistor at greater voltages invites premature failure from internal dielectric breakdown. How to apply this information to transient overvoltage applications is not clear. A conservative design can be done by specifying resistors with steady-state voltage ratings equal to or greater than the worst-case transient situation.

The reader is cautioned that there is another maximum voltage given in resistor specifications, called *dielectric withstanding voltage*. This value is the maximum voltage between either terminal and the exterior of the case of the resistor. The value of the dielectric withstanding voltage is important when the resistor is to be mounted next to a conducting metal chassis.

Real resistors have parasitic shunt capacitance and series inductance. Woody (1983) measured values between 1 and 2 pF for 17 different resistors of various types with a typical value of 1.6 pF. At 10 MHz, a capacitance of 1.6 pF has a reactance of about 10 kΩ, which is much larger than the resistance values of series resistors that are commonly used in circuits for protection against transient overvoltages. Thus, the parasitic capacitance of resistors is not expected to be a problem in overvoltage protection applications.

Woody (1983) measured the parasitic inductance of 15 different resistors. He found about 20 nH inductance in an Allen Bradley 56 Ω, 0.25 W carbon composition resistor. Wire-wound resistors had much larger parasitic inductance: values ranged from about 0.3 μH to more than 1 μH. Because transient overvoltages are a high-frequency phenomenon, the inclusion of parasitic inductance in the real resistor increases the magnitude of the impedance during transients. Since resistors are included in transient protection circuits to increase the impedance between two points (and thus limit the current), the parasitic inductance will usually be helpful in transient protection applications. For example, a parasitic inductance of 0.3 μH has a reactance of about 19 Ω at 10 MHz. If this inductance is in series with a 10 Ω resistance, the magnitude of the impedance will be about 21 Ω, more than twice the resistance.

The idea of placing a resistor in series with a node to be protected is so simple that it is often ignored as a protective technique. A resistor alone is probably the least expensive transient protection circuit, although it has limited ability to protect vulnerable devices. Van Keuren (1975) showed that the minimum failure level for a particular integrated circuit line driver increased from 38 V to more than 100 V when a 270 Ω resistor was inserted in series with the input terminal. Burger (1974) briefly discussed protection of semiconductor junctions with a series resistor.

B. POSITIVE TEMPERATURE COEFFICIENT DEVICES

Nearly all nonlinear devices that are used for transient protection of electronics have a smaller resistance for greater voltages across the device. Such devices are suitable for shunt elements in protection circuits. It would be attractive to also have a device whose resistance increases as the voltage across it increaes. This kind of device would be suitable for series insertion and would decrease the power dissipated in shunt protective elements that are located downstream from the nonlinear series element. Unlike the shunt elements, the nonlinear series element does not necessarily require a capability to carry a large current or absorb a large energy pulse. However, the series element must be able to tolerate large transient voltages across the device without breakdown or flashover.

There is a remarkable family of devices known as *positive temperature coefficient* (PTC) *resistors*. The PTC resistors have an approximately constant resistance at temperatures below their switch temperature, T_S. At internal device temperatures above T_S, the resistance increases dramatically. The normalized temperature coefficient, $(1/R)(dR/dT)$, often has values between 30 and 60%/°C. The maximum resistance is typically a factor of 10^4 greater than the resistance below T_S, the so-called "cold" resistance. Because of this dramatic change in resistance, PTC resistors are used as resettable fuses to protect motors or transformers from overheating or from damage by sustained excessive currents.

Most PTC resistors are fabricated from a bulk barium titanate semiconductor, $BaTiO_3$ (Andrich, 1966; Saburi and Wakino, 1963). One company fabricates PTC resistors from conductive polymers. The polymer device has carbon black dispersed in a plastic matrix. Upon reaching the switch temperature, the plastic has a phase change, expands, and disrupts the conducting paths of carbon (Doljack, 1981). When the plastic cools, the polymer PTC device recovers its original cold resistance.

Typical values of T_S are between about 40 and 120°C, depending on the materials in the PTC resistor. Values of cold resistance range from about 22 to 330 Ω for barium titanate devices and from about 0.01 to 22 Ω for conductive polymer devices.

The barium titanate PTC devices exhibit a negative temperature coefficient behavior when sufficiently large voltages are applied across the device. If such large voltages are maintained for a time on the order of 1 second or more, thermal runaway will occur with consequent degradation or destruction of the device. The designer of transient protection circuits should check prototypes of barium titanate PTC devices to be certain that after switching they maintain suitably large resistances at all voltage levels that could be encountered in the application. PTC devices that are rated for 120 V rms (or more) maximum voltage appear to be suitable for insertion between a spark gap with a dc firing voltage of 90 to 150 V and an avalanche diode with a smaller conduction voltage.

The conductive polymer PTC devices appear to be able to withstand very large voltages across the terminals without damage, and they do not exhibit the negative temperature coefficient of barium titanate devices. Moreover, the polymer PTC devices have a larger temperature coefficient than barium titanate devices, so the switching action of the polymer PTC devices is more abrupt.

Manufacturers of PTC resistors are listed in Appendix B.

C. INDUCTORS

1. Choke Coils

Because an inductor has a voltage drop that is proportional to the rate of change of current in it, the inductor is an attractive device for transient protection circuits. To be suitable for transient protection applications, an inductor should meet all of the following criteria:

1. Thick insulation on wire in the coil to prevent dielectric breakdown between adjacent turns. Sherman (1975, p. 108) recommends magnet wire with a thick varnish insulation. Regular hook-up wire with polyvinyl chloride or polyethylene insulation with a thickness of at least 0.4 mm (and 1 kV rating) would probably be even more suitable; however, it also has a larger volume.
2. The high-permeability core to the inductor (e.g., iron or ferrite) should not saturate at small values of transient current.
3. The inductor core should be insulated from the coil with tetra-fluoroethylene (TFE) tape to prevent both dielectric breakdown and removal of insulation from the wire by mechanical stress (Sherman, 1975, p. 108). One might also consider wire with a nylon jacket over polyvinyl chloride insulation.
4. The coil should have a single-layer construction. This minimizes the parasitic capacitance shunted across the inductor. Also, the large voltage that may appear between wires in adjacent layers may lead to insulation breakdown during a transient overvoltage.

5. Wire with a diameter of at least 0.6 mm should be used in the inductor. During surges with large values of current, the coil will encounter large mechanical stresses due to magnetic forces. If the coil is not properly designed, such forces can cause the wire to break. miniature coils that are designed for use in radio frequency circuits with steady-state currents of the order of 10 mA are *not* suitable for transient overvoltage applications.

For circuits that have a normal current on the order of 10 mA or less, the *IR* drop across a series resistor of 10 to 50 Ω is not particularly objectionable. Most analog and digital signals are in this range of currents. Since resistors are less expensive than inductors, most circuit designers will specify a resistor rather an an inductor. For these reasons, inductors have been more commonly used in transient protection circuits for power lines (dc to 400 Hz normal operation) than for analog or digital data lines. The large currents in power circuits make appreciable series resistance undesirable, so an inductor is commonly used as a series device in power circuits. The construction and use of LC filters is discussed in Chapter 13; specific design techniques for filters for the ac supply mains is in Chapter 19.

2. Ferrite

The inductance of a coil with a given shape and number of turns can be increased by placing a high-permeability core inside the coil or surrounding the conductor. For attenuation of transients and noise, ferrite material is generally recommended for the core. Ferrites are a ceramic material with a large permeability and small conductivity. At frequencies above about 1 MHz, the effect of a ferrite core on a circuit can be modeled as a resistance in series with an inductance. The presence of the resistive (or lossy) part of the impedance will damp any oscillations caused by the interaction of the inductance with capacitance elsewhere in the circuit. In addition, the resistive part of the impedance will act to dissipate transients rather than just reflect them (Sherman, 1975, pp. 106–107).

Small cylinders of ferrite material with an axial hole for conductor(s) are called *ferrite beads* and are widely used to cure electromagnetic interference problems. Placing a typical ferrite bead on a wire adds about 5 μH of inductance.

Large surge currents can make intense magnetic fields inside the core of inductors. When ferromagnetic materials, such as ferrite, are placed inside the core to increase the inductance, the magnetic field may saturate during surges. This will make the inductance decrease. In fact, the inductance can vanish when the core is completely saturated.

D. COMMON-MODE CHOKE

A common-mode choke is a four-terminal component that is useful for inserting a large inductance in series with common-mode sources but has

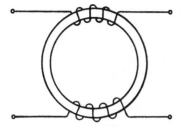

Fig. 12-1. *Common-mode choke.*

negligible inductance in series with differential-mode sources. In this way, the use of a common-mode choke on a balanced line has a minimal effect on desired signals, which are differential-mode, but offers substantial impedance to common-mode noise. All common-mode chokes have two independent coils of the same size and number of turns. These coils are commonly wound on a toroidal core of ferrite, as shown in Fig. 12-1. The continuous toroidal core structure is particularly desirable, because it has no air gap to radiate magnetic field and increase ambient electromagnetic noise levels. The use of a ferrite core introduces an appreciable resistive (lossy) component to the reactance, which is a desirable feature for noise suppression and damping of resonance in inductor-capacitor filters. The design and use of common-mode chokes have been discussed by Kübel (1975).

Use of common-mode chokes if one of the oldest transient protection techniques. A relatively ancient reference to common-mode chokes is in a patent by Rovere et al. (1934), who were interested in protecting carrier telegraph and telephone circuits from lightning.

A typical common-mode choke for use on the ac supply mains has an inductance between about 1 and 30 mH to common-mode signals. The inductance for differential-mode signals is of the order of 30 μH (Standler, 1984). The small differential-mode inductance prevents saturation of the magnetic core by steady-state currents in power circuits, which are often more than 1 A. For this reason, this component is sometimes called a *current-compensated choke*.

E. CAPACITORS

Capacitors are commonly used as shunt elements in low-pass filters and as bypass applications on dc supply lines. Capacitors may also be used as series elements when the signal frequency is greater than about 10 MHz, as in radio frequency applications. The important properties of capacitors in such applications are capacitance, dc voltage rating, parasitic inductance of the real capacitor, and failure mechanism when the voltage rating of the capacitor is exceeded.

An ideal capacitor has an impedance that is proportional to $1/f$, where f is frequency. However, for all lumped-element capacitors (the type used in circuits below a few hundred megahertz), there is a resonance frequency f_0 caused by the internal inductance of the capacitor. At frequencies greater than f_0, the impedance of the capacitor is dominated by the inductance term. The impedance is then proportional to f, instead of the $1/f$ proportionality that one would expect of an ideal capacitor. The value of f_0 is usually between 1 and 100 MHz for nonelectrolytic capacitors that are commonly used in electronic circuits. Therefore a capacitor in a circuit may behave like an inductor when transients with appreciable high-frequency content are present. Plots of impedance versus frequency and techniques for reducing the effects of inductance in capacitors on circuit performance are discussed in Chapter 15 on parasitic inductance.

Tasca (1981) described tests of various capacitors when charged with a pulse of constant current. Breakdown voltages of glass, mica, ceramic, plastic, and paper dielectrics are independent of the rise time of voltage over the interval between 0.01 and 20 μs. For capacitors with a dc breakdown voltage rating of 100 to 600 V dc, the actual breakdown voltage is usually about a factor of 10 to 15 greater than the rated voltage. The ratio of actual breakdown voltage to the rated voltage gives a safety factor. About 0.3% of the capacitors have a safety factor as small as 2. There is a comfortable margin of safety between the rated voltage and the actual breakdown voltage.

Tasca (1981) found that tantalum electrolytic capacitors, when pulsed with a polarity that was the same as the normal working polarity, exhibited a breakdown behavior similar to an avalanche diode. The breakdown voltage across the tantalum capacitor was approximately constant at a value between 2 and 10 times the maximum rated working voltage.

When capacitors are used in applications where the voltage rating may be exceeded, it is important to specify "self-healing" capacitors. Such capacitors are commonly made with thin metal electrodes that are deposited on a sheet of dielectric material (Maylandt and Sträb, 1956; Netherwood, 1956, 1965). The thickness of the electrode is usually between about 0.03 and 3 μm. If the dielectric should break down, the resulting arc evaporates the thin electrode in the region of the arc. The arc then extinguishes automatically when the radius between the dielectric puncture site and the metal electrode becomes too great to maintain the arc. This process is known as *self-healing*. After a number of internal arcs there may be changes in the capacitance and insulation resistance, which degrade the capacitor. The degradation can be minimized by careful design of the capacitor.

Self-healing capacitors with a dielectric, for example, of polyester are known as *metalized* polyester types. Inexpensive plastic dielectric materials tend to be flammable.

Early self-healing capacitors had a paper dielectric. Paper dielectric capacitors are obsolete in essentially all applications except bypassing the ac supply mains, an application that is relevant to transient protection. However, it is possible to make an inexpensive capacitor with a paper dielectric that is impregnated with nonflammable materials so that the capacitor will not cause a fire if the capacitor suffers a severe internal dielectric breakdown.

It may seem plausible to use a shunt capacitor to store the charge that is transferred during electrical overstress, thus reducing the voltage. This technique can be quite effective for electrostatic discharge (ESD), where the total charge transfer is usually less than a few microcoulombs. However, when large transients from lightning, NEMP, or switching reactive loads are encountered, diverting the transient with a shunt capacitor is uneconomical. For example, the standard $8/20\,\mu s$ test waveform with a peak current of $3\,kA$ transfers $0.05\,C$ of charge. A capacitance of about $5000\,\mu F$ would be required to reduce the overvoltage to $10\,V$. This large value of capacitance is economically obtainable only with electrolytic capacitors. Such capacitors have relatively large internal series inductance and are not effective for transient suppression or pulse current applications. Even if one could obtain a suitable capacitor, the drastic reduction in system bandwidth and the increase of output current drive requirements resulting from insertion of such a large capacitance would probably be unacceptable except on dc power buses.

F. FUSES AND CIRCUIT BREAKERS

The conventional wisdom is that fuses and circuit breakers are too slow to be effective in transient overvoltage protection circuits. Nevertheless, there are several reasons to discuss them. First, generalizations that are usually valid have a way of inhibiting progress by restricting designers from breaking the rule in exceptional cases. Second, fuses continually appear in overvoltage protection circuits, often in ways that may do more harm than good. Third, there are valid reasons to include fuses in the output of dc power supplies and to isolate defective loads from an ac power line. Fuses might be included in an overvoltage protective circuit, with resulting increase in reliability of the system.

There are many different types of fuses and circuit breakers, each of which is available in a variety of different current ratings. A fuse is an expendable component that is designed to become a permanent open circuit when exposed to excessive current. A circuit breaker is a device that opens a circuit when excessive current is detected but that can be reset manually (or automatically on some models) to restore the circuit continuity.

The common model of "fast" fuse used in electronic equipment is a glass cylinder 1.25 inches in length, 0.25 inch in diameter (known as type 3AG or AGC). Data in technical literature from fuse manufacturers state that the

1 A rated model typically takes 1 second to open at a 2 A current (maximum of 5 seconds at 2 A). At very large overloads the fuse is much quicker: at 10 A, the 1 A fuse typically takes between 7 and 20 ms to open.

Circuit breakers tend to be slower than fuses, although they are satisfactory to prevent fires from excessive current in wires.

The value of the action integral, $I^2 \Delta t$, required to open a fuse was essentially constant from 0.01 s to about 10 μs (Martzloff, 1985). A "fast" 10 A fuse had $I^2 \Delta t \approx 200 \, A^2 \, s$. To open this fuse during an 8/20 μs waveform, a peak current of 9 kA was necessary. Martzloff (1985) also presented evidence of mechanical fatigue of fuse wire caused by magnetic force during large current surges.

Figure 12-2 shows three possible ways to use a fuse with a shunt protective device, such as a varistor. Before the circuits shown in Fig. 12-2 are discussed, it is worthwhile to explicitly mention that the fuse shown in Fig. 12-2 is symbolic of any device that normally has a low impedance but

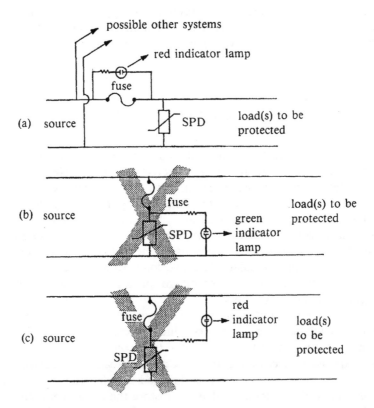

Fig. 12-2. *Three ways to connect a fuse and a surge protective device. The disadvantages of methods (b) and (c) are discussed in the text.*

changes to a very large impedance when excessive current passes through it. In particular, the fuse could be replaced by a PTC resistor or a circuit breaker. Likewise, the varistor shown in Fig. 12-2 is symbolic of any device that normally has a large impedance but changes to a very small impedance during electrical overstress. In particular, the varistor could be replaced by a spark gap, an avalanche diode, or a triac.

There are several possible hazards that could result in damage to the source in Fig. 12-2: (1) the load may fail and draw excessive current; (2) a crewman or repairman may make a mistake while connecting cables and inadvertently short-circuit the source; (3) the insulation on the wires may be damaged and short-circuit the source; or (4) the varistor might fail as a short circuit. The circuit in Fig. 12-2a protects against all of these hazards to the source, whereas the circuits in Fig. 12-2b and c *only* protect the source from a shorted varistor. The circuits in Fig. 12-2b and c do not provide comprehensive protection for the source. Therefore, one *must* use the circuit of Fig. 12-2a when *two or more critical and independent systems* are to be operated from the same source.

The circuit of Fig. 12-2a cannot be used in some critical systems where it is intolerable to risk failure of the system due to interruption of power by a blown fuse. The circuit of Fig. 12-2b is often recommended, because it cannot interrupt power to the critical load. This circuit also has the advantage that it will jettison the varistor if the varistor draws excessive current and threatens the power supply or load. Although many engineers have endorsed the circuit shown in Fig. 12-2b, it has some major shortcomings, many of which are not obvious.

First, the resistance of the fuse will increase the clamping voltage across the power circuit and thus partly defeats the purpose of the varistor. This is not a trivial issue: a 1 A fuse has a resistance between about 0.1 and 0.3 Ω (depending on ambient temperature and current). At a transient current of 1 kA, this resistance of the fuse contributes an extra 100 to 300 V of overvoltage across the load.

Second, the parasitic inductance of the fuse can increase the clamping voltage across the power circuit. The standard glass fuse and a minimal length of connecting wire have a length of about 5 cm, which has inductance of the order of 0.1 μH. At a value of dI/dt of 1 kA/μs, the extra overstress across the load due to the inductance of the fuse is about 100 V.

Third, if the fuse blows during a transient when the current is rapidly changing, the sudden interruption of the transient current can create a very large overvoltage upstream from the fuse owing to the inductance in the wiring upstream from the fuse. This effect has been observed by Smith (1973, pp. 5, 9). Smith commented that the overvoltage that was produced by the opening of the fuse was more severe than the overvoltage that was applied to test the protection circuit. Meissen (1983) studied the production of overvoltages by fuse opening.

Fourth, once the fuse opens, the system is without overvoltage protection. During a thunderstorm or nuclear war, the system might survive for only a few tens of seconds before the next overvoltage destroys the unprotected system.

If reliability of the system is indeed paramount, as was asserted during the rejection of Fig. 12-2a, one should ask, "Do overvoltages really threaten the system?" If overvoltages are a threat, then provide a varistor that is more than adequate to withstand the worst-case overvoltage overstress and omit the fuse. If overvoltages are not a threat, then the varistor and the fuse are both superfluous (but the varistor alone might be included "just in case").

If the circuit of Fig. 12-2b is to be used in spite of its shortcomings, it would be prudent to include a lamp as shown in Fig. 12-2b to indicate that the varistor is still connected to the circuit and providing protection. Because the lamp is on during a "good" condition and off during a "warning" or "failure" condition, the lamp should have a green color, not a red or orange color. Miniature fluorescent glow lamps are available in green; long-life tungsten lamps (with a green lens) or green LEDs are also possibilities. A disadvantage of this scheme is that a lamp that signals failure by turning off does not naturally attract attention and is therefore poor human factors engineering. Further, in applications where the varistor is required to be installed in a remote location, extra cabling will be needed to route current to the lamp, since the lamp must be located in an area where it is readily observable. Replacement of the fuse in such a situation might be a major maintenance task. If the fuse is located in a convenient place, far from the varistor, the inductance in the cable to the fuse can vitiate the overvoltage protection offered by the varistor.

The circuit of Fig. 12-2a, however, can use a neon lamp and a series resistor to display a bright red light to signal failure of the fuse. This lamp may be mounted far from the fuse without compromising the operation of the circuit.

The circuit of Fig. 12-2c is a particularly bad one. This circuit has the same undesirable features as Fig. 12-2b, since the fuse and SPD are connected in the same way. The only difference between Fig. 12-2b and 12-2c is the way the indicator lamp is connected. In the circuit of Fig. 12-2c the indicator lamp is connected. In the circuit of Fig. 12-2c the indicator lamp is lit when the fuse has opened *and* the SPD has shorted. This would seem to be a good feature, since the lamp calls attention to the defective protection circuit. However, suppose the overstress opens the fuse but does not short the SPD. Then the SPD will not be connected across the line (the resistance and inductance of the tungsten lamp vitiate the protection), but the lamp will not indicate this failure state. Alternatively, suppose the circuit operates "properly" when a large overstress shorts the SPD and blows the fuse. The light then comes on. However, the next large overvoltage may cause the lamp to become an open circuit, since there is no longer any overvoltage protection in the circuit. If someone did not notice (and remember) that the lamp *was* on briefly and repaired the circuit, the system would remain without overvoltage protection. This circuit has been developed for use by the military (Reynolds, 1972). It would be prudent to avoid this circuit.

G. SUPERCONDUCTORS

Recently there has been an enthusiastic quest for compounds that will be superconductors at room temperature. When such compounds are available for commercial use, they will have obvious applications in design of high-efficiency electric power distribution, electromagnets, and motors. A possible early application of these room-temperature superconductors is to fabricate a series overvoltage protection device. When the current density, J, in a superconducting material exceeds a critical value, the superconductivity ceases, and the material reverts to its normal behavior. The critical current density depends on the type of material and the external magnetic field. The critical current is smaller as the external magnetic field increases.

These properties might make superconductors a useful series protection device. When the current in the superconductor reaches the critical limit, the material becomes an insulator and blocks the overvoltage. The superconducting series protection device would be applied in the same way as the PTC resistors described earlier in this chapter. However, the superconducting devices would not have the time delay that is characteristic of temperature-responsive devices such as the PTC resistors.

The major disadvantage of superconducting series protection devices is that once either the critical current density or the critical magnetic field is exceeded, the material becomes an insulator. To restore the superconductivity, the material must be annealed, something that is difficult to do with a component in an electronic circuit.

13

Filters

A. INTRODUCTION

Low-pass filters are commonly connected in series with the power cord for electronic equipment to achieve electromagnetic compatibility. These filters attenuate high-frequency noise that is generated inside the chassis and conducted on the power cord out of the chassis and into the environment. This application of filters is well-known and is covered in many standards and regulations. Low-pass filters may also protect equipment from some disturbances on the mains, such as high-frequency noise and transient overvoltages.

Filters are also connected to data and signal lines to attenuate unwanted voltages that would otherwise cause upset. Some of these filters are inadvertent: they arise from parasitic capacitance of shunt surge protective devices, such as avalanche diodes.

Is it reasonable to use a linear device, such as a filter, *alone* to protect a circuit or system from transient overvoltages? The answer is often no, as will be explained later in this chapter. However, filters can be very useful to attenuate the remnant that propagates downstream from a nonlinear device such as a spark gap or metal oxide varistor. The nonlinear device can protect both the load and the filter from damage and degradation, and the filter gives some additional protection to the load. In this way nonlinear devices and filters can be complementary, not antagonistic.

Protecting systems from damage and upset by transient overvoltages and attenuating conducted noise to acceptable levels is a difficult engineering problem. Reliable operation of an electronic system from the ac supply mains will probably require *both* nonlinear devices, such as metal oxide varistors, *and* low-pass filters connected to the mains.

The discussion of filters is begun with a review of the ordinary use of filters to attenuate conducted noise. Then the problems of using filters alone to provide protection against overvoltages are discussed. Finally, several brief examples of using filters with overvoltage protection devices are given. A more extensive treatment of low-pass filters and surge protective devices for applications on the mains is given in Chapter 19.

B. CUSTOMARY USE OF FILTERS

Prefabricated filters are one of the principal components available to attenuate high-frequency (e.g., greater than about 100 kHz) noise conducted from systems into mains and into communication cables, or to prevent such noise from entering a system on the same conductors. For optimal benefit, these filters must be used in conjunction with an effective shielding technique.

To save space and maintain the integrity of shield topology, filters are being included in pins inside connectors. The mating connector is attached to a conducting plane, commonly a bulkhead or chassis. The noise removal must be accomplished in a feed-through arrangement as shown in Fig. 13-1. Figure 13-2 illustrates two other ways to connect a filter, both of which are undesirable. No other good choices exist, given a single-layer shield.

1. Applications on the Mains

High-frequency noise is transferred from the mains through step-down transformers by parasitic capacitance between the primary and secondary coils (Bull, 1968, p. 15). Therefore, the high-frequency noise across the

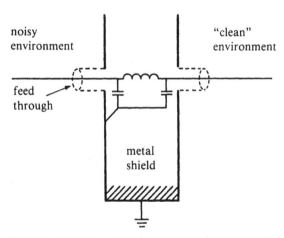

Fig. 13-1. Use of low-pass filters with an electrostatic shield.

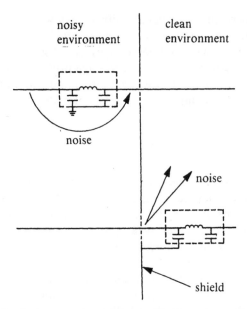

Fig. 13-2. *Two bad ways to use a low-pass filter to remove noise from a line.*

secondary coil may be larger than would be expected on the basis of the turns ratio and the amplitude of the noise on the mains. One cannot rely on common dc voltage regulator circuits to attenuate high-frequency noise or transient overvoltages, because these circuits do not reject high-frequency noise at the same degree as the ripple voltage at twice the mains frequency on account of limitations imposed by the gain-bandwidth product. Electrolytic filter capacitors that reduce the ripple in dc power supplies have appreciable series inductance which makes them ineffective for bypassing noise to ground at frequencies of about 1 MHz or greater. Therefore, high-frequency noise or transient overvoltages on the mains may be able to propagate through a dc power supply and upset the operation of an electronic circuit.

Common low-pass filter modules for commercial service on the mains provide approximately 30 to 70 dB of attenuation in a 50 Ω system at frequencies between 0.15 and 30 MHz. Owing to limitations on the physical size of inductors and capacitors, most filters have little attenuation at frequencies below 0.15 MHz. However, mains filters for switching power supplies often have significant attenuation at frequencies as low as 10 kHz. Above about 30 MHz, most noise is transferred in or out of a chassis by radiation through space rather than by conduction along wires.

2. Simple Filter Circuit

Many power and signal paths use three or more conductors, so that both common and differential modes are present. The simple π filter, shown in

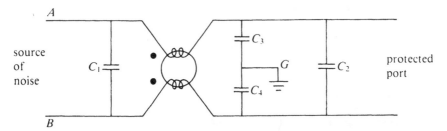

Fig. 13-3. *Simple filter for attenuation of both common- and differential-mode noise.*

Figs. 13-1 and 13-2, can be used only on a two-conductor system. A simple filter is shown in Fig. 13-3 for use on a system with two conductors, A and B, and a third conductor, G, for grounding. The inductor is a common-mode choke, which was described in Chapter 12. This inductor presents a minimal series impedance to differential-mode signals that propagate in opposite directions on conductors A and B and have no current in the grounding conductor. This inductor can present a large series impedance to common-mode signals that propagate in the same direction on conductors A and B and return through the grounding conductor.

There are two practical ways to connect capacitors in a low-pass filter shown in Fig. 13-3. One or more capacitors, such as C_1 and C_2 in Fig. 13-3, can be shunted between the signal or mains conductors to attenuate differential-mode noise. A pair of capacitors, such as C_3 and C_4 in Fig. 13-3, can be shunted between the grounding conductor, G, and each single conductor to attenuate common-mode noise. The capacitance values of C_3 and C_4 are nominally identical, so common-mode noise is not converted to an appreciable differential-mode voltage by unbalanced impedances.

Series impedance upstream from the filter acts with capacitor C_1 to provide attenuation (voltage division) of high-frequency noise. The parasitic differential-mode inductance of the choke forms a voltage divider with capacitor C_2 to further attenuate differential-mode noise. The large common-mode inductance of the choke forms a voltage divider with C_3 and C_4 to attenuate common-mode noise.

For filters on the mains the capacitance of C_1 and C_2 is usually between about 0.1 and 0.5 μF, and it can be made larger. For filters on the mains, the capacitances of C_3 and C_4 must be small for safety reasons; values of the order of 2 nF are typical. If the ground connection were open, an electric shock hazard would exist if the capacitance between the hot conductor in the mains and the chassis ground is large.

The maximum capacitance between the mains and ground is determined by the allowable leakage current. The leakage current is measured between the conducting case (which must normally be connected to circuit ground) and the mains neutral conductor when the filter is connected to the mains without a ground wire. The maximum leakage current and the maximum capacitance for use on 120 V rms, 60 Hz mains are given in Table 13-1.

TABLE 13-1. Maximum Leakage Current and Capacitance*

Environment	Maximum rms leakage current	Maximum capacitance line to ground 120 V rms, 60 Hz Service
Medical equipment:		
patient care area	0.1 mA	2.2 nF
Residential and		
nonpatient-care medical	0.5 mA	11 nF
Commercial	3.5 mA	77 nF

*Maximum leakage current specifications are contained in Section 27 of Underwriters Laboratory Standard 544 (for medical equipment) and in Section 24 of Underwriters Standard 1283 (for mains filters).

C. PROBLEMS WITH THE USE OF FILTERS AS A SURGE PROTECTIVE DEVICE

1. General

There are several reasons why filters *alone* are not a good surge protective device:

1. The spectrum of the overvoltage may overlap the spectrum of the desired signal.
2. To use a linear device for overvoltage protection is expensive and bulky.

Further problems arise because filters are normally characterized with a sinusoidal voltage with an amplitude of only a few volts in a test fixture with an impedance of 50 Ω. Attenuation data from these measurements are not relevant to performance during high-voltage transients for the following reasons:

3. Insulation of components inside the filter can break down during overvoltages, which acts as an unexpected source inside the filter.
4. The 50 Ω impedance of the test fixture damps resonances inside the filter and obscures the possibility of resonance when the filter is used with reactive source or load impedances.
5. The inductance of chokes typically decreases during large currents owing to partial or complete saturation of ferromagnetic core material. Abnormally large currents may flow in chokes during high-voltage transients.

These considerations are discussed in the following paragraphs.

2. Spectrum of Surges and Linear Devices

As discussed in Chapter 5, the frequency spectrum of various overvoltages extends from dc to at least 1 kHz. The spectrum of the desired signal often has a wide bandwidth that overlaps the spectrum of the overvoltage, so it is not possible to discriminate between the overvoltage and desired signal with a filter.

It is difficult for any linear circuit, such as a filter, to provide adequate attenuation of high-voltage transients. Consider a situation where the incident wave has a peak of 5 kV and it is required to have a maximum output voltage of 10 V. The filter would be required to have an attenuation of about 50 dB across the spectrum of the transient overvoltage. It is difficult to design filters with such a large attenuation at frequencies below about 10 kHz, a frequency range where overvoltages have appreciable energy. If a filter can meet the required attenuation specification, it is likely to be expensive and occupy a large volume.

3. Insulation Breakdown

The very large voltages and currents in severe transient overvoltages can overwhelm filters. The input capacitor, C_1 in Fig. 13-3, is particularly vulnerable to damage by transient overvoltages on the mains. Consider the common 8/20 μs current waveform with a peak of 500 A that is used for testing surge-protective devices. This test waveform transfers 8.7 mC of charge. If we consider a typical π filter with two 0.5 μF capacitors and a 30 μH choke, the potential difference across the capacitors will have a peak value of about 9.6 kV. This large voltage across the capacitor could cause dielectric breakdown, followed by an arc inside the capacitor. The peak voltage of 9.6 kV that is predicted by calculation would probably not occur in a realistic situation: common electrical outlets flash over at a potential of about 6 kV. Whatever the peak voltage may be during a severe overstress, it still threatens the dielectric in shunt capacitors.

To avoid destruction of shunt capacitors in unprotected filters during operation on the mains, specifications commonly address behaviour during high-voltage stresses. For example, Underwriters Laboratory Standard 1283 requires filters that are rated for 250 V ac or less to withstand a potential difference of 1414 V dc between the hot and neutral terminals and a potential difference of 2121 V dc between ground and either the hot or neutral conductor. However, at larger overvoltages the shunt capacitors may break down. If the breakdown can be confined to "self-healing" capacitors, which were discussed in Chapter 12, the breakdown *may* have no serious consequences for the reliability of the filter module. However, dielectric breakdown of shunt capacitors acts as an unexpected source, which has a large value of dV/dt, that will cause the actual attenuation of the filter to be much less than the expected value.

Large voltage stresses across series inductors may cause insulation breakdown and shunt the inductor with an arc. When this occurs, large surge currents can be passed directly to the output port (see Fig. 13-3) with catastrophic effects on the "protected" load.

4. 50 Ω Impedance in Test Fixture

Attenuation data in manufacturer's specifications are measured with a 50 Ω resistive load. Although this is a reproducible test method, it does not correspond to actual operating conditions (Schlicke and Weidmann, 1967). This is particularly insidious. The 50 Ω resistive load in the standard test method damps resonances that may be presented with the LC elements inside the filter. Furthermore, a reactive load during real-world operation of the filter may cause resonance with the output port of a low-pass filter and produce voltage gain (Ricketts et al., 1976). Using a test fixture with only a purely resistive impedance avoids determining behavior of the filter with reactive source or load impedances.

D. USE OF FILTERS WITH NONLINEAR SURGE PROTECTIVE DEVICES

Laboratory tests have shown that some prefabricated filter modules suffer arcing between an input terminal and ground at voltages between 5 and 10 kV. A spark in air at the external terminals of a filter module has approximately the same effect on the circuit as a discharge inside a gas tube. Actually the gas tube is better; its properties are more predictable, and the gas tube may respond more quickly owing to radioactive preionization and other features. Aboard a military aircraft, for example, the air pressure, and hence the breakdown voltage, changes dramatically with altitude. The gas tube, which was discussed in Chapter 7, is a sealed system that is free from effects of dust, dirt, and ambient air pressure.

Given that filters fail by either internal or external arcing, one may as well combine gas tubes or varistors with filters that may be exposed to transient overvoltages. In that way the current is carried in a predetermined path and does not degrade the filter's future performance. This recommendation has also been made in U.S. DoD DCPA TR-61A (1972). Hays and Bodle (1958) recommended that filters have an inductive input and be preceded by a spark gap, as shown in Fig. 13-4. The input inductor of the T filter shown in Fig. 13-4 produces a large voltage across the spark gap during a transient that has a large value of dI/dt. This large voltage across the spark gap causes the spark gap to conduct more rapidly. The filter shown in Fig. 13-4 is not suitable for use on the mains, owing to the power follow current in the spark gap. Specific designs of low-pass filters with overvoltage protection for the ac supply mains are discussed in Chapter 19.

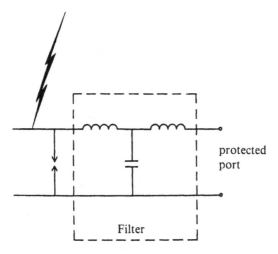

Fig. 13-4. *Use of a spark gap with a low-pass filter.*

The concept of combining varistors and ferrite core inductors to form a π filter, as shown in Fig. 13-5, has been investigated and endorsed for filter modules for signal lines (Campi, 1977). The parasitic capacitance of the varistors forms the capacitors in the filter. Campi's filter was to be a pin in a multipin electrical connector. In this way, the transient protection was conveniently included in the system along with proper shielding.

The performance of linear devices, such as filters alone, is quite properly characterized with gain or loss in decibels. When nonlinear devices are combined with filters, it is *not* desirable to specify the attenuation of transient overvoltages with gain or loss in decibels. Nonlinear devices are appropriately characterized by a clamping voltage at a specified surge current and waveshape, not by a gain.

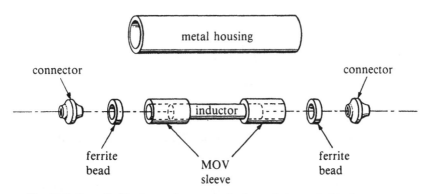

Fig. 13-5. *Campi's filter/overvoltage protector for a pin in an electrical connector.*

E. LOSSY LINE

Most surge protective devices (SPD) operate by shunting the transient to ground through an element with a very small impedance (e.g., spark gap, varistor, diode, or shunt capacitor in a low-pass filter). When the interaction of the line and the transient protection circuit is viewed from transmission line theory (see Chapter 4), an appreciable part of a rapidly rising transient will be reflected by the SPD, which approximates a short-circuit termination. Other types of transient protection circuits, such as the inductor input of a T filter shown in Fig. 13-4, may approximate an open-circuit termination, again producing a reflection of an appreciable part of a rapidly rising transient.

It would be desirable to attenuate these reflections. Otherwise, the reflections may radiate energy into other systems and cause problems there. Also the reflections may produce overstress on the cable and cause insulation to fail. Ultimately, the energy contained in the transient overvoltage must be converted to heat. This conversion should be done in a controlled way so that neither damage nor upset occurs anywhere. One might use ferrite core inductors or resistors to accomplish this controlled attenuation.

There are several ways in which the cable itself can be converted into a filter to attenuate transients that propagate along the cable. The simplest way is to use wire with a large resistance per unit length. This is the solution adopted by manufacturers of oscilloscope probes for use with a ($1\,M\Omega \parallel 20\,pF$) input impedance. The coaxial cable in these probes has a center conductor of nickel-chromium alloy and a resistance per unit length between about 130 and $350\,\Omega/m$. Unfortunately for transient protection applications, this resistance wire has a diameter only of the order of 75 to $100\,\mu m$. During a large surge current, this thin resistance wire might melt.

A more robust coaxial cable with a center conductor of resistance wire is available as military type RG-222 (formerly RG-21A). The center conductor is Nichrome wire with a diameter of 1.4 mm, the outer conductor is a double shield of silver-plated copper braid. This cable has the following parameters:

$$C/\ell = 95\,pF/m$$

$$L/\ell = 0.24\,\mu H/m$$

$$R/\ell = 0.65\,\Omega/m$$

$$Z_0 = 50\,\Omega$$

The attenuation per unit length is $0.14\,dB/m$ at $10\,MHz$ and $0.9\,dB/m$ at $400\,MHz$.

Another approach is to use a magnetic lossy medium between the two conductors in a transmission line (Mayer, 1986). This material can introduce appreciable attenuation at frequencies greater than about 10 MHz. One can fabricate a coaxial line with lossy material between the inner and outer conductors that is reported to have an attenuation of about 60 dB/m at 100 MHz (Capcon Inc. data sheet). Much greater attenuations can be obtained by using a helical center conductor with lossy material both inside the helix and between the helix and the outer conducting of the transmission line.

If interference above 50 MHz is likely, lossy line should be included downstream from a conventional low-pass filter module. This is suggested because conventional lumped-element filters do not work well at very high frequencies. In this situation, the lossy line should penetrate the bulkhead of a shielded enclosure, so that radiation does not shunt the lossy line.

14

Isolation Devices

A. INTRODUCTION

An isolation device is an element that is placed in series between the source and load, as shown in Fig. 14-1. An isolation device has no conductive path between input and output ports; this property is called *isolation*. It may seem strange to discuss a component that is used in an electrical series circuit but has no conductive path. Yet such devices can be quite useful to block common-mode voltages, shown as V_C in Fig. 14-1, from appearing across a load. With an ideal isolation device, the voltage appearing across the load is a function only of V_D. As discussed in Chapter 4, the common-mode voltage is often undesirable, whereas the differential-mode voltage is desirable. Thus, isolation devices can be used to separate desirable and undesirable signals.

There are two common ways that an electrical signal can be coupled from the input to output port of the isolation device: magnetic field and light beam. A device that uses a magnetic field to couple the input port to the output port is called an *isolation transformer*; a device that uses light is called an *optical isolator*. A special class of optical isolators is the use of fiber optics communications. Isolation transformers will be discussed first, then optical isolators and fiber optics.

B. ISOLATION TRANSFORMERS

In elementary theory, a transformer is a differential device: the output (secondary) voltage is only a function of the voltage across the input

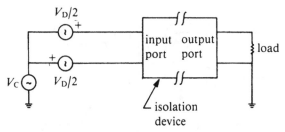

Fig. 14-1. *General use of an isolation device.*

(primary) terminals. In reality some capacitance exists between the primary and secondary coils, as shown as C_1 in Fig. 14-2. C_1 is drawn with dashed lines in Fig. 14-2 to illustrate that it is a parasitic element, not a deliberately included component. The common-mode voltage at the input is coupled through C_1 to the load. By interposing one or more electrostatic shields between the primary and secondary coils, as shown in Fig. 14-3, the effective capacitance between the input and output ports can be reduced to negligible values. All of the shields must be bonded to earth ground in order to block common-mode noise. When this is done, an isolation transformer is obtained. A good isolation transformer will have less than 0.005 pF capacitance between the input and output terminals. If a pair of capacitors is connected between the output port of the isolation transformer and ground, as shown by C_2 in Fig. 14-4, common-mode voltages can be attenuated to very small values by the capacitive voltage division. If C_1 is 0.005 pF and C_2 is 0.01 μF, a common-mode attenuation of 126 dB is obtained.

Isolation transformers are most commonly used to block common-mode voltages on the mains, an application that will be discussed in Chapter 20 on

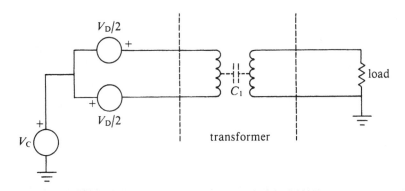

Fig. 14-2. *Parasitic capacitance, C_1, between primary and secondary coils of a transformer couples common-mode voltage, V_C, to load.*

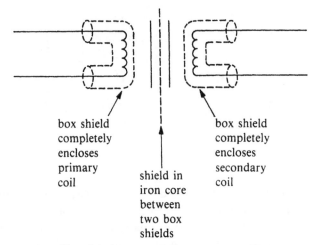

Fig. 14-3. *Isolation transformer construction.*

ac power applications. Note that a simple isolation transformer provides *no differential-mode attenuation.* Therefore, a differential-mode voltage transient that appears at the input will be transmitted to the output side (Martzloff, 1983). Also note that a simple isolation transformer provides no voltage regulation. If the input voltage decreases from 120 to 95 V rms, so will the output voltage. There are techniques to add differential-mode filtering and regulation of steady-state voltages to an isolation transformer. When this is done, the resulting product is no longer just an isolation transformer but is a *line conditioner.* Line conditioners are described in Chapter 20.

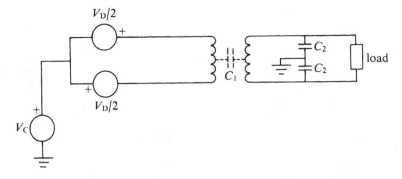

Fig. 14-4. *Use of bypass capacitors, C_2, to attenuate common-mode signals passed through a nonisolation transformer.*

C. OPTICAL ISOLATORS

An optical isolator is an electronic component that contains a light source and a photodetector, with no electrical connection between the two. A light beam transfers information from the input to the output. Electrical insulation between the light source and detector is provided by a piece of transparent glass or plastic. This insulation can typically withstand a steady-state voltage of several kilovolts.

The light source in nearly all modern optical isolators is an infrared light-emitting diode (LED). The photodetector in modern optical isolators is usually a silicon phototransistor, which has a response time of the order of $1\,\mu s$. Another photodetector, the photo-Darlington, is quite slow: response times of 0.1 to 1 ms are common. This inherent slowness provides immunity to transients that have too brief a duration to affect the output state. This slow response is useful to prevent upset of protected systems.

A silicon photo-diode detector provides the fastest response in an optical isolator. This is useful for wide-band and communication systems, such as digital local area networks. When the detector in an optical isolator is a light-activated SCR or triac, power systems can be controlled by the optical isolator. Optical isolators for use in analog signaling systems will be discussed later in this chapter.

The capacitance between input and output ports of an optical isolator in a six-pin dual-in-line package (DIP) is typically between 0.3 and 2.5 pF. This capacitance and the shunt capacitance across the output port determine the common-mode rejection of the optical isolator. Although an input-output capacitance of 1 pF may seem small, at a $1\,kV/\mu s$ rate of change of common-mode voltage, a current of 1 mA can pass from the input to output through this parasitic capacitance. Any attempt to intercept this current with a guard ring will inevitably decrease the spacing between the input and output circuit connections and thus reduce the isolation voltage. When large rates of change of common-mode input voltage are anticipated, avoid optical isolators with the base terminal of the phototransistor connected to a pin on the package, since the base terminal is sensitive to noise injected by currents in the parasitic capacitance.

When a sufficiently large potential difference is present between two adjacent conductors on a printed circuit board, a spark forms on the surface of the board between the conductors. The heat from the spark can form a blackened conductive path, called an *arc track*. To prevent this degradation of the insulation, specify a large distance between conductors. However, the DIP of common optical isolators has only 7.6 mm (0.3 inches) between input and output pins, so one cannot conveniently increase the spacing beyond this distance. To obtain greater isolation, a slot about 2 mm in width can be milled in the printed circuit board between the input and output pins.

A disadvantage of optical isolators is that they have a finite lifetime. The brightness of the LED source decreases to about half of its initial level during a time of the order of 10^5 hours (10^5 hours = 11 years). The rate of degradation is greater at larger values of current in the LED. This phenomenon is not widely discussed in the literature; the note by Lopez et al. (1977) is one of the few exceptions.

The LED in an optical isolator generally requires protection from being driven into reverse breakdown. Figure 14-5a shows a standard circuit that protects the LED from operation in reverse breakdown. The optical isolator is shown inside a dashed box in Fig. 14-5 to illustrate that it is a single module that contains a LED and a photodetector. Diode D_1 will conduct at about -0.6 V across the LED. The reverse breakdown voltage of the LED is typically between -5 and -20 V, so D_1 provides adequate protection. A switching diode (e.g., model 1N4447) is quite satisfactory for D_1.

The circuit in Fig. 14-5b provides comprehensive protection for the LED in the optical isolator (Standler, 1984). When V_{IN} is positive, the avalanche diode protects the LED from operation with excessive current when the LED is forward-biased. The minimum value of R_2 is calculated from

$$R_2 \geq (V_Z - V_D)/\max(i_D)$$

where V_Z is the breakdown voltage of the avalanche diode, and V_D is the voltage drop across the LED when it is operating at its maximum rated current, $\max(i_D)$. The voltage drop across the LED at normal operating

(a)

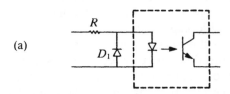

(b)

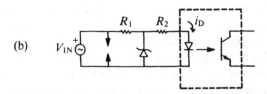

Fig. 14-5. *Basic circuits to protect an optical isolator from overstresses.*

currents (e.g., 10 mA) could be used for the value of V_D in order to obtain a safety margin. In the absence of other design criteria, V_Z should be about 6.8 V, since the zener or avalanche diodes have a minimum incremental impedance, $\Delta v / \Delta i$, at this value of V_Z. When V_{IN} is negative, the avalanche diode is forward-biased and protects the LED in the same way as diode D_1 in Fig. 14-5a.

The value of R_1 can be obtained from

$$R_1 + R_2 = (V_{IN} - V_D)/i_D$$

where V_{IN} is the maximum voltage across the source during *normal* (no overstress) operation of the circuit, V_D is the voltage across the LED, and i_D is the current through the LED when the LED is on. The resistance R_1 should be as large as possible to protect the avalanche diode, so the value of R_2 is chosen to be slightly greater than its minimum value. The value of R_1 can be much larger than the 20 to 50 Ω that is commonly included between a spark gap and avalanche diode.

Example:

Given: $\max(i_D) = 50 \text{ mA}$ $V_Z = 6.8 \text{ V}$

$V_D = 1.2 \text{ V}$ at $i_D = 10 \text{ mA}$

$V_{IN} = 5 \text{ V}$

Solution: $R_2 \geq 110 \ \Omega$

$R_1 + R_2 = 380 \ \Omega$

Let $R_2 = 110 \ \Omega$; then $R_1 = 270 \ \Omega$.

Notice that if the avalanche diode has a steady-state power rating of 5 W, the maximum current in the reverse-biased avalanche diode, 735 mA, occurs when there is 205 V across the spark gap in the circuit shown in Fig. 14-5b. This is more than enough to operate a spark gap with a dc breakdown voltage of 150 V, so this circuit is well coordinated and will withstand a *continuous* overstress of either polarity.

The optical isolator requires a power supply for the output side of the device. This is a serious disadvantage when the optical isolator is used as a transmitter, since it demands an independent power supply with an insulation level at least as great as that of the optical isolator. An independent power supply is required, since one dares not use the same power supply to transmit signals into an environment that contains overvoltages *and* to operate vulnerable electronic devices, since the overvoltages could be conducted directly to vulnerable devices through the power supply.

Another possible disadvantage to optical isolators is that they are inherently unidirectional: information can pass only from the LED side to the photodetector side. Thus a single pair of wires cannot be used for bidirectional communications when optical isolators are connected at the end of the cable. This restriction could increase the number and the total mass of the wires that are required for communication, a serious disadvantage aboard aircraft and missiles.

Unlike the isolation transformer, the optical isolator can transfer dc signals. This inherent capability is a particular advantage with relatively slow logic communications (e.g., 300 baud), since a transformer would distort these signals owing to inadequate low-frequency response.

Optical isolators are usually reserved for digital interfaces owing to distortion of analog data with simple optical isolator circuits that have a switching device at the output port. There are a few optical isolators that are designed for linear operation in analog circuits. The General Electric H11F1 optical isolator has a photosensitive JFET at the output port. The value of V_{DS}/I_D (the output port resistance) is a linear function of the LED current. The Motorola MOC5010 optical isolator has a linear amplifier output for converting input current variations to output voltage variations.

D. FIBER OPTICS

Often it is very difficult to use adequate shielding and protection devices to provide adequate protection against damage by lightning and EMP from nuclear weapons, and even then, upset can occur. In addition to concern about damage and upset from transient overvoltages, one must often be concerned about illumination of cables by high-power radio or radar transmitters. Noise will be injected by these transmitters if the electromagnetic energy penetrates the shield of the cable in electronic systems. These problems are difficult to solve when conventional conductive cables are used for data communications. Use of fiber optics instead of conductive cables has several advantages compared to conductive cables:

1. smaller diameter, less massive
2. inherent immunity to noise pickup in strong electromagnetic fields
3. electrical isolation
4. wider bandwidth
5. less attenuation with distance at high frequencies

The smaller size and mass of fiber optic cable is a major advantage in miniature electronic equipment aboard aircraft and missiles. Because fiber optic cable is a dielectric, it will not act as an antenna for current or voltage pulses from EMP, lightning, or radio waves. Fiber optic cable does not pick up noise. Because there is no conducting path, fiber optic cable cannot form a ground loop, nor does it radiate noise into adjacent cables. In this way, the electrical overstress and upset problems that plague systems that are connected with conducting cables are avoided. These advantages are sufficient to make fiber optic cables attractive for many electronic systems.

However, it is possible that transient overvoltages or noise in the electronic transmitting circuit may corrupt the signal that is sent over the fiber optic cable.

15

Parasitic Inductance

A. INTRODUCTION

When current passes through a conductor, a magnetic field is created (Ampere's law), so *all* conductors have self-inductance. When the inductance is gratuitous (or not deliberately included), it is called "parasitic inductance."

The amount of parasitic inductance is a critical issue when surge currents are diverted by nonlinear components, such as spark gaps, varistors, avalanche diodes, etc. Parasitic inductance can greatly increase the clamping voltage of varistors and avalanche diodes and cause damage or upset in systems that are supposedly protected.

The same physics and engineering involved in understanding and minimizing parasitic inductance for overvoltage protection is also applicable to bypassing power supplies with capacitors and design of filters that are effective at high frequencies. Although this book is concerned with overvoltages and not with noise, it seems a shame not to show how the same principles and techniques can solve both overvoltage and noise problems. This chapter concludes with a brief coverage of bypassing capacitors on dc supply buses.

The inductance per unit length of a straight conductor is of the order of $1\,\mu H/m$ for conductors with a diameter between about 0.5 and 2.0 mm. At higher frequencies the inductance is slightly smaller owing to the skin effect.

B. PARASITIC INDUCTANCE AND SHUNT SPDs

A representative circuit is shown in Fig. 15-1. The voltage drop across the shunt path consists of two terms: (1) the voltage due to the parasitic

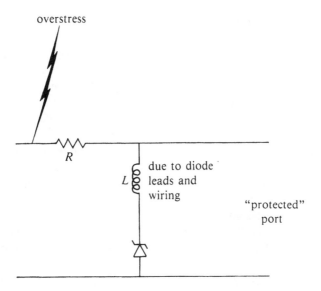

Fig. 15-1. *Schematic diagram showing effect of parasitic inductance of leads and package in a surge protective device.*

inductance, $L\, dI/dt$; and (2) the voltage due to the current in the nonlinear device. In order to get the smallest clamping voltage and best protection, the parasitic inductance that is in series with shunt SPDs must be minimized.

There are two situations in which minimization of such parasitic inductance is particularly critical. The first SPD in a hybrid protection circuit usually has the largest surge current during an overstress. Since this makes dI/dt particularly large, it is important to minimize L for this first device, which is usually a spark gap or varistor. The other critical situation is the final SPD in a hybrid protection circuit. The voltage across the final shunt element determines the clamping voltage across the protected port. To minimize the magnitude of the clamping voltage, one must minimize L for the final device.

In many practical situations the voltage resulting from the parasitic inductance is greater than the voltage corresponding to the current in the nonlinear device. In fact the expected benefit from the nonlinear device in the circuit may be vitiated by excessive parasitic inductance. Sherwood (1977) says a "gas discharge device with 0.5 inch leads on either side has lost a large percentage of its effectiveness under fast rise time surge conditions." This point can be emphasized with an example.

Suppose a spark gap is connected with a 10 mm length of copper wire with a diameter of 1 mm. This short piece of wire will have an inductance of about 5 nH. When dI/dt is 20 kA/μs, 100 V will be developed along this "short circuit" by the parasitic inductance. The 100 V owing to the inductance should be compared to the approximately 20 V across the spark gap due to the arc voltage.

Clark and Winters (1973, pp. 25, 50) give a value of "approximately" 10 nH for the parasitic inductance of a DO-13 case with no leads. Clark (1975) and Clark and Pizzicaroli (1979) evaluated the effect of lead length on clamping voltage of avalanche diodes and emphasized that the minimum length of leads is desirable. Fisher (1978) discussed the effects of parasitic inductance in applications of metal oxide varistors. This work emphasizes the importance of keeping the leads as short as possible. He showed that long leads are undesirable for two reasons: the series inductance shown in Fig. 15-1, and the mutual inductance, which is discussed later in this chapter under the heading "inadvertent transformer effect."

Electronic circuit designers often work with currents of the order of 10 mA. Therefore, it is difficult for them to appreciate the large values of dI/dt that occur in transients, such as 10^{11} A/s during a lightning return stroke. To obtain a peak value of dI/dt equal to 10^{11} A/s from a sinusoidal waveform with a 10 mA peak current, a frequency of 1.6 GHz would be required. (This frequency has a free-space wavelength of about 0.2 mm.) Such a large frequency is far beyond the range of frequencies that are familiar to circuit designers, so such values of dI/dt will not be encountered in their low-current work.

Clearly, one must minimize the length of conductors in shunt paths to minimize the parasitic inductance and maintain the good voltage clamping properties of the nonlinear device. What is not obvious is how to do this. Specific suggestions for practical situations are given later in this chapter.

Besides the increase in clamping voltage owing to parasitic inductance in shunt paths, the parasitic inductance has another undesirable effect. The parasitic inductance acts to increase the apparent response time of the shunt SPD, because high frequencies see the inductance as a large impedance. When there is appreciable inductance in the shunt path, one would expect a large remnant to travel downstream from the shunt SPD. In this view the parasitic inductance acts as a delay line.

For example, a diode in a DO-35 package has about 3 nH of parasitic inductance even when its leads have *zero length*. In practice, there is more inductance than 3 nH, because the lead length is not zero. A particularly insidious practice occurs when mounting components on a printed circuit board. To avoid mechanical stress on the seal between the component lead and case, about 2 mm of straight lead is needed before forming each bend. The extra 4 mm of lead is associated with increased parasitic inductance. Each lead will have one right-angle bend, so there will be two right-angle bends for each component case in the shunt path. These right-angle bends further increase parasitic inductance.

Another undesirable practice is to use components that have crimped leads to maintain proper spacing between the printed circuit board and the component. Crimped leads are commonly available for capacitors and varistors. The bends in each crimp, as well as the extra lead length, add parasitic inductance.

Printed circuit boards themselves are not undesirable. A printed circuit board can force all of the production modules to have the same geometry. Therefore, a properly designed board can ensure that the shunt SPD components have minimal parasitic inductance.

However, printed circuit boards are not desirable for construction of surge protective circuits that will be exposed to large currents (e.g., more than 1 kA for an 8/20 μs waveshape). Large surge currents should be routed through conductors with a larger cross-sectional area than is possible on conventional printed circuit boards.

C. TECHNIQUES FOR MINIMIZING INDUCTANCE IN SHUNT DEVICES

The first rule when using conventional packages is to *make the shunt path as short as possible,* as shown in Fig. 15-2. This means not only to make the leads of the SPD as short as possible, but also to bring the signal path in the printed circuit board etched traces right up to the SPD, as shown in Fig. 15-3. In Fig. 15-3a, an arrangement is shown that has short leads for the SPD but that retains the effect of long leads by making an unnecessarily long shunt path on the printed circuit board. A better way to arrange the printed circuit board is shown in Fig. 15-3b. Before the surge current can reach the terminals of devices downstream from the SPD, the surge current first touches the terminals of the SPD package in Fig. 15-3b.

With conventional packages there appears to be nothing else that one can do to minimize the effect of the parasitic inductance. However, the technique of making the surge current touch the terminals of the shunt SPD before the surge can propagate downstream has led to the development of novel packages that are discussed below.

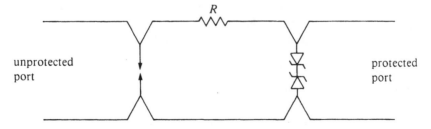

Fig. 15-2. *Circuit diagram emphasizing need to minimize parasitic inductance in shunt elements.*

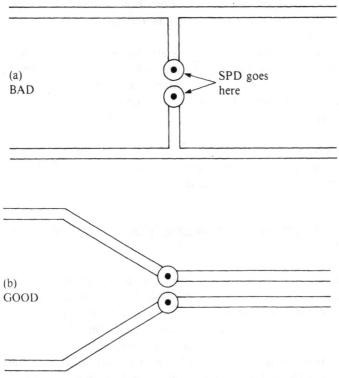

Fig. 15-3. *Mounting of surge protective device on printed circuit boards.*

1. Four-Terminal Structure

In four-terminal structure, the inherent inductance in conductors is rearranged to be in series with the surge current, as shown in Fig. 15-4. The four terminals of the shunt SPD in Fig. 15-4 are indicated by the four black dots. The shunt SPD is shown in Fig. 15-4 as an avalanche diode but could be a spark gap or varistor or any other type of shunt SPD. Although inductance in shunt elements is *not* desirable, inductance in series elements is desirable, except when the signal frequency is greater than the highest significant frequency in the spectrum of the transient.

The parasitic inductance in a shunt path can be reduced to less than 1 nH by using the four-terminal construction that was patented by Clark (1982) for avalanche diodes, as shown in Fig. 15-5. Clark mounted the diodes on a strip of material as shown in Fig. 15-5a. A small piece of bent metal was used to connect the upper terminal of the diodes to the other strip, as shown in Fig. 15-5b. The four pins were then bent at a right angle to the plane of the diode and two strips, and the assembly was molded in a plastic package, as shown in Fig. 15-5c.

The four-terminal structure is commonly known as a "zero-inductance"

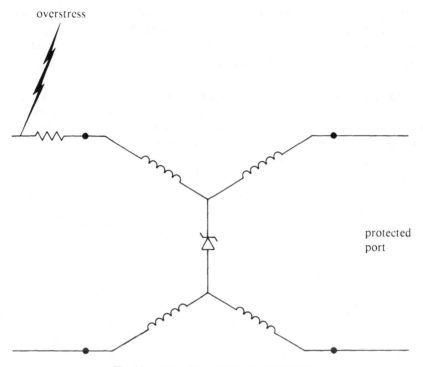

Fig. 15-4. Use of four-terminal construction.

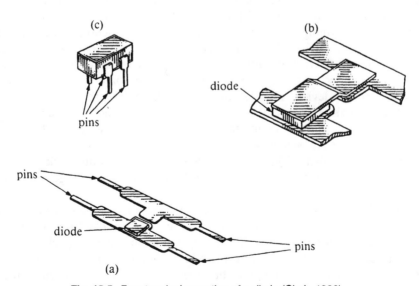

Fig. 15-5. Four-terminal mounting of a diode (Clark, 1982).

package, but this is inappropriate. The inductance that is naturally present in any conductor has been *relocated in series* with the transient (and signal) path by the four-terminal construction, as shown in Fig. 15-4. The inductance is still present, but it has been moved to a location where it does no harm.

Four-terminal construction is available on some models of spark gaps. Use of spark gaps in a four-lead configuration is not popular, although it deserves to be considered. The large magnitude of dI/dt in the spark gap when it changes from the insulating to a highly conducting state can cause problems, as discussed in Chapter 7. By reducing the parasitic inductance of the spark gap, the clamping voltage can be reduced during large magnitudes of dI/dt. By including inductance upstream from the spark gap, the magnitude of dI/dt can be reduced transient overvoltages.

Four-terminal construction is also used in electrolytic capacitors for use at frequencies between about 10 kHz and 1 MHz (Lambert and Peterson, 1969; Bowling, 1979). These four-terminal electrolytic capacitors are used in switching power supplies. They may be useful in filters for a dc voltage regulator that is located a long distance from the transformer/rectifier module. However, the parallel combination of an ordinary two-terminal electrolytic capacitor and two ceramic capacitors, as explained later in this chapter, will give a smaller impedance over a wider range of frequencies than one four-terminal electrolytic capacitor, and at less cost.

2. Surface Mount Devices

In order to build smaller electronic circuits, surface mount devices were introduced around the year 1980. A circuit board for surface mount devices has no holes for leads, and automatic assembly is easier, because there is neither lead forming nor cutting, which makes production of circuits with surface mount devices less expensive than with conventional packages. Surface mount devices are attached to the circuit board with adhesive, and electrical connections are then made with wave soldering. Alternatively, surface mount devices may be attached with solder paste, which is then melted to make a permanent electrical and mechanical connection. The small size of the surface mount devices makes manual assembly difficult. For several years only semiconductor devices for operation at steady-state currents of less than 0.25 A were available in surface mount packages. Such devices were too fragile to be useful in most overvoltage protection applications.

In 1985 metal oxide varistors became available in leadless surface mount packages. In 1986, avalanche diodes that are characterized for suppression of transient overvoltages also became available in this new package. SPDs in surface mount packages were introduced because customers were using the new packaging technology to build smaller circuits, not because surface mount packages offer smaller inductance values than conventional packages.

There is no question that the leadless surface mount packages for SPDs offer opportunities to decrease the parasitic inductance of shunt SPDs. This is critical for applications in which transient overvoltages have small rise times—for example, electrostatic discharge, high-power microwaves, and EMP from nuclear weapons. Designers are only beginning to exploit the opportunities that these new leadless packages offer, and no applications information was available in 1988, when this book was written.

D. INADVERTENT TRANSFORMER EFFECT

1. Physics

A very insidious effect occurs when diverting large currents through SPDs on a printed circuit board on in other confined geometries. Consider Fig. 15-6, which is a bird's-eye view of the layout of a printed circuit. The schematic symbols are shown for clarity instead of component packages. However, Fig. 15-6 shows the geometrical arrangement of a circuit and is not just a schematic diagram. The spark gap is symbolic of any SPD that can tolerate large surge currents; in some situations a varistor might be used in place of the spark gap. The avalanche diode is symbolic of any SPD that is suitable for the final clamping element; in some situations an ordinary diode or even a metal oxide varistor might be used as the final SPD.

The surge current, I_S, travels down the cable from point A to point B, through the spark gap to point C, and back down the cable to point D. This current is associated with a magnetic field that travels outward. There is a loop area formed by the final SPD between points E and H and the protected device that is connected between points F and G. The magnetic field, $\vec{B}$, from current I_S intersects the area formed by points $EFGH$. There will be a voltage induced in this loop that is given by

$$\int (d\vec{B}/dt) \cdot d\vec{S}$$

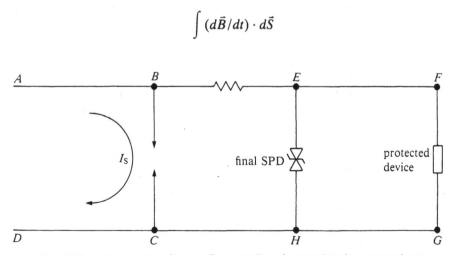

Fig. 15-6. Inadvertent transformer effect: coupling of surges into the protected port.

where $d\vec{S}$ is the differential area and the integral is taken over the surface bounded by points *EFGH*. For most practical situations, exact computation of the voltage from the magnetic field is complicated.

An alternative view of this same situation is that a parasitic transformer has been inadvertently inserted into the circuit. The current in the primary of this transformer is I_S. The secondary of this transformer is in series with the protected device.

2. Ways to Minimize the Inadvertent Transformer Effect

With either view, it is clear that the tight clamping voltage offered by the avalanche diode has been negated by the magnetic field from the surge current. Using a four-terminal package for the final SPD does *not* remove the inadvertent transformer effect. However, there are a number of practical things that can been done to reduce the magnitude of this effect.

1. Relocate the spark gap (or other components that have current I_S) as far as possible from the loop *EFGH*. This suggestion works because the magnetic field decreases with distance from the source current. When the surge current comes from a source outside the building, the spark gaps may be placed in a vault at the point of entry of the conductors into a building, and the remainder of the protection circuit can be placed in another room that contains the protected device or equipment to be protected.

2. Minimize the area of loop *EFGH*. The final SPD should be as near the protected device as possible. If the final SPD is not near the protected device, twisted-pair wire should be used to connect the final SPD to the protected device. The random orientation of the twisted pair wire makes the net area $\int d\vec{S}$ be near zero.

3. Minimize the area of loop *ABCD*. The best solution for critical applications may be to use coaxial cable for the conductor that carries I_S and install the spark gaps in a coaxial mounting.

4. If the spark gap must be near the final SPD, consideration should be given to orienting the path of I_S so that the magnetic field associated with the surge current, I_S, is parallel to the plane of *EFGH*.

5. Put an impedance, preferably an inductor, upstream from the spark gap to decrease the magnitude of dI_S/dt.

3. Mounting Techniques for Spark Gaps

Standler (1986) recommended that spark gaps be mounted with their unbent leads in a plane that intersects the plane of the circuit board at a right angle, as shown in Fig. 15-7. Removing the right-angle bend in each spark gap lead, which would be present in a conventional mounting, and minimizing

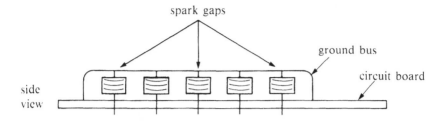

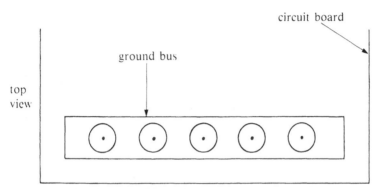

Fig. 15-7. *Mounting of spark gaps on printed circuit board.*

the length of the lead minimizes the inductance in series with the spark gap. It also orients the surge current in the spark gap so that the associated magnetic field is parallel to the plane of the printed circuit board, in order to minimize inadvertent transformer effects. The ground bus is a strip of brass or copper that can be easily soldered. A flexible copper-braid ground strap, which is not shown in Fig. 15-7, is soldered to the ground bus that is common to all of the spark gaps.

4. Point-to-Point Wiring of SPDs

The same principles stated above for mounting SPDs on printed circuit boards also apply for point-to-point wiring of SPDs. Figure 15-8 shows three variations for connecting an SPD to protect a specific device. A varistor is shown as the SPD in Fig. 15-8, but the same principles apply for any type of SPD.

Figure 15-8a shows a particularly bad way to connect the SPD. For overstresses with a short rise time, the overstress may damage the "protected" device before the SPD can clamp the voltage. Furthermore, the voltage across the protected device will have a large component owing to

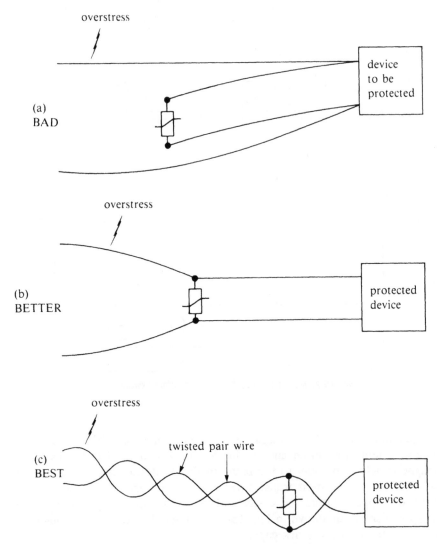

Fig. 15-8. *Point-to-point wiring of a surge protective device.*

the inductance in the long wires to the SPD. This bad method is often used to install SPDs in retrofits, because it is easy to do. Although the SPD does provide some protection for slowly rising transient overvoltages, this method of installing the SPD should be condemned.

The method shown in Fig. 15-8b is much better. The long distance between the SPD and protected device does, however, invite the inadvertent transformer effect.

The inadvertent transformer effect is suppressed by using twisted-pair wires on both sides of the SPD, as shown in Fig. 15-8c. This method of connecting an SPD is recommended when point-to-point wiring is to be used.

E. PARASITIC INDUCTANCE OF CAPACITORS

1. General

Figure 15-9 shows the magnitude of impedance of a capacitor as a function of frequency, f. When the impedance is proportional to $1/f$, the capacitor is behaving in the ideal way. However, for all lumped-element capacitors (the kind used in circuits at frequencies below a few hundred megahertz), there is a resonance frequency f_0 due to internal inductance of the capacitor. At frequencies greater than f_0, the impedance of the capacitor is dominated by the inductance term, and the impedance is then proportional to f. The internal inductance of the capacitor can be estimated from knowledge of the resonance frequency and the capacitance value, C,

$$L = 1/(\omega_0^2 C)$$

where $\omega_0 = 2\pi f_0$.

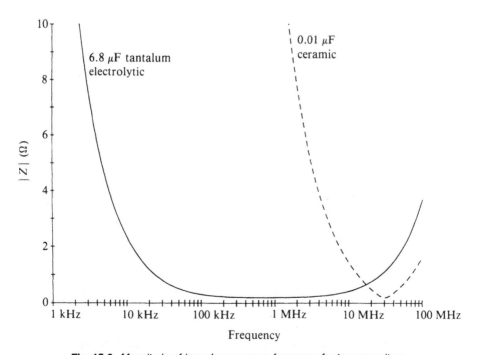

Fig. 15-9. *Magnitude of impedance versus frequency for two capacitors.*

Clearly, a lumped-element capacitor is not useful as a capacitor at frequencies that are much greater than f_0. Typical values of f_0 for nonpolar lumped-element capacitors are between 1 and 100 MHz; typical values of parasitic inductance are between 1 and 50 nH. Woody (1983) measured values of parasitic inductance between 21 and 37 nH for 29 different capacitors.

Notice in Fig. 15-9 that for frequencies above 20 MHz, the 0.01 μF capacitor has a smaller impedance than a 6.8 μF capacitor. This observation leads to a simple, practical way to avoid the effects of parasitic inductance in capacitors. To get a small impedance for a wide range of frequencies, connect several different capacitors in parallel. Figure 15-10 shows the magnitude of reactance versus frequency for the parallel combination of three capacitors with the following properties:

	C	L	Series Resistance
Tantalum electrolytic	6.8 μF	5.8 nH	0.2 Ω
Ceramic disk	0.1 μF	7.0 nH	0.2 Ω
Ceramic CK06 style	0.01 μF	2.8 nH	0.2 Ω

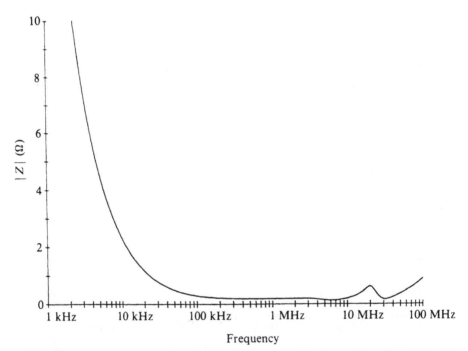

Fig. 15-10. *Magnitude of impedance for parallel combination of 6.8 μF tantalum electrolytic and ceramic capacitors: 0.1 and 0.01 μF.*

The 0.1 μF ceramic disk capacitor is not essential, but it does reduce the impedance between 10 MHz and 20 MHz by about a factor of 2.

This network of capacitors is useful to bypass noise and remnants of transient overvoltages to ground, as discussed in the next section.

2. Bypassing DC Power Supplies

Noise on dc power supply lines can cause unsatisfactory operation of the system during normal conditions. The manner in which noise appears on power supply lines and how one removes it are related to parasitic inductance. It is therefore appropriate to treat this subject here. Noise is generated by varying currents in the wires that are used to connect the power supply to the load. The load is composed of the various amplifiers and other devices that require power from the supply. The wires that connect the power supply to the load have a small amount of resistance and inductance, as shown in Fig. 15-11.

To analyze the effect of the inherent resistance, R, and inherent inductance, L, in the power supply connections, consider a numerical example with values that are from a realistic situation. Suppose that each wire that connects a power supply to amplifier A_2 is a 50 cm length of copper wire with a diameter of 0.5 mm (24 AWG). Then R will be 0.042 Ω. The inductance, L, is difficult to predict, but measurements show that it is about 1 μH.

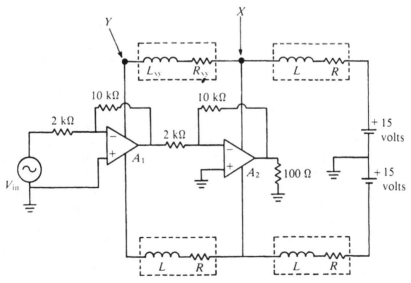

Fig. 15-11. *Parasitic resistance and inductance in power supply bus of a simple operational amplifier circuit.*

Suppose that the voltage source is sinusoidal:

$$V_{in} = 0.4 \sin(2\pi\, 10^5\, t)\ V$$

The output current of amplifier A_2 is sinusoidal with an amplitude of 101 mA (the output voltage has an amplitude of 10 V; the load is the parallel combination of 100 and 10 kΩ). This output current must come from the power supply lines. The total current drawn by amplifier A_2 from the power supply is approximately the quiescent current plus the output current. The quiescent current is the current in the power supply terminals when V_{out} is zero. Let us use 5 mA for the quiescent current. The total power supply current for amplifier A_2 is

$$5\ mA + 101\ mA \sin(2\pi\, 10^5\, t)$$

Similar analysis tells us that amplifier A_1 has a total power supply current of

$$5\ mA - 1.2\ mA \sin(2\pi\, 10^5\, t)$$

The total current at point X in Fig. 15-11 is about 10 mA + 100 mA $\sin(2\pi\, 10^5\, t)$. This current, I, causes a voltage drop between the +15 V supply and point X of $IR + L(dI/dt)$. The IR term has a peak value of about 4.6 mV. The inductive voltage drop has a peak value of 63 mV, much larger than the IR voltage drop. The peak of the IR and $L(dI/dt)$ voltage drops are not in phase, since dI/dt is a maximum (or a minimum) when I is zero.

What is the value of the power supply noise at point Y, where the power supply is connected to amplifier A_1? Point Y will see all of the noise that is at point X, plus an additional amount due to the resistance, R_{xy}, and inductance, L_{xy}, between points X and Y. If the wire between X and Y is short, and the current to A_1 is small compared to the current to A_2, the noise on the power supply has approximately the same amplitude at points X and Y. Then most of the noise at amplifier A_1 (point Y in Fig. 15-11) is due to current drawn by amplifier A_2, which is an important lesson.

Amplifier A_2 can "talk" to amplifier A_1 via the power supply connections. Voltage fluctuations at the power supply terminals of an amplifier will result in some change in output voltage of that amplifier. Logic devices with edge-triggered inputs (e.g., edge-triggered flip-flops) are quite sensitive to noise. Noise on power supply lines during normal operation of the system can produce unsatisfactory operation of the system.

The standard cure for noise on the power supply lines is to use *bypass capacitors*. Bypass capacitors are connected between each power supply terminal and ground. These capacitors must be located near the sensitive device (load), as shown in Fig. 15-12. The bypass capacitors with the smallest capacitance should be located nearest to the sensitive device (load).

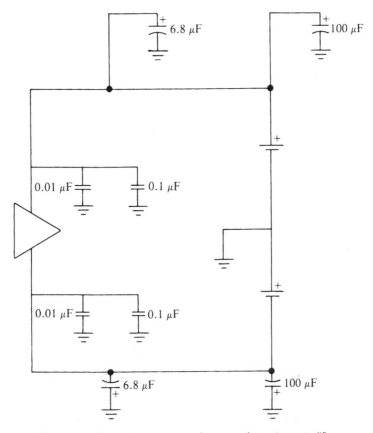

Fig. 15-12. *Bypassing power supply connections at an amplifier.*

The function of bypass capacitors can be understood from either of two perspectives.

1. The parasitic series inductance in the power supply lines increases the output impedance of the power supply at high frequencies. The bypass capacitors provide a source of charge (or current) that has a much smaller output impedance, owing to the shorter distance (and therefore less parasitic inductance) between the capacitor and load.

2. We can think of an ideal dc power supply conductor as an incremental ground (because its voltage does not change with time). The parasitic inductance (and resistance) in the power supply wiring inserts an impedance between the ideal dc voltage source and the power supply terminals of the load. Using bypass capacitors at the power supply terminals of the load provides a low impedance path to ground and removes voltage fluctuations at that point. In this view, power supply bypass capacitors have the same function as the capacitor across the

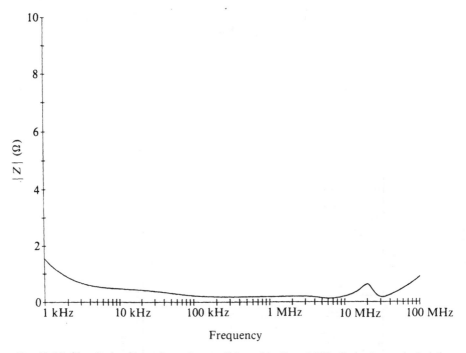

Fig. 15-13. Magnitude of impedance for parallel combination of 100 μF aluminum electrolytic, 6.8 μF tantalum electrolytic, and ceramic capacitors: 0.1 and 0.01 μF.

emitter or source resistor in the self-bias common-emitter or common-source transistor amplifiers.

Because real capacitors have parasitic inductance, it is standard practice to place several different kinds of capacitors in parallel to form the bypass element. A typical selection would be a tantalum electrolytic capacitor (1 to 10 μF) in parallel with one or more ceramic capacitors (0.01 to 0.1 μF). The exact value of the bypass capacitance is not critical; it is often difficult to see the effect of a factor of two change in capacitance. What is critical is that the bypass capacitors maintain a suitably low impedance at all frequencies of interest.

At frequencies below about 30 kHz, the impedance of a parallel combination of 6.8, 0.1, and 0.01 μF capacitors is not necessarily small (see Fig. 15-10). However, the parasitic inductance of the power supply lines is usually not a problem at these low frequencies. If it is, it is simple to add a 100 μF aluminum electrolytic capacitor to the parallel network to get the impedance versus frequency plot shown in Fig. 15-13. Because the 100 μF capacitor is only effective at low frequencies, probably only one 100 μF capacitor will be needed for each dc supply voltage on each printed circuit board.

Part 3

Applications of Protective Devices

16

Overview of Applications

In this chapter consideration is given to the following questions:

1. What should be protected?
2. How should the design of overvoltage protection be approached?
3. Should overvoltage protection be included in a chassis?
4. Is it economical to protect against overvoltages?
5. What are desirable properties for protective circuits?
6. Where should overvoltage protection be installed?

The issues in these questions are all related, so the discussion of each issue continues throughout this chapter.

A. LESSONS FOR MANAGERS

Probably the most important principle in obtaining reliable system operation is not a technical electrical engineering decision. Managers must decide that reliability, including protection from overvoltages, must be *considered* during the initial design of the system.

It is all too common to see a system designed that works well during tests in a gentle environment in a laboratory but fails mysteriously at some field sites. When overvoltages are suspected, protective circuitry may be ordered in a panic retrofit. Such panic retrofits are more expensive than including the protection in the initial design. Retrofits are often bulky, because space was not provided in the equipment for overvoltage protection components, so these components are placed in an additional box that sticks out like a sore thumb.

The user is not free of responsibility in this matter. The user should query the manufacturer about the level of overvoltage protection inside the system and about what surge protective devices should be installed upstream from the system.

The system should not be operated until appropriate surge protective devices have been installed. Degradation during surges applied to an unprotected system can cause latent failures that reduce the reliability of the system after appropriate surge protection has been installed.

B. DESIGN STRATEGIES

There are three basic strategies for designing overvoltage protection.

1. Given a certain amount of money for components and labor in assembly and testing, design the "best" protection circuit for this ceiling on costs.
2. Design the lowest cost protection circuit that will allow a given product to survive a specific test waveform without damage.
3. Do not add overvoltage protection to a product.

1. Three Strategies

The first strategy is a good choice for designing add-on hardware that is to be used to protect unknown products. The first strategy is also useful when little money is available for protection, which translates to "give me all the protection that I can afford," which may be less than what is needed for reliable operation in all environments. The word "best" is explained later in this chapter, in the section on desirable properties of protective circuits.

The second strategy is a good choice when protective devices are installed inside a specific piece of hardware.

One might expect the author of a book on overvoltage protection to condemn the absence of overvoltage protection in a product, the third strategy. However, the absence of protection is not such a terrible thing, provided the manufacturer informs or reminds the user that external overvoltage protection is desirable for reliable, long-term operation. If the manufacturer makes certain assumptions about the environment, those assumptions should be communicated to the user. This is commonly done for assumptions about ambient temperature, relative humidity, steady-state rms mains voltage, and mains frequency.

It is not possible to suggest that one strategy is always better than another. Different situations require different approaches and treatments.

2. Should Overvoltage Protection Be Included Inside a Chassis?

The third strategy above deserves a more detailed examination. Consider protecting a product from damage by transient overvoltages on the mains. Although the protection will probably be accomplished by using varistors, the broader term "surge protective devices" (SPDs) will be used to keep this discussion general.

It is very convenient for the user if each chassis contains SPDs. This convenience to the user is the major benefit of placing SPDs in every chassis. However, such internal protection is difficult to coordinate with other SPDs in the environment. The problem is simple: the designer does not know what SPDs, if any, are present in the other equipment in the user's environment. If every designer makes the "worst-case" assumption that there are no other SPDs in the environment, each chassis will need to contain heavy-duty SPDs, which are more expensive than moderate-duty or light-duty SPDs. The system may be well-protected, but the use of redundant or excessive SPDs increases costs. However, if the designer assumes that other equipment will divert some of the surge current when, in fact, the other equipment has no SPDs, catastrophic failure may occur during an overvoltage.

This situation could be resolved by use of mandatory standards by regulatory agencies. Such standards would specify minimum acceptable levels of protection to be provided by SPDs in each chassis as well as SPDs elsewhere in the building, such as a lightning arrester at the distribution panel. At the time this book was written (1988), the author saw no hope of such standards being adopted in the United States for overvoltages on the mains.

If SPDs are *not* included in each chassis by the manufacturer, then the user is responsible for all overvoltage protection. In most situations the user is not an electrical engineer. Even if the user is an electrical engineer, the user probably lacks the specialized knowledge to design an appropriate overvoltage protection for the equipment or system. If the user installs SPDs inside of chassis or outlet strips, the user voids both the warranty and the safety approvals of those products. The user can purchase a surge protection module that attenuates overvoltages that would otherwise reach vulnerable load(s). These modules come in several forms, which are discussed in Chapter 19.

The situation is quite different for protection against damage by overvoltages on data communication cables from the situation for overvoltages on the mains. Protection of telephone systems is regulated by governmental agencies. A spark gap (carbon block with air, or a sealed gas tube) is often required near the point of entry of telephone lines into a building. Telephone receivers are required in the United States to survive exposure to the standard overstress waveforms in part 68 of the Code of Federal Regulations, which were described in Chapter 5. In addition there are various industry standards from Bellcore, IEEE, and ANSI.

Overvoltage protection for computer peripherals (e.g., CRT terminals, printers, plotters) may be unnecessary. Protection may be a luxury in applications where the peripheral is connected to a computer by a cable that is no more than a few meters in length and where both the computer and peripheral use the same protected mains. Modems already have overvoltage protection on the telephone port because they must meet the same survival specifications as telephone receivers. Computer peripherals that have long data cables, and especially those which cables go between buildings, should have overvoltage protection on the data lines, as described in Chapter 17. Such protection can be added by the user when the situation requires it. Guidelines for deciding where to add overvoltage protection are given later in this chapter.

There is another issue in the debate between SPDs that are internal to the chassis versus added on by the user. In situations where large transient currents could occur (e.g., direct lightning strikes to overhead power or telephone conductors), it would be *un*desirable for the sole SPD to be located inside the chassis of vulnerable equipment. The intense magnetic field from the passage of large current would likely cause problems, as described in Chapter 15. Lightning arresters for the mains and spark gaps for telephone lines should be near the point of entry of those conductors into the building, *not* inside electronic equipment.

Some hardness can be obtained without SPDs by increasing the specifications of parts that are already required, such as transformers, voltage regulators, and filter capacitors in the power supply. The cost and benefits of this approach to hardness should be compared with the cost and benefits of other approaches, such as adding SPDs to the power supply as discussed in Chapters 18 and 19.

3. Suppression at Source

Nearly all of the discussion of electrical disturbances focuses on the effect of disturbances on vulnerable or susceptible equipment and how to protect that equipment. This is quite understandable from the point of view of the user of the equipment, who does not care where the disturbance came from (unless he is going to sue for damages in court!). It definitely makes good sense to place protective devices between expensive or critical equipment and the rest of the world. However, it also makes sense to identify sources of disturbances and try to suppress the disturbances at the source. In the author's opinion, engineers who design equipment that is a source of transient overvoltages have an ethical responsibility either (1) to install internal protective devices to attenuate their product's overvoltages, or (2) to warn the customer that external protective devices may be required to prevent damage to vulnerable electronic systems.

An analogy might be made to discharge of raw sewage into a river. Such discharge ought to be prohibited by law to protect people downstream. But at the very least, one should post "no swimming" signs downstream from the point where sewage is injected into the river.

4. Graceful Failure

When it is not economical to provide protection against damage or upset by transient overvoltages, one might try to provide some features to make failure graceful rather than catastrophic.

Graceful failure of a computer system might be accomplished in several ways. A crowbar circuit could be placed on the dc supply bus inside the computer. If the dc power supply failed for any reason, including overvoltage on the ac supply mains, the crowbar circuit would protect the logic circuits (e.g., microprocessor, floating point processor, memory, etc.) from damage. The dc power supply would be considered expendable, since it is relatively inexpensive, compared to the logic circuits, and easily replaced as a unit.

Graceful failure of a computer can also be accomplished by software engineering. For example, a text editor could store a backup copy of the document on a disk drive after 50 keystrokes had been entered by the operator. If the computer operation is upset for any reason, including a transient overvoltage, the operator can reboot the computer and get back to work with a loss of less than 50 keystrokes.

Another example of graceful failure would be an automatic teller machine for a bank that simply ceases to function rather than blithely and inappropriately dispense $20 bills.

5. What to Protect

There are several situations in which comprehensive protection from disturbances may be desirable, and even economically defensible. Some equipment may perform critical functions where long-term continuous operation is essential. Examples include life-support equipment for medical patients, telephone network, and military defense systems. Other equipment may be too expensive to risk damage or degradation by electrical overstresses. Examples may include equipment in a supercomputer facility, CAT and MRI scanners in hospitals, and military defense systems.

In other situations, overvoltage protection may be required, since the expected incidence or severity of overvoltages may be large. In the author's opinion, the mains should always have overvoltage protection. Switching reactive loads on the mains is too common to ignore. Data lines should be protected from overvoltages if they either (1) are illuminated by exterior electromagnetic fields or exposed to direct injection of current by lightning,

(2) carry data of a particularly critical nature, or (3) have relatively lengthy transmission lines.

A particular example of the third condition is the case where data lines go between different buildings. If one of the buildings is struck by lightning, one can expect substantial currents in the conductors that go to other buildings.

C. ECONOMIC ISSUES

A scientific approach to protecting a system is to calculate the expected cost of damage and upset. To do this requires the knowledge of the expected number of disturbance of the ith type in a year, N_i, and the average cost of the damage or upset that results from that disturbance, C_i. One would then expect to spend an amount K each year to recover from these disturbances:

$$K = \sum N_i C_i$$

If the amount K is sufficiently large, one may be immediately motivated to obtain protection from at least the most significant disturbances. The cost of removing a disturbance depends on the type of disturbance. For example, overvoltages may be suppressed with metal oxide varistors that cost about $1 each, whereas operation of equipment during an outage may require an uninterruptible power supply that costs more than $1000.

If the cost of removing a disturbance of the ith type is R_i, then it makes sense to remove the disturbance if

$$R_i < N_i C_i$$

since the cost of removing the disturbance will be recovered in reduced operating costs during the first year. If this inequality is almost satisfied, then it may still be justifiable to remove the disturbance, but it will require an investment that pays off over several years.

Although all of this is logical and scientific, there are several practical problems with using this calculation as a rigorous basis for making decisions. The number of disturbances per year at a specific site, N_i, is unknown. Knowledge of N_i implies that one can predict the future! The average cost of a particular type of disturbance, C_i, is also difficult to determine. There are many types of disturbances, and each of them comes in many sizes. The best way to determine C_i would be to generate the disturbances under laboratory conditions and keep track of the repair bills, but the results of this kind of research are rarely known. It is clear that an exact calculation of $N_i C_i$ is not possible. However, one should not denigrate the utility of a crude estimate as a guide to help determine when it is less expensive to tolerate the effects of disturbances than to remove the disturbances.

D. DESIGN GOALS: WORST-CASE OVERSTRESS AND ABSOLUTE PROTECTION

All too often, managers and engineers want to set a goal of absolutely certain adequate performance during a worst-cast overstress. There are three troublesome aspects to this goal: certainty is an unreasonable goal, the term "worst-case" can have many different meanings, and trying to obtain this goal tends to increase costs to unacceptable levels.

Requiring absolute certainty is a bad idea. Even if one is willing to spend large amounts of money for an extremely heavy-duty SPD, one should realize that there are always events—many of them unexpected when the SPD was specified and designed—that will cause damage or upset to the "protected system" and possibly also damage the SPD. These unanticipated events make guarantees of 100% perfection an illusion. Examples of surprises include radiated magnetic fields from large values of dI/dt in the surge, corroded or missing grounding conductors, surges that enter the system from unexpected paths, and systems that have been made more vulnerable by heating, aging, or "improvements," etc.

What does "worst case" really mean? Is it some level, determined by calculations from the laws of physics, that is theoretically impossible to exceed? Is it the largest magnitude measured and reported in the technical literature? Is it a maximum value in an engineering standard? Before one specifies "worst case," one ought to determine what it means and why it is relevant to the engineering project at hand. In many cases the customer and manufacturer should agree on a *reasonable* worse-case stress that is expected during the lifetime of the product or system. However, in some cases it may be appropriate to attempt to design protection so that the system will be neither damaged nor upset by rare, extremely severe stresses.

There is a risk to requiring super protective circuits that tend to be expensive. It is possible to increase vulnerability by placing a super protection module on *some* conductors or nodes, rather than a moderate protection module on *all* conductors that interface to the system or nodes with a vulnerable device. Thus, a lower cost module might provide better protection, because more of them can be installed for a given cost. Alternatively, the money saved by specifying less expensive overvoltage protection might be reallocated to increase the reliability of the system in another way. After all, a chain is only as strong as its weakest link: it makes no sense to harden a system against events that are expected once in a 100 years when the system remains vulnerable to common events.

Once it is decided to include overvoltage protection in a product, there is a question of what level of protection to include in the product. Consider a small product (e.g., personal computer or stereo receiver) that is to be used inside a building. It would be ludicrous to design this product so that it could withstand an 8/20 μs waveshape with a peak current of 20 kA, which is typical of a direct lightning strike. There is no way that lightning could strike just that one product. There will be other devices in the building, each of which would also be exposed to a fraction of the lightning current. From an engineering viewpoint, it makes sense to place a lightning arrester at the point of entry of the mains into the building, as described in Chapter 19. Additional protective devices could be placed in vulnerable appliances or on branch circuits that supply vulnerable appliances. A reasonable design criterion for protection from transient overvoltages on the mains for a single, high-quality appliance might be to survive an 8/20 µs waveshape with a peak current of 500 A and a 0.5 µs–100 kHz ring waveform with a peak open-circuit voltage of 3 kV.

The choice of test waveforms for validation of overvoltage protective measures is discussed in Chapter 22. It is reasonable to use survival of standard test waveforms as a design goal. Validation of protective measures then involves testing with these same waveforms.

Abrahams (1988), in an editorial on specifications for software, states that the "guiding principles ought to be:

"1. Ask only for what is realistically needed.

"2. If you don't know what you want, you aren't ready to specify it."

This same advice applies to writing specifications for overvoltage protection. However, in many situations there is inadequate information to allow a precise, scientific justification of all of the specifications for overvoltage protection. Some arbitrary choices, compromises, and informed guesses will need to be made in the absence of adequate information. More research will be helpful in reducing the uncertainty, but it can never be completely removed.

E. DESIRABLE PROPERTIES OF PROTECTIVE CIRCUITS

There are a number of desirable properties of circuits for protection against overvoltages. These are listed in summary form in Table 16-1 and are discussed below.

Probably the most important property of an overvoltage protection circuit is adherence to the doctrine of *primum non nocere*—first do no harm.

1. *The protection circuit should have negligible effect on the system during normal operation.* In particular, the shunt resistance should be suitably large, and both the series resistance and shunt capacitance should be suitably small. After all, if the protection circuit makes the normal operation of the system unsatisfactory, then there is no reason to use the system. The engineer who designs the protection

TABLE 16-1. Desirable Properties of SPDs

1. Negligible effect on the system during normal operation
2. Good clamping of voltage across the protected port
3. Exposure to expected overvoltages will not destroy SPD
4. Fast responding
5. Minimal or no routine maintenance
6. Small volume
7. Packaged for proper environment
8. Protection against continuous overvoltages sometimes desirable
9. Minimal cost
10. Protect all conductors
11. "Fail-safe" design
12. Coordinate with other SPDs
13. Length and diameter of wire leads or type of connector
14. Small leakage currents
15. Long lifetime
16. Small remnant

circuit must resist the temptation to modify the performance requirements of the system in order to make it easier to protect.

2. *The protection circuit should have good clamping: the voltage across the protected port during a large transient current should be near the maximum operating voltage of the system.* The word "near" might mean a factor of 1.0 to 3.0 times the maximum operating voltage of the system, with the specific factor dependent on the vulnerability of a specific system. For example, components connected to the ac supply mains may be more tolerant of transient overvoltages than semiconductors for applications at frequencies above about 10 MHz.

3. *The protection circuit should be able to survive expected transient overvoltages without being destroyed.* The specific worst case for a given environment or application may be listed in relevant standards or in the specifications in the purchase contract. It is important that overvoltages not be inflated too much, otherwise the cost of the protection circuit may be unacceptably high.

4. *The protection circuit should be fast responding.* A response time of less than 1 ns is desirable if electrostatic discharge (ESD) or high-power microwaves (HPM) is expected. The protection circuit should be coordinated with other measures, such as shielding. For a system with bandwidth limited by a low-pass filter downstream, a response time of 1 μs might be adequate.

Several other desirable properties of circuits for protection against electrical overstress are:

5. *The protection circuit should require minimal or no routine maintenance.* Consumable components, such as fuses, should have in-

dicator lamps to signal the need for replacement. Requiring routine maintenance increases the cost of the protection circuits, although the money comes from a different budget.

6. *The protection circuit should require a small volume.* In retrofit applications, the volume available for additional modules is limited. Where overvoltage protection is included in the original design, space is still valuable.

7. *The protection circuit must be packaged for the proper environment.* The relevant environmental variables (temperature, humidity, corrosion resistance, air pressure, etc.) must be considered. Temperature is probably the most critical environmental parameter for electronic components. For example, the conduction of metal oxide varistors increases as temperature increases. Varistors that have an acceptably small steady-state conduction current at a temperature of 25°C, may be unacceptable in an environment where ambient temperatures can reach 75°C. Environmental variables such as humidity and corrosion resistance are usually addressed by specifying an appropriate package for the components or the entire protection module. Air pressure is critical for spark gaps that use ambient air, because the firing voltage is a function of air pressure. However, most spark gaps used to protect electronic systems (other than telephones) are sealed units. Air pressure is also relevant for cooling of equipment. Cooling of equipment is not a critical issue for overvoltage protection circuits that are designed only for exposure to nonrepetitive events (perhaps 60 seconds or more between overvoltages).

8. *It would be desirable if the circuit for protection against transient overstresses could also protect against sustained or continuous electrical overstress without damage to the protection circuit.* Nonrepetitive transient overvoltages are not the only threat to equipment. The ultimate goal is to increase the reliability of equipment in adverse environments and not just to solve the narrower problem of transient overvoltages. In some situations it may be possible to enhance a circuit that protects against transient overvoltages and also obtain protection against repetitive or even continuous overvoltages. Examples of these other threats include (1) overvoltages caused by switching of inductive loads on each cycle by a thyristor, and (2) overvoltages caused by accidental connection of the ac supply mains or a dc supply bus to a communication line.

9. *The cost of the protection circuit should be minimal.* Cost is an engineering parameter too. The cost of protecting a system often exceeds the cost of what is being protected. For example, an SPD module that sells for $50 may be necessary to protect an integrated circuit that costs less than $1. An uninterruptible power supply that sells for $1000 and a $400 line conditioner may be required to protect

a $50 power supply in a desktop computer from upset. Spending more money on the protection than the cost of the device being protected is often reasonable. The *value* of the device being protected is not just its cost; there are also the value of the reliability of the system that contains it, the cost of labor to repair, the loss of function while the system is awaiting repair or being repaired, etc. If minimal cost is critical, protection should be designed into the product and not added on. An SPD module that sells for $50 may contain components and assembly labor that cost less than $5. The remaing $45 goes for engineering design, testing, packaging, advertising, shipping, and profit. Some of these extra expenses can be avoided or minimized by including the SPD in the same package as the vulnerable equipment.

10. *What conductors should be protected?* For example, in a single-phase mains application, is one varistor between the hot and neutral conductors adequate? Or should one consider common-mode overvoltages that appear between the hot and grounding conductors and also between the neutral and grounding conductors? Complete, but more expensive, protection requires three varistors, as discussed in Chapter 19.

11. *Is an inherent "fail-safe" design desirable?* The term fail-safe is ambiguous. The word "fail-safe" is dependent on the application and not just a function of the protection circuit. Martzloff (1982) advocated replacing the single term "fail-safe" with a choice of two terms for *shunt* SPDs: "fail-short" or "fail-open." In fail-short, when the SPD fails, the equipment is protected from damage but is upset until the SPD is replaced (because the SPD shorted the line). In fail-open, when the SPD fails, the SPD is jettisoned, and the system continues to operate without overvoltage protection. These two choices were illustrated in the discussion of how to use fuses in Chapter 12. A fusible metal, such as tin-lead solder, can be used to provide a short circuit across the electrodes of a fail-short spark gap or diode. Fail-open circuits usually have a fuse, circuit breaker, thermal cut-out, or explosive device (which is often the SPD itself!) to remove the failed shunt SPD. It might appear on first glance that fail-short is more conservative. Fail-short is preferred when protection from damage by overvoltages is more important than continuous operation. However, there are situations where the importance of continuing the mission overrides any concern about operating the system without an appropriate SPD: fail-open is desirable for such situations.

12. *Coordination with other protection components and circuits.* In practical situations this is often difficult to do, because the system designer does not know what other items are in the customer's

environment. The user does not know what protective components are hidden inside of each chassis in his or her environment. However, neglect of coordination can lead to money wasted on overvoltage protection that is ineffective, as well as possibly produce catastrophic failure.

13. *Minimum diameter of wire in arrester, length of leads, or type of connector.* Relatively small-diameter wire can be used to carry large transient currents. For example, spark gaps that are rated for 10 kA peak currents with an 8/20 μs waveshape often have leads with a diameter of 0.8 mm (20 American wire gauge). It is important, as discussed in Chapter 15, that the length of leads on shunt SPDs be minimized. This can present practical problems when the SPDs are to be installed in existing fixtures that were designed without considering transient overvoltages. Technicians learn by experience not to cut leads to the minimum length, because such short leads may create difficulties at a later time, when components need to be rearranged. By using longer than minimum leads, splices (which tend to be less reliable than continuous wire) can be avoided in the future. These concerns are normally admirable, but long leads are definitely undesirable in shunt SPDs. Perhaps the most elegant solution is to design SPDs with a male and female conductor attached, so that the SPD can be conveniently installed in series with the present power or data lines. Because the user does not need to connect any wires, the possibility of making mistakes is avoided. However, this forces the SPD to use existing grounding conductors (e.g., the green wire in the mains cable, the shield in data cables) that are often long and tortuous and which therefore have more inductance than is desirable for shunting transient currents away from vulnerable objects.

14. *SPDs should have small leakage currents.* There are limits on line-to-ground capacitive current for equipment connected to the mains (see Chapter 13). The steady-state power dissipation of the SPD during normal operation should be small.

15. *SPDs should have a long lifetime.* It would certainly be desirable that the expected lifetime of the SPD equal or exceed the lifetime of the equipment that the SPD is to protect. The limits on lifetime are often imposed by exposure to severe overstresses (e.g., waveforms with larger peak currents, larger charge transfer, or longer duration than were assumed in the design of the SPD). Other limits are inherent in degradation mechanisms of SPD components: decay of radioactive material or erosion of electrodes can both limit the life of spark gaps; metal oxide varistors are degraded as described in Chapter 8.

16. *The voltage and energy in the remnant should be small.* In a sense, this property is a combination of properties Nrs. 2 and 4, small clamping voltage and fast response. *If* the SPD were to respond

instantly and have a small clamping voltage, then the voltage in the remnant would be small. However, for SPDs that use *only* crowbar devices (e.g., spark gap or SCR), the response time can be appreciable. Before these crowbar devices switch, the remnant is essentially the entire overvoltage, which is called the *let-through voltage*. The let-through voltage can be minimized by adding additional devices downstream from the crowbar, as in a hybrid protection circuit. The let-through voltage can also be minimized by selecting a fast-responding crowbar device.

F. WHERE SHOULD SPDs BE INSTALLED?

There are three routes by which components in a chassis may be damaged or upset by electrical overstresses: (1) electromagnetic radiation from outside the chassis, (2) overvoltages conducted on cables going into the chassis, and (3) overvoltages generated inside the chassis.

The first route can be closed by shielding the chassis. The second route may be closed by an appropriate combination of hardening power supplies and interfaces, installing SPDs and filters. The third route is closed by installing SPDs at the source of internally generated overvoltages.

The protective devices should be installed *between* the source of the overvoltage and the equipment to be protected, as shown in Fig. 16-1.

Consider another way of installing an SPD, as shown in Fig. 16-2. Vulnerable load *A* is protected against the overvoltage injected at point *X* in Fig. 16-2. However, the overvoltage can travel to loads *B* and *C*, and possibly damage them, before the one SPD can protect them. An SPD that has multistate (hybrid) construction may have a voltage clamping level across the input port that is too great to protect devices *B* and *C*.

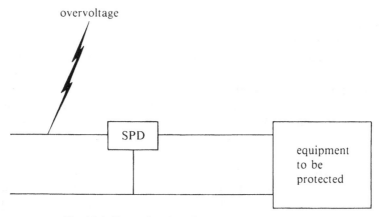

Fig. 16-1. *Proper location of a surge protective device.*

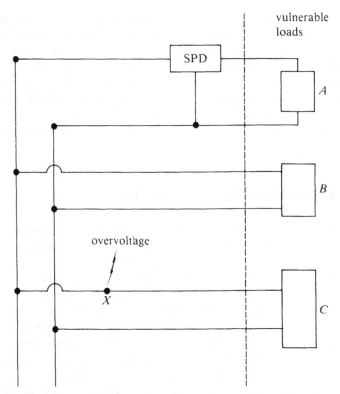

Fig. 16-2. *Inadequately protected system: loads B and C may be damaged by surge that originates at point X.*

There are two good locations for overvoltage protection in most applications. One is near the point of entry of a cable into a building or room. The other is at the point where the cable is to be connected to the equipment that is to be protected.

The advantage to placing the protection near the point of entry is that this minimizes electromagnetic radiation by transient overvoltages on cables inside the building. However, protection at the point of entry does not protect equipment from electrostatic discharge (ESD) or switching transients that are generated within the building.

It is good practice to place heavy-duty SPDs that might conduct transient currents of more than 1 kA at the point of entry. Examples of such devices are lightning arresters on the mains and spark gaps on telephone lines.

It is also good practice to place light-duty SPDs such as avalanche diodes or small-diameter metal oxide varistors near or inside equipment to protect the equipment from locally generated overvoltages and from the remnant that travels downstream from the heavy-duty SPD at the point of entry.

After it is decided where the SPD is located, the relation question of how "ground" is to be obtained can be answered. SPDs at the point of entry are usually connected to a copper water pipe that is bonded to the building ground. SPDs in equipment are usually connected to the green wire in the mains cord, which is connected to the building ground at the distribution panel. When a transient with a large value of dI/dt occurs, the two ground points are not at the same potential owing to inductance of the conductors.

It should be noted that if the grounding wire is disconnected, an SPD will provide *no* protection against common-mode overvoltages. The grounding conductor is more than a safety feature: it is absolutely essential for the protection against common-mode overvoltages.

17

Applications in Signal Circuits

A. INTRODUCTION

This chapter covers applications of overvoltage protection to signal circuits. *Signal circuits* are defined here as those in which the magnitude of the voltage is less than about 15 V, and the magnitude of the current is less than about 50 mA. The bandwidth of a signal circuit may extend from dc to more than 1 MHz.

Line receivers have their inputs connected to a signal line, and they receive data from a distant source. Line drivers have their outputs connected to a signal line, and they transmit data to a distant receiver. Line receivers and line drivers are examples of *interface modules.* Most of the overvoltage protection problems on a signal lines arise during protection of interface modules in a system, since the overstresses that are conducted on long cables from the exterior of a system tend to be much more severe than internally generated overstresses.

Some basic circuits are discussed that can be used on almost any signal line, regardless of whether it is analog or digital. Then we shall discuss some specific examples for analog circuits, followed by specific examples for digital circuits. The discussion of digital circuits will emphasize computer data lines.

B. SPARK GAP AND AVALANCHE DIODE CIRCUIT

1. General

Bodle and Hays (1957) invented the basic circuit that combines a spark gap, series impedance, and avalanche diodes, as shown in Fig. 17-1, to provide

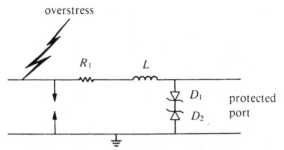

Fig. 17-1. *Basic protection circuit for signal lines.*

comprehensive protection from electrical overstresses. This circuit has been described by many other authors, including Popp (1968), Greenwood (1971, p. 318), Erickson (1972), Knox (1973), Clark (1975; 1976), and Smithson (1977) among others.

The load is protected by an avalanche diode absolute value clipping circuit, D_1 and D_2, which will be discussed in detail later. The resistor, R_1, provides a large voltage across the spark gap when the current in the avalanche diodes is large. If the voltage across the spark gap is sufficiently large, the spark gap will conduct and shunt current away from the avalanche diodes. The avalanche diodes protect the load; the spark gap and resistor protect the avalanche diode.

When the spark gap is conducting, it acts in two ways to protect the circuit. First, it dissipates some of the energy of the overvoltage as heat. Second, its low impedance at the end of a transmission line causes some of the transient overvoltage to be reflected. The reflection phenomenon is useful for delaying the dissipation of transient with large values of dI/dt. If we have 500 m of line that has a short at each end, owing to conducting spark gaps, the round trip time delay is about 5 μs. After multiple reflections the energy of the transient overvoltage must be dissipated in the protection circuit and the resistance of the line.

2. Specifications for Spark Gap

A spark gap will be nonconductive until the voltage across the gap exceeds the dc firing potential. Clearly, the dc firing voltage of the spark gap must be greater than the maximum voltage across it during normal operation of the system. This is an easy requirement to meet in signal circuits that often operate with potential differences of no more than ±20 V. The dc firing voltage is not the only important specification in overvoltage protection, owing to the time delay between the application of a transient voltage and the onset of appreciable conduction in the spark gap. The value of the time delay is complicated to predict and is a function of the rate of rise of the potential as well as the spark gap parameters. For example, if 300 V is applied with a rise time of less than $0.05\,\mu s$ to a spark gap with a 150 V dc firing voltage, the gap will typically remain nonconducting for about 0.5 to $1.0\,\mu s$, then operate in the glow region for about 0.5 to $1.5\,\mu s$, and then operate in the arc region. The point to be made is that a spark gap can be (briefly) nonconducting with a potential across it that is several times its dc firing potential. Furthermore, this high potential can remain across the spark gap for durations on the order of a microsecond. This is long enough to damage vulnerable electronic circuits if other protective devices (e.g., avalanche diodes) are not used. The main reason for including spark gaps in overvoltage protection circuits is that they excel at shunting currents of the order of 5 to 10 kA away from vulnerable circuits.

The author usually specifies spark gaps with a nominal dc firing voltage between 150 and 250 V for use in circuits that protect signal lines with normal voltages of less than ± 50 V. These spark gaps are faster responding at large values of dV/dt than gaps with dc firing voltages between 70 and 100 V (see Chapter 7).

In typical signal circuits there is no need to worry about the possibility of follow current in the spark gap. Most signal sources cannot supply adequate current to maintain an arc (or a voltage of at least 60 V so the spark gap could operate continuously in the glow regime). However, if follow current is possible in a particular application, then a current-limiting device must be included between the source of the follow current and the spark gap. Possible current-limiting devices include a fuse, a positive temperature coefficient (PTC) resistor, or a circuit breaker.

3. Specifications for Avalanche Diodes

The reverse breakdown voltage, V_z, of the avalanche diodes is usually specified so that it will not conduct during normal operation of the system. This implies that the avalanche diode should have a large value of V_z. However, there are two reasons why the diode should have the smallest possible breakdown voltage. First, small values of breakdown voltage allow the protection circuit to clamp at a lower level and thus provide less stress on the devices to be protected. Second, avalanche diodes can be destroyed by large amounts of power or energy during overstresses. For two diodes of the same type, the one with smaller value of V_z can tolerate larger surge currents, since power or energy are approximately proportional to V_z. The value of V_z will usually not be more than a few volts greater than the supply voltage for the active circuit that transmits the signals.

If noise is critical, then the normal system operation should not put the avalanche diode in a weakly conducting state, because some weakly conducting avalanche diodes can be potent sources of noise (see Chapter 9). However, if noise is not particularly critical *and* the current during normal system operation is limited, one may specify a value of V_z that is *less* than the normal system voltage. For example, the nominal values of V_z for D_1

and D_2 might be 11 V when connected to the output terminal of an operational amplifier that operates from ±15 V supplies. This is not a common practice, but it may be useful in particular applications.

The effects of tolerance should be considered when specifying the value of V_z. There is often a very small cost increase (e.g., $0.10 in an item with a cost of $1.50) associated with diodes with a 5% tolerance, when compared to diodes with a 10% or 20% tolerance. In many situations the smaller tolerance devices will be worth the extra cost, since the nominal value of V_z can be smaller without interfering with the normal operation of the system.

The decision to use special transient suppressor diodes or an "ordinary" avalanche diode is largely dependent on the value of R_1 in a particular application. In some applications, the resistance upstream from the diode can be 1 kΩ or even larger. In these situations, a diode with a steady-state power rating as small as 1 W might be used without exceeding the diode's *steady-state* power rating (except for a few microseconds when the input voltage is greater than V_f, an overstress that the diode can easily withstand). In other applications, the resistance upstream from the diode will be as small as 3 to 10 Ω. In these situations, a special transient suppression diode is highly desirable, because the steady-state power rating of the diode will be greatly exceeded during an overstress. Examples of both types of applications are given later in this chapter.

There are several advantages of using "ordinary" avalanche diodes, instead of transient suppression diodes: (1) ordinary diodes likely have lower cost; (2) ordinary diodes have smaller size, less mass; and (3) ordinary diodes are likely to have smaller parasitic capacitance. The lower cost is always welcome. The smaller size can be critical in portable equipment, aircraft, missiles, and satellites. The smaller parasitic capacitance is critical in high-frequency applications.

Transient suppression diodes are available in bipolar models, for which D_1 and D_2 are combined into a single package. This reduces the cost of components and circuit assembly time. Moreover, the bipolar diodes permit a shorter distance on the circuit board than the series connection of two unipolar diodes. This shorter distance reduces the parasitic inductance that is in series with the shunt path through the bipolar avalanche diodes.

It is common for D_1 and D_2 to have the same nominal value of V_z. However, this is not required. If the signal voltage always has the same polarity, the forward-biased diode may be omitted. For example, to protect a digital line that has a potential between 0 and +5 V, D_1 may be omitted and D_2 may have a value of V_z of 6.8 V. In other applications, D_1 and D_2 may have different values of V_z.

4. Specifications for Resistor

The series resistor, R_1, limits the current through the avalanche diodes, D_1 and D_2, to a safe value. The resistance value of R_1 is calculated by the following conservative procedure and Eq. 1:

$$R_1 = [V_f - \max(V_z)] \max(V_z)/P \tag{1}$$

P is the maximum *steady-state* avalanche diode power rating specified by the manufacturer, $\max(V_z)$ is the *maximum* reverse breakdown voltage of the avalanche diode, and V_f is the *maximum* value of the dc firing voltage of the spark gap, including tolerances.

The value of $\max(V_z)$ in Eq. 1 may be estimated by multiplying the *nominal* avalanche voltage, V_z, by a factor of about 1.2 to compensate for the increase in magnitude of the breakdown voltage at large currents (owing to the incremental resistance of the diode and the temperature coefficient) and for the 5% tolerance of the avalanche breakdown voltage. The value of V_z at small currents may be used in Eq. 1 for crude estimates.

During transient overvoltages with a rapid rate of rise, the spark gap will conduct at a voltage that can be several times greater than V_f. During the microseconds that the spark gap is nonconducting, the current in the avalanche diode will be greater than its steady-state maximum value. This will not damage the avalanche diode, because the overload is relatively small, and the time of the overload is brief.

Using Eq. 1 to determine the value of resistance R_1 is perhaps too conservative. A constant voltage source of magnitude V_f, which is typically 150 V dc, is uncommon in modern industrial environments. It is therefore unlikely that the overvoltage protection circuit would be connected for an indefinite time to such a large dc voltage. However, it is conceivable that the circuit could accidentally be connected to a sinusoidal voltage source (e.g., mains with a nominal voltage of 120 V rms). To recalculate the value of R_1, two simplifying assumptions are made: (1) that the avalanche diode has a constant voltage, V_z, across it when it is reverse-biased and conducting, and (2) that V_f is much greater than V_z, so that each avalanche diode is reverse-biased and conducting during essentially half of the period of the sinusoidal voltage. Equation 2 is then obtained for the minimum value of R_1.

$$R_1 = [(V_f/\pi) - V_z](V_z/P) \tag{2}$$

If both avalanche diodes are in the same package, then V_f is divided by $\pi/2$, not π, in Eq. 2. This correction is necessary, since power is dissipated in the diode package during each half-cycle, rather than allowing the package to cool during alternate half-cycles.

The conventional method for calculating the value of R_1 uses a standard model for the transient overstress. Let

V_S = maximum expected transient voltage (e.g., peak voltage of a 10/1000 μs waveform)

I_S = manufacturer's rated peak surge current of diode for same waveform as used for V_S

V_c = maximum clamping voltage specified by manufacturer of diode at surge current I_S

Then use Eq. 3 to determine the minimum value of R_1.

$$\min(R_1) = (V_S - V_c)/I_S \tag{3}$$

The value of $\min(R_1)$ should be multiplied by a factor for tolerance to obtain the nominal value; for example, multiply by 1.05 to convert $\min(R_1)$ to R_1 for a 5% tolerance resistor. A more conservative calculation for R_1 would be to use the nominal value of V_z in place of V_c in Eq. 3 (Huddleston and Bush, 1975).

The author objects to this conventional method for determination of R_1, because the method requires the designer to specify a maximum value of surge voltage and a surge waveshape. Surges with larger peak voltage or longer duration threaten to destroy the avalanche diode(s). Also the effect of the spark gap is ignored in the derivation of Eq. 3, which appears unreasonable. Data sheets for transient suppressor diodes typically list maximum parameters for the $10/1000\,\mu s$ waveform but not for other waveforms. This makes it difficult to do design calculations for other waveforms that may be more appropriate for a particular environment.

Larger values of R_1 should be used if they are tolerable in a particular application, because increasing the value of R_1 increases the amount of overvoltage protection in an economical way. The value of R_1 that is obtained from Eqs. 1 to 3 will probably be a nonstandard value (e.g., 19 Ω instead of 20 Ω), so it should be increased to the next-larger standard stock value.

The series component between the spark gap and avalanche diodes is not required to be purely resistive. Inductance is desirable if surges with short rise times are expected. One might use a wire-wound resistor for R_1 to obtain both resistance and inductance in the same package. Or a choke may be substituted for R_1. Calculation of the proper value of inductance is complicated and depends on the waveshape of the surge, but values of the order of $30\,\mu H$ are commonly used. Using inductance alone has the disadvantage of offering little or no protection from long-duration over-stresses. Also, the core of the inductor may saturate during large surge currents, which makes the inductance vanish.

The resistor should be physically large to prevent flashover. The author usually uses a carbon composition resistor with a steady-state power rating of 2 W. A wire-wound resistor with a steady-state power rating between about 5 and 12 W is also suitable for R_1. If long-duration overstresses are expected that may overheat R_1, then a positive temperature (PTC) device should be considered for R_1. Metal film and carbon film resistors are *not* suitable for limiting large surge currents, as discussed in Chapter 12.

5. Other Considerations on Resistance Value

A resistance value that is determined by using Eqs. 1 to 3 may be so large that it interferes with the normal operation of the circuit.

Resistor R_1 forms a voltage divider with the load resistance, R_L. In systems with a small value of source impedance, the voltage across the load is reduced by more than 5% when

$$R_1 \geq 0.053 R_L$$

If high-frequency signals are to be transmitted, one should consider the parasitic low-pass filter in the circuit of Fig. 17-1. If the protection circuit is connected to the receiving or load end, a low-pass filter is formed by R_1 and the parasitic capacitance of the avalanche diodes. If the protection circuit is connected to the transmitting (source of normal signals) end of a long line, the shunt capacitance between signal and common conductors is down-stream from R_1 and forms a low-pass filter.

Various ways to reduce the effective parasitic capacitance of avalanche diodes were presented in Chapter 9. These methods may be useful to extend the bandwidth of a protection circuit.

When propagation of high-frequency signals is required, the value of R_1 should be as small as possible. On the other hand, R_1 should be as large as possible in order to limit the current in the avalanche diodes. Clearly there must be a compromise for the value of R_1. Inspection of Eq. 1 shows that R_1 is a maximum when V_z is $V_f/2$. Because V_f is usually greater than twice V_z, smaller values of V_z will give smaller values of R_1. Inspection of Eq. 1 also shows that R_1 is smaller when the power rating, P, is larger. However, larger power ratings may be achieved by increasing the cross-sectional area of the diode, which increases the parasitic capacitance of the diode.

A positive temperature coefficient resistor can be used for R_1, so that the value of resistance may be small during normal operation but increases to large values during long-duration overstresses.

Another approach to the design of the circuit in Fig. 17-1 is to make R_1 as large as possible without interfering with the normal operation of the system and use transient suppressor diodes that are specified to survive relatively large surge currents. Typical values of R_1 in this approach are often between 3 and 10 Ω. When R_1 is too small, there is danger that the avalanche diode will clamp the voltage across the spark gap and prevent the gap from conducting. The spark gap is superfluous when R_1 is too small. If this method of design is used, one should be careful to also include long-duration overvoltage waveforms with peak voltages near V_f in the testing of the circuit (see discussion of "blind spot" in Chapter 22).

6. Assembly Considerations

This circuit is usually assembled on a printed circuit board in order to obtain miniature size and to reduce production costs. The length of the leads of the spark gap and diodes should be as small as possible (see Chapter 15).

Because transient overvoltages can have high voltages, the circuit designer should take precautions to avoid an arc path that could shunt resistor R_1. To intercept flashover, should it occur, a wide ground band should be etched on both sides of the printed circuit board underneath resistor R_1.

The mounting of the spark gaps in a vertical plane minimizes the length of the spark gap leads and removes the right-angle bend in each lead that would otherwise be present. This minimizes the inductance in series with the spark gaps and helps avoid the inadvertent transformer effect, as explained in Chapter 15.

An interesting phenomenon was noted by the author during surge testing of this type of circuit. When the printed circuit board foil evaporated during the test, the trace always disappeared where it was connected to the "donut pad" for the spark gap. The intense peak currents in these particular tests, which were of the order of 10 kA, create large magnetic fields near the spark gaps. The magnetic field pushes the conduction electrons toward the top or bottom surface of the printed circuit board trace and concentrates the current in a thin layer. If the current density is sufficiently great, the local heating will evaporate the conductor on the printed circuit board. The result is complete removal of the printed circuit board conductor for a gap of 1 to 4 mm. This weakness in the printed circuit board can be avoided by bending the spark gap lead after it passes through the circuit board and routing it for about 6 mm along the printed circuit trace. The connection of the spark gap lead to both the pad and the straight line trace is made with solder in the usual way. The wire lead and solder provide a heat sink as well as a reduced parasitic resistance, so that the traces are less likely to be evaporated.

Further, a piece of solid bus-bar wire (22 to 16 AWG) is soldered in contact with the common (ground) on the printed circuit foil and to the common terminal of all of the spark gaps. The bus-bar reinforces the printed circuit foil in the region of the board where large transient currents could flow. The signal conductors that are downstream from the spark gaps are not reinforced with bus-bar wire, because evaporation of these signal conductors is unlikely and poses no hazard to the equipment to be protected.

The typical grounding conductor for the mains does not have a small value of inductance because of its relatively long path and bends in cable. However, most alternative grounding practices are unsafe.

C. DIODE CLAMPS TO POWER SUPPLY

The other basic overvoltage protection circuit is to use diodes to clamp the signal line to the dc power supply line, as shown in Fig. 17-2. The device to be protected in Fig. 17-2 is U_1, which might be an amplifier, logic device, or other integrated circuit. The two dc power supplies for the U_1 are V_{CC} and V_{EE}. Each of these two supplies has a typical magnitude of 15 V for analog circuits. Diode D_1 prevents the input terminal of U_1 from having a potential greater than $V_{CC} + 0.7$ V. Similarly, diode D_2 prevents the input terminal of U_1 from having a potential less than $-(V_{EE} + 0.7$ V$)$. Since *most* integrated circuits can continually withstand any potential at the input terminal that is between V_{CC} and $-V_{EE}$, this circuit provides good protection for the input terminal of U_1. The circuit of Fig. 17-2 must not be used if the input terminal cannot withstand application of the power supply voltages.

Core-switching diodes, which were discussed in Chapter 10, are commonly used for D_1 and D_2. These diodes are normally reverse-biased in the circuit of Fig. 17-2, and their leakage current, which is typically less than a few tens of picoamperes, would not be expected to cause problems with U_1 in most applications. The shunt capacitance of D_1 and D_2 might couple noise from the power supply to the input of U_1. However, the capacitance of core-switching diodes is only a few picofarads. If the input impedance of U_1 is much smaller than the impedance of the diode's capacitance, the voltage divider that is formed by these two impedances will substantially attenuate the noise injected from the power supply. Power supply bypassing is necessary for other reasons, but this bypassing will also attenuate changes in voltages on the power supply lines caused by injection of transient currents through D_1 or D_2.

The small parasitic shunt capacitance of this protection method makes it suitable for protecting high-frequency circuits. Of the available overvoltage protection components, only spark gaps and core-switching diodes have a parasitic capacitance of the order of picofarads. All other common components have a larger capacitance.

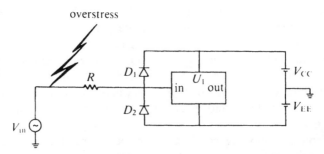

Fig. 17-2. *Protection circuit that diverts surge current to power supply bus and away from input of circuit U_1.*

The circuit in Fig. 17-2 has a serious problem. The overstress that threatened the input of U_1 was shunted to one of the power supply lines. It is possible that the overstress could damage U_1, or other devices that are connected to these dc power supply lines, by injecting the overstress into the power supply terminals. *If* the overstress has a very small charge transport, e.g., less than $0.1 \mu C$, then it can be shunted to ground through bypass capacitors without causing any distress, provided that the bypass capacitors are located near U_1. However, overstresses with a charge transport of substantially more than $0.1 \mu C$ should not be diverted with the circuit of Fig. 17-2. The value of $0.1 \mu C$ comes from an arbitrary decision to limit transient overvoltages on the dc power supply to less than 0.1 V and use a $1 \mu F$ bypass capacitance. The circuit in Fig. 17-2 can offer excellent protection of the input terminals from electrostatic discharge, a transient with a relatively small charge transfer.

The value of the resistor R in Fig. 17-2 limits the current in diodes D_1 and D_2. If the value of R is made no larger than 1% of the input impedance of U_1, appreciable attenuation of the input signal can be avoided. Core-switching diodes can tolerate transient currents of the order of 4 A for $1 \mu s$, which is a $4 \mu C$ charge transfer. Because of concern about injecting noise on the power supply lines, the circuit in Fig. 17-2 was restricted to situations in which the charge transfer was less than $0.1 \mu C$. The core-switching diodes are robust enough to survive these small transient currents without a series resistance.

The general circuit concept of Fig. 17-2 is used to protect complementary metal oxide semiconductor (CMOS) logic integrated circuits. The input circuit of a basic CMOS cell is shown in Fig. 17-3, along with the overvoltage protection circuit. What is particularly interesting in Fig. 17-3 is that the overvoltage protection circuit is fabricated as part of the integrated

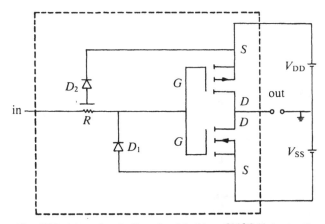

Fig. 17-3. *Internal protection circuit inside a CMOS logic circuit.*

circuit. The circuit has been described by Cergel (1974) and Pujol (1977). The resistor, R, has a value between 100 and 2500 Ω, depending on the manufacturer and the particular model of CMOS. Diode D_2 is distributed along the length of the series resistor R. The reverse breakdown voltage of the protection diodes is greater than the maximum value of $(V_{DD} - V_{SS})$, so the diodes will not operate in reverse breakdown during sustained overvoltages. The maximum steady-state input terminal current in a CMOS integrated circuit is 10 mA.

If the CMOS circuit is to be connected to a voltage source that has a level greater than V_{DD} or less than $-V_{SS}$, an external series resistor should be connected to the input terminal to reduce the current in the protection diodes to a safe level. If the external resistor has a value of 1 MΩ, the CMOS device can be connected to a continuous source of 1 kV without stressing the protection diodes. The external resistance and input capacitance form a low-pass filter. The typical CMOS input capacitance is about 5 pF. With a 1 MΩ external resistance, a time constant of 5 μs or a bandwidth of 32 kHz is obtained. This is much slower than the time constant of a few nanoseconds that is produced by the internal resistor R and the input capacitance. Unfortunately, nearly all robust protection circuits reduce the signal bandwidth.

D. BALANCED LINE APPLICATIONS

Many long signal lines are balanced in order to reject noise, which often appears as a common-mode signal. These balanced signal lines require special consideration for overvoltage protection. A basic spark gap and avalanche diode circuit for a balanced transmission line is shown in Fig. 17-4.

The spark gap should be a three-electrode model, which was described in Chapter 7. A pair of two-electrode spark gaps can pass a large differential-

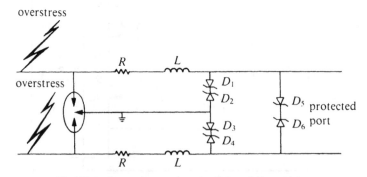

Fig. 17-4. *General protection circuit for a balanced line.*

mode voltage downstream if one, but not the other, of the two electrode gaps conducts. Therefore, three-electrode spark gaps are preferable for protecting a balanced line.

The same value of R and L must be present in each side of the circuit to preserve the balanced configuration. The value of R can be determined in the same way as for the unbalanced circuit, which was presented in Fig. 17-1, by considering one-half of the balanced circuit and ignoring D_5 and D_6. In Eqs. 1 through 3 above, V_z becomes the voltage for the reverse-biased diode D_1 or D_2; V_f is the dc firing voltage of the spark gap from either line to ground.

The inclusion of inductance L is particularly desirable in the balanced circuit. All of the commercially available three-electrode spark gaps have dc firing voltages of at least 300 V, because they are most commonly used for protecting telephone lines. This dc firing voltage is considerably larger than the 150 V models that are commonly used on unbalanced signal lines. Placing series inductance between the 300 V dc spark gap and the avalanche diodes helps create a large voltage across the spark gap during the initial part of an overvoltage without increasing the series impedance during normal operation to an unacceptably large value. If the signal bandwidth extends to relatively high frequencies, the pair of inductors should be a bifilar choke, which was described in Chapter 12. A bifilar choke inserts a series inductance for common-mode signals but has a negligibly small inductance for differential-mode signals.

Avalanche diodes D_1 and D_2 clamp the upper signal conductor in Fig. 17-4 to a potential between $\pm V_z$ of ground. Diodes D_3 and D_4 clamp the lower signal conductor to a potential between $\pm V_z$ of ground. If diodes D_5 and D_6 are absent, the maximum differential-mode output voltage could be as large as $2 \cdot V_z$. This situation occurs, for example, when the upper signal conductor is at a potential $+V_z$ from ground and the lower conductor is at a potential $-V_z$ from ground. Inclusion of diodes D_5 and D_6 can limit the maximum differential-mode output voltage. In most cases, diodes D_1, D_2, D_3, D_4, D_5, and D_6 will have identical specifications, including breakdown voltage, V_z. However, situations are possible where D_5 and D_6 might have a different conduction voltage, probably smaller, than the other four diodes.

During normal operation of a balanced line, the upper signal conductor will be a potential of $V_d/2$ with respect to ground, and the lower signal conductor will be at a potential of $-V_d/2$ with respect to ground. The value of V_d may be either positive or negative. In this way there is no common-mode voltage present during normal operation. This requires that the line drivers for a balanced line have two power supplies, one positive and one negative.

The RS-422 specification for computer data exchange on a balanced line, which is widely used, does not require a zero common-mode voltage (although it is acceptable to use one). Most integrated circuit line drivers that conform to RS-422 specifications use only a single supply, which is usually +5 V. The voltage between one signal conductor and ground is typically either 0.2 or 3 V, depending on whether the line is at the low or high state. This implies that there will be a dc common-mode signal of 1.6 V, which is about 60% of the differential-mode signal. This use of a single power supply decreases the cost of the line driver system, since including a negative power supply would substantially increase the cost owing to an extra secondary winding on the power transformer, extra filter capacitor, and extra voltage regulator. The presence of a dc common-mode signal does *not* preclude the use of a common-mode choke.

Smithson (1977) developed a protection circuit for a balanced computer data line that is shown in Fig. 17-5. His circuit is interesting, because he used three stages of protection—spark gaps, metal oxide varistors, and avalanche diodes. The use of varistors is significant. The varistors decrease the stress on the diodes when the overvoltage does not cause the spark gap to conduct (e.g., when the transient overvoltage has a peak magnitude of less than 300 V). Moreover, the varistor also decreases the stress on the diodes during the initial portion of transient overvoltages before the spark gap fires.

It is interesting to estimate the varistor current and voltage when the voltage across the spark gap is just equal to its 300 V dc firing voltage. This condition occurs when there is about 30 A in the varistor and about 53 V across the varistor. (This information is from the maximum clamping voltage graphs in the 1978 edition of the General Electric varistor data sheets.) When 30 A passes through the 8 Ω resistor between the spark gap and varistor, 240 V will be dropped across the resistor. The sum of these

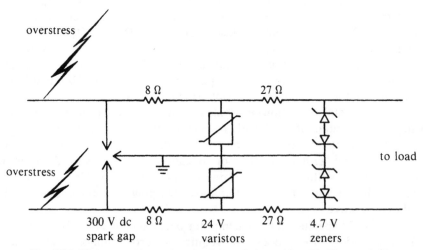

Fig. 17-5. *Three-stage protection circuit for a balanced line (Smithson, 1977).*

two voltage drops is 293 V, essentially the dc firing voltage of the spark gap. That about 80% of the voltage across the spark gap comes from the drop across the resistor emphasizes the importance of placing a series impedance between shunt nonlinear devices. Without the series resistor, the spark gap would never conduct.

If this example is continued, a current of about 1.8 A is found in the avalanche diodes when there is 293 V across the spark gap. This corresponds to a worst-case power dissipation of about 8.5 W in the reverse-biased avalanche diode. There is reasonable coordination between the varistor and avalanche diodes.

The 1N3825 avalanche diodes chosen by Smithson have a shunt capacitance of about 2 nF. This is in addition to the 8.5 nF shunt capacitance of each varistor. These large shunt capacitances will degrade the rise time of digital signals and reduce the system bandwidth. The output of such a transient overvoltage protection circuit on a digital signal line should be connected to an interface circuit with a Schmitt-trigger input to restore the proper logic waveform.

Smithson (1977) placed the avalanche diodes on the same circuit board that contained the load to be protected. This prevented ground impedance from impairing the clamping voltage. The other protection components were located at the point of entry for the cable into the building. This takes care of the inadvertent transformer effect that was described in Chapter 15.

E. ANALOG APPLICATIONS

Integrated circuit operational amplifiers are inexpensive, versatile analog "building blocks," which can be used to solve most analog signal conditioning problems for which the highest frequency of interest is less than about 1 MHz. Operational amplifiers are commonly found connected to analog inputs and outputs. Protection of these inputs and outputs condenses to the problem of protecting the operational amplifiers. Protection of inverting and noninverting voltage amplifier inputs is considered first, then protection of the output port.

1. Inverting Voltage Amplifier

When the operational amplifier is used as an inverting voltage amplifier, electrical overstress protection for the input terminals is provided by the circuit shown in Fig. 17-6. The voltage gain of this circuit is $-R_f/R_{in}$. The values of R_f and R_{in} are usually between 1 kΩ and 1 MΩ, although R_f is sometimes larger than 1 MΩ.

The two diodes, D_1 and D_2, in Fig. 17-6 prevent the magnitude of the differential-mode input voltage from exceeding about 0.7 V. Since the

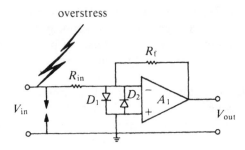

Fig. 17-6. Protection circuit for input of operational amplifier that is connected as an inverting voltage amplifier.

noninverting input is directly connected to ground, these diodes also limit the common-mode input voltage. All commercially available operational amplifiers that do not have internal diode clamping can easily withstand ± 0.7 V across the two input terminals. In fact, *most* modern integrated circuit operational amplifier input terminals can be continuously connected to any potential between the power supply voltages (which are typically $+15$ and -15 V) without damage to the operational amplifier.

During normal operation of the operational amplifier, there is no more than a few millivolts potential difference across D_1 and D_2 so the diodes will be nonconducting, and effects of the diode's capacitance will be negligible. There is no hazard to D_1 and D_2 from operation in the reverse breakdown region, since one of these diodes is always forward-biased and will protect the other diode from operation at a large reverse-bias potential difference.

Switching diodes (e.g., 1N4447, 1N4148 types) are commonly used for D_1 and D_2, because these diodes have fast response and are inexpensive. These diodes can tolerate steady-state currents of 500 mA without damage. This current corresponds to an input voltage of 500 V if R_{in} is 1 kΩ, and an even larger input voltage if R_{in} is greater. If the spark gap shown in Fig. 17-6 has a dc firing voltage between 90 and 250 V, the spark gap and resistor R_{in} will protect diodes D_1 and D_2 from excessive currents.

The weakest link in the circuit of Fig. 17-6 may be resistor R_{in}. It should be rated to withstand large transient voltages across it without degradation. (It is unacceptable to allow R_{in} to degrade, since its value determines the voltage gain of the amplifier.) A carbon composition resistor with a steady-state power rating of 2 W is generally suitable, and so is a wire-wound resistor with a steady-state power rating of 5 W. Metal film or carbon film resistors should be avoided for R_{in}, because these types of resistors are less able to withstand transient overvoltages. If continuous overstresses are anticipated, a PTC device should be inserted in series with an ordinary resistor to make the total value R_{in}. The PTC device will protect the ordinary resistor from damage by excessive power dissipation.

If the operational amplifier has a JFET, MOSFET, or other input stage with input terminal currents of the order of a picoampere or less, then standard switching diodes (e.g., 1N4447 type) are inappropriate for D_1 and D_2 if the very large input impedance of the amplifier is to be maintained. In this situation special low-leakage diodes for D_1 and D_2 should be used. Such diodes can be fabricated from the gate to channel junction of a Si n-channel JFET or GaAs LEDs, as described in Chapter 10. To maintain the low-leakage conditions, the inverting input terminal should be connected to a PTFE (Teflon, registered trademark of DuPont) insulated standoff. Low-leakage diodes have maximum steady-state currents between 10 and about 100 mA. If steady-state overload conditions are anticipated, the value of R_{in} should be more than 1 kΩ, and the spark gap should have a dc firing voltage of about 150 V.

2. Noninverting Voltage Amplifier

The noninverting voltage amplifier, shown in Fig. 17-7a, is another basic operational amplifier circuit. Resistors R_1 and R_2 determine the voltage gain; typical values are:

$$10 \, \Omega \leq R_1 \leq 1 \, k\Omega$$

$$1 \, k\Omega \leq R_2 \leq 10 \, k\Omega$$

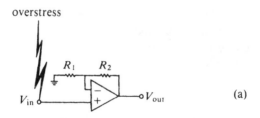

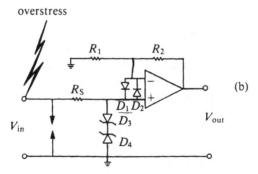

Fig. 17-7. Protection circuit for input of operational amplifier that is connected as a noninverting voltage amplifier.

Since this circuit cannot have a voltage gain of less than unity, the largest useful input voltage is equal to the maximum output voltage of the operational amplifier. This is typically about 12 V if ±15 V power supplies are used. If the voltage gain is greater than one or if the power supplies are less than ±15 V, the maximum useful input voltage is less than 12 V. For example, if the gain is 10, then the maximum useful input voltage is only about 1.2 V if ±15 V power supplies are used and only about 0.4 V if ±5 V power supplies are used.

A recommended protection circuit is shown in Fig. 17-7b. Diodes D_3 and D_4 are avalanche diodes that are connected to form an absolute value clamp circuit. Diodes D_3 and D_4 conduct when the absolute value of the voltage across them exceeds $V_z + 0.6$ V. This conduction voltage, $V_z + 0.6$ V, should be slightly greater than the maximum useful input voltage, so that D_3 and D_4 do not interfere with the normal operation of the circuit. If D_3 and D_4 were to conduct during normal operation, they would act as a voltage divider with resistor R_s and might also become a broad band noise source.

However, values of V_z greater than 12 V are not recommended. It is desirable to keep the maximum voltage at the noninverting input terminal of the operational amplifier *less* than the positive power supply voltage. When ±15 V supplies are used, this means the noninverting input terminal should be kept at less than 15 V from ground. There are several terms that can cause the clamping voltage to increase from the nominal value:

ΔV due to large current in reverse-biased diode	>2 V
Effect of 5% tolerance on 12 V diode	0.6 V
Forward-biased diode in D_3, D_4 pair	0.7 V
Parasitic inductance in the diode path	?

The value of ΔV due to large currents, and the forward-biased voltage drop, were discussed in detail in Chapter 10. Because of these terms, one cannot use diodes with $V_z = 15$ V to clamp the input at 15 V. There must be an interval of at least a few volts between V_z and the maximum clamping voltage.

When the maximum useful signal is between 1 and 6 V, one may wish to specify $V_z = 6.8$ V for D_3, D_4. Diodes with values of V_z less than about 6 V use the Zener mechanism rather than the avalanche mechanism and have much larger values of ΔV at large currents than avalanche diodes.

Diodes D_1 and D_2 prevent large differential-mode input voltages. Such voltages can arise if a rapidly changing signal or transient overvoltage waveform is applied to the noninverting input terminal so that the output voltage of the operational amplifier is slew-rate limited. These two diodes have no other function in this circuit.

The resistor R_s is included to limit the current to diodes D_3 and D_4 (and also D_1 and D_2 if R_1 is small). The input resistance presented by the operational amplifier circuit, r_{in}, is approximately

$$r_{in} = A_0 r_i R_2 / (R_1 + R_2)$$

where A_0 is the open-loop voltage gain of the operational amplifier itself, and r_i is the incremental resistance between the inverting and noninverting input terminals. The value of A_0 is typically at least 10^5, r_i is typically at least $1\,M\Omega$, and $R_2/(R_1 + R_2)$ is usually between 1 and 10^{-3}. This makes the input resistance of this circuit at least $10^8\,\Omega$. Therefore, the value of R_s has negligible effect on the voltage gain for low-frequency signals when R_s is a few kilohms. However, significant attenuation of high-frequency signals can occur owing to the low-pass filter that is composed of R_s and the parasitic capacitance of avalanche diodes D_3 and D_4. The time constant of this filter is usually of the order of a few microseconds, because $R_s \approx 1\,k\Omega$ and $C \approx 3\,nF$. This may not be a serious limitation, since common, inexpensive operational amplifiers (e.g., 741 and 307 types) also act as a low-pass filter with a time constant of $1\,\mu s$ due to the gain-bandwidth product of the amplifier.

To reduce the value of the RC time constant formed by the protective circuit, one could use a PTC device in series with (or in place of) an ordinary resistance. During normal operation, the value of R could be as small as $22\,\Omega$. The capacitance of the clamping circuit (D_3, D_4 in Fig. 17-7b) could be reduced by using a series string of switching diodes in place of the avalanche diodes, as described in Chapter 10, or in series with the avalanche diodes, as described in Chapter 9.

3. Protection of Output

Destructive overvoltages can also reach the operational amplifier through the output port. The circuit shown in Fig. 17-8 is suggested for protection of the operational amplifier from overstresses on an output cable. Diodes D_5 and D_6 can be avalanche diodes with $V_z = 12\,V$ when the operational amplifier is operated from $\pm 15\,V$ supplies. When smaller supply voltages are used, the value of V_z should be chosen to be near the maximum output voltage of the operational amplifier. Notice that no resistance is included between the operational amplifier output terminal and D_5, D_6. If these diodes conduct when the magnitude of the output voltage is large, internal output current limiting in the operational amplifier (which is also called *short-circuit protection*) is relied on to protect the operational amplifier, as well as diodes D_5 and D_6. If the maximum output current from the operational amplifier could harm D_5, D_6, or if there is no output current limit inside the operational amplifier, then a PTC resistor should be included between the operational amplifier output and D_5.

The resistor R_s limits the current in D_5 and D_6 from overstresses that appear on the output cable. Because the output current from the opera-

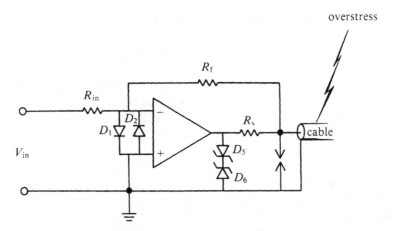

Fig. 17-8. *Protection circuit for output of operational amplifier.*

tional amplifier must also pass through R_s, we can not automatically specify a large resistance (e.g., $1 \, \text{k}\Omega$) at this point in the circuit. The maximum output current of a typical integrated circuit operational amplifier is about $15 \, \text{mA}$. If the maximum acceptable voltage drop across R_s during normal operation is $1 \, \text{V}$, R_s must be less than $68 \, \Omega$. A PTC resistor is preferable, instead of an ordinary resistor, for R_s. If an ordinary resistance is used for R_s, a value of $330 \, \Omega$ is suggested in order to protect the avalanche diodes.

Notice in Fig. 17-8 that the feedback point is not taken at the output terminal of the operational amplifier. The circuit shown in Fig. 17-8 has a small output impedance when viewed from the cable, despite the large value of R_s. This surprising result is produced by the negative feedback and the choice of the feedback point shown in Fig. 17-8.

Because it is possible for overvoltage on the output cable to reach the inverting input terminal through resistor R_f, the input must also be protected. Comprehensive input protection is furnished by switching diodes D_1 and D_2 when R_f is at least $1 \, \text{k}\Omega$. Typical values of R_f in this circuit would be of the order of $10 \, \text{k}\Omega$, since smaller values of R_f shunt output current from the amplifier away from the output cable to the input circuit.

F. DIGITAL APPLICATIONS

Protection of input and output terminals of integrated circuit logic devices is particularly challenging. Faster logic devices are *more* vulnerable to damage from electrical overstresses, for reasons specified in Chapter 1. Yet circuit designers, for good reasons, have gone away from the slow, robust DTL and ordinary TTL logic families, and tended to use various versions of high-speed devices which are much more vulnerable to damage.

Logic circuits commonly use a single +5 V power supply, although some CMOS circuits can operate with power supply voltages as small as +3 V. Protection devices that will clamp a line at 3 to 5 V during an overvoltage are difficult to obtain. True zener diodes have a relatively large value of $\Delta V/\Delta I$ and do not offer tight clamping. Avalanche diodes have a minimum value of V_z of about 6.8 V. A series string of core-switching diodes can provide clamping levels of a few volts.

G. PASSIVE SPD FOR RS-232 COMPUTER DATA LINES

1. General

Computer terminals and desktop computers are common in manufacturing, office, engineering, and university environments. To exchange data between users, the terminals or desktop computers are connected together. The most common way to exchange data is defined in EIA Standard RS-232. Lightning can cause a large transient overvoltage on computer data lines, which can damage the data communications circuit (Clark, 1981; Tetreault and Martzloff, 1985). Continuous overvoltages, which are discussed later, are an additional hazard. The following material describes a protective circuit that can be installed in series with a RS-232 serial interface near the computer or terminal. The circuit is designed for minimum expense and is purely passive. Optionally, it can protect against continuous overvoltages.

It is common to use four-conductor telephone cable for RS-232 communications between buildings. These wires are not connected to the telephone switching network but are merely used as a convenient direct line. Therefore modems are unnecessary. If the lines are sufficiently long, capacitance between the wires can be a problem. For example, the author measured a capacitance of $0.07\,\mu\text{F}$ between an unused pair of telephone conductors that runs between two buildings at one university, which were separated by a straight line distance of about 1 km. The limited output current of a common integrated circuit driver (e.g., 10 mA for model 1488), together with the large capacitance of long lines, makes high-speed communications impossible. However, experience at that university shows that communication can be achieved at speeds of up to 2400 baud without additional equipment, such as buffer amplifiers.

The basic spark gap and avalanche diode overvoltage protection circuit of Fig. 17-1 is commonly used to protect receivers and transmitters in the RS-232 computer interface. A separate protection circuit is required for each end of the cable, as shown in Fig. 17-9. The circuit shown in Fig. 17-9 is duplicated for each signal conductor. A typical installation would need three of these circuits: one for each of pins 2, 3, and either 6 or 20 on the 25-pin "subminiature D" connector at each equipment interface. Pin 7 of the connector, signal ground, is common to all of these circuits. In addition,

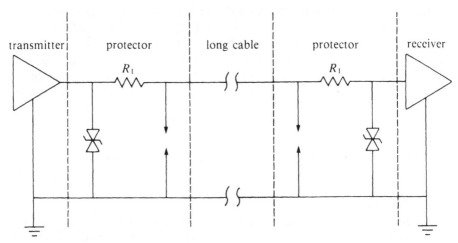

Fig. 17-9. *Passive protection circuit for RS-232 computer data line.*

pin 1, chassis ground, may be bonded to pin 7 at each end of the cable in order to reduce the series resistance and inductance in the ground conductor.

2. Selection of Avalanche Diodes

Conventional RS-232 data line protection modules in the commercial market use avalanche diodes with a breakdown voltage, V_z, of about 30 V. Clark (1981) recommended bipolar avalanche diodes "selected for a 25 V operating level with a maximum clamping level of 40 V" during a surge. The apparent justification for this choice is that RS-232C specifies the maximum *open-circuit* transmitter output to be 25 V. The 30 V diodes are specified to have less than 5 μA leakage current at 25 V, and, hence, they certainly do not affect the normal operation of the RS-232 interface.

However, a smaller value of V_z, such as 18 V, is also acceptable, as explained below, and has two advantages. First, for a given steady-state power or pulse energy rating and two diodes of the same type, the diode with the smaller value of V_z can safely divert a larger surge current. Second, the smaller clamping voltage during surges places less stress on the interface to be protected.

The maximum *open-circuit* voltage in RS-232 is 25 V. However, *when a load is connected* to the transmitter, section 2.6 of RS-232C specifies that the maximum magnitude of voltage at the interface point be *only* 15 V. Since RS-232 specifies the load resistance to be between 3 and 7 kΩ, a maximum current of 5 mA is available from the transmitter when the output voltage of the transmitter is greater than 15 V. This small current will not produce significant heating in an avalanche diode with a steady-state power rating of 5 W and a breakdown voltage between 15 and 25 V.

Common RS-232 transmitters (or drivers) use a type 1488 integrated circuit. These devices have an "absolute maximum" rating specified by their manufacturers of 15 V on the power supply pins and on the output terminal. A 25 V open-circuit output voltage is not possible with these devices when they are used within their manufacturer's specifications. If another type of transmitter is used that has a larger output voltage than 15 V, this larger output voltage will be experienced only when the interface is not connected to a receiver, which is certainly not the usual way to operate an interface.

Therefore, the avalanche diodes can be specified to have a value of V_z as small as 18 V $\pm$ 5% without concern about affecting the normal operation of the RS-232 interface. Nor will there be concern about damaging the interface or diodes by using 18 V avalanche diodes in this application.

The avalanche diodes are usually specified to be bipolar transient suppressor diodes.

Appropriate spark gaps would be models with a dc firing voltage between about 150 and 250 V. In the absence of other criteria, the author usually specifies a value of 150 V for V_f.

3. Selection of R_1 for Passive RS-232 Protection

If Eq. 1 is used to obtain a value for the resistance when V_f is 150 V, V_z is 18 V, and P is 5 W, then R_1 is about 470 Ω. This value is much too large if high-speed data communications are to be used, since R_1 forms a low-pass filter with the shunt capacitance of the cable and the parasitic capacitance of the avalanche diodes at the receiver. It is probably not reasonable to protect the RS-232 interfaces against continuous overstresses in this way.

Usual practice is to make the value of R_1 be between 10 and 33 Ω. A carbon composition resistor with a steady-state power rating of 2 W is suitable for R_1.

4. Protection of RS-232 Interface Against Continuous Overstress

There are several reasons why it is desirable to include protection from sustained overvoltages rather than consider only transient overvoltages:

1. Cloud-to-ground lightning can have "continuing currents" of the order of 100 A for a duration between about 0.04 and 0.5 seconds (see Chapter 2).
2. Overhead power lines can sag or fall and touch telephone lines. This will inject sustained overvoltages into a telephone line, which may enter the computer data lines.
3. Telephone line voltages may be accidentally connected to hardwired computer data lines if the telephone cables are used for hardwired computer communications.

4. There is also the possibility of malicious damage or sabotage by connection of a potent voltage or current source to a computer data line. The author is aware of one instance in which students connected 120 V rms mains to a RS-232 data line in order to have an excuse for not having their computer assignment finished!

There are two considerations that will make the circuit in Fig. 17-9 able to withstand continuous overvoltages. First, specify the dc firing voltage of the spark gaps to be greater than the maximum magnitude of expected continuous voltages. Second, use a positive temperature coefficient resistor (PTC) for R_1. The author has had good experience with a polymer PTC that has a resistance of 22 Ω when at room temperature and switches to more than 3 kΩ when hot. These two changes are inexpensive and extend the range of threats that will damage neither the interface nor the protective circuit.

5. Use of Connectors on RS-232 SPD

Many commercial transient protection circuits have screw or solder terminals. The user must connect cables to these terminals. One of the problems with requiring the user to properly connect input, output, and ground wires is that it may not be done correctly. Consider a circuit that protects four signal conductors from voltages that are more than about ±30 V from common (ground). Suppose that the user connected pins 2, 3, 7, and 20 to the four signal conductors. Then one could obtain output voltages as large as 60 V between pin 7 (signal common) and the other three wires. Such a condition is not likely to protect electronic equipment. This is not just a theoretical possibility: the author has seen installations where someone had "protected" the signal common by treating it as just another signal line.

For unskilled users, it is better to have prewired protection circuits with the 25-pin D connectors permanently connected and attached to the protection modules. Pin 7 on the unprotected side and pins 1 and 7 on the protected side of the circuit should be permanently connected to the common conductor in the protection circuit.

When the author surveyed the market for passive RS-232 protection modules in 1983, several companies had protection for only two to four pins and had wires connected directly between input and output on all of the other pins of the interface connector to "conveniently maintain continuity." This defeats the purpose of the protection module. If an overvoltage propagates down a cable and bypasses the protection module, then the supposedly protected equipment may be damaged. These modules may be unsatisfactory even when the unprotected cable does not have conductors attached to the unprotected pins of the input connector. If a spark forms between pins of the input connector, hazardous voltages may be coupled to pins on the output conductor. It is definitely preferable to avoid any transient protection module that has a direct connection between the input and output connectors.

H. ACTIVE OVERVOLTAGE-PROTECTED RECEIVER FOR RS-232 COMPUTER DATA LINES

In this section the design of a protection circuit for a RS-232 receiver is described that uses an optical isolator to reject large common-mode voltages.

The circuits described above can be used to protect either line drivers or line receivers, since there is nothing in the protection circuit to restrict the flow of information to a single direction. This situation changes when optical isolators are used. The information flow must pass from the LED side to the photodetector side. This introduces the complication that separate circuits must be designed for receivers and transmitters. In addition, the photo-detector output usually requires an active circuit to condition the signal.

The RS-232 data communications standard was never intended for use on cables with a length greater than about 15 m (50 ft). The RS-232 system is "single-ended"; that is, there is one common conductor for all of the bidirectional signal conductors. This presents a problem when the two ends of the cable are in different buildings, where there may be a difference of potential between "ground" at the two buildings. The major symptom of this ground loop is a quasisinusoidal 60 Hz waveform superimposed on the digital data transmission. Such a problem has been avoided in the newer RS-423 data communication standard by requiring receivers with differential inputs and a separate common conductor for signals that propagate each way. The RS-422 data communication standard takes an even more aggressive approach by using a balanced pair of wires for every digital signal, so that there is no common conductor, and also requires receivers with differential inputs. However, advantages of these newer standards are not readily available to users of computer terminals and peripherals, since continued use of RS-232 makes it easy to interface older equipment to newer equipment. By incorporating isolation into the active protection circuit, the ground loop that is inherent in RS-232 communication circuits can be interrupted.

Figure 17-10 shows a schematic of an improved overvoltage protection circuit that, in addition to the features of the circuit shown in Fig. 17-1, also used isolation to provide protection against very large common-mode transient overvoltages. Values of components are given in Table 17-1. The common-mode protection is provided by two optical isolators (D_3 and Q_1; D_4 and Q_2) that combine light-emitting diodes (LEDs) and phototransistors in the same package. This optical isolator should be rated by the manufacturer to withstand at least 3 kV between the LED and the phototransistor, be relatively fast, and have a large phototransistor gain.

The part of the circuit in Fig. 17-10 that is composed of a spark gap,

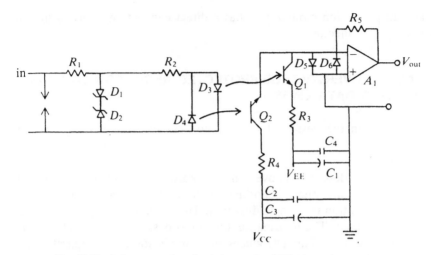

Fig. 17-10. *Active protection circuit for received RS-232 computer data.*

resistors R_1 and R_2, and avalanche diodes D_1 and D_2 is similar to that in Figure 17-1. The series combination of resistors R_1 and R_2 provides 3 kΩ impedance, the minimum allowed by RS-232C, and limits the current in the LEDs. The maximum current in the LEDs will be no more than about 4.7 mA, since the infrared LED has a drop of about 1 V, and the maximum input signal is 15 V. The minimum nonzero LED current, 1.3 mA, occurs when the voltage on the RS-232 line is 5 V, the minimum specified magnitude. This minimum current should generate sufficient light in the LED to cause the phototransistor to conduct. The light-emitting diodes are

TABLE 17-1. Components for Optical Isolator Received Data Circuit

G_1	150 V dc spark gap in ceramic case
D_1, D_2	1N4736 (6.8 V, 1 W avalanche diode)
D_3, D_4	Part of CNY17-IV optical isolator
D_5, D_6	1N4447 switching diode
Q_1, Q_2	Part of CNY17-IV optical isolator,
	breakdown voltage $V_{CEO} > 70$ V,
	about 5 µs switching time between active and cutoff regions,
	current transfer ratio > 160%
A_1	LF356 operational amplifier
R_1	2 kΩ, 2 W carbon composition
R_2	1 kΩ, 0.5 W carbon composition
R_3, R_4	2.2 kΩ, 0.5 W carbon composition
R_5	100 kΩ, 0.25 W
C_1, C_2	100 µF, 25 V
C_3, C_4	0.1 µF ceramic

connected in antiparallel so that one LED is always forward-biased. The forward voltage drop (about 1 V) of one LED prevents the other LED from being operated in the reverse breakdown region (more than 3 V). Because the RS-232 signal is bipolar, two optical isolators are required: one for each polarity.

The maximum magnitude of an RS-232 signal is 15 V when the output is terminated with a resistance between 3 and 7 kΩ. The largest signal voltage that will appear across avalanche diodes D_1 and D_2 is 5.7 V (notice that R_1 and R_2 form a voltage divider, and recall that the LED has a 1 V forward drop). The value of R_1 in Fig. 17-10 can be larger than the value of R_1 in Fig. 17-1, because the Fig. 17-10 circuit terminates a RS-232 line, which requires a minimum resistance of 3 kΩ. Therefore there is no particular need for avalanche diodes D_1 and D_2 in Fig. 17-10 to have special construction for absorption of transient overvoltages; ordinary models with a steady-state power rating of 1 W are sufficient. One manufacturer rates its 1N4736 avalanche diodes, which have a steady-state power rating of 1 W, for a maximum power dissipation of 60 W for a 15 μs nonrepetitive rectangular pulse (Motorola Zener Diode Manual, 1980, p. 11-56). This is adequate overload capability, since 2 W dissipation in the avalanche diode will produce more than 550 V across the spark gap (which is assumed to be nonconducting initially) when R_1 is 2 kΩ. The spark gap will quickly conduct, and the power dissipated in the avalanche diodes will then be much less than their maximum steady-state value. During a transient overvoltage, avalanche diodes D_1 and D_2 will limit the magnitude of the voltage to about 10 V. Resistor R_2 will then limit the current in the forward-biased LED to about 10 mA, well below the 90 mA maximum continuous current specified by the manufacturer.

The output of the optical isolators is recombined with an operational amplifier. When either of the phototransistors is on, the operational amplifier is in a nonlinear mode, and the inverting input is not a virtual ground. Larger values of R_3 and R_4 than those given in Table 17-1 increase the switching times of phototransistors Q_1 and Q_2. The operational amplifier, A_1, should be moderately fast (slew rate between 6 and 30 V/μs to satisfy Sections 1.3, 2.3, and 2.7 of the specifications in RS-232C). Diodes D_5 and D_6 allow the phototransistor collector current to flow to ground. If the optical isolators should fail, these diodes may protect the operational amplifier and the load.

The operational amplifier has internal protection against output short circuits to any voltage between the +15, −15 volt supplies, including ground. The operational amplifier alone provides a power-off source impedance of much more than the 300 ohms that is specified in Section 2.5 of RS-232C.

Bypass capacitors C_1, C_2, C_3, and C_4 are included to reduce the effect of inductance in the power supply lines.

I. PROTECTION OF RADIO FREQUENCY SIGNAL CIRCUITS

There is a broad class of radio frequency (rf) circuits that require protection. The distinguishing feature of this class of circuits is the presence of rf signals during normal operation, usually at least 10 MHz and perhaps up to 1 GHz. Because of the high frequencies involved, the signal path is a high-quality transmission line, usually coaxial. Specific examples include computer local area networks, inputs to radio receivers, cable television systems, etc.

One cannot protect rf circuits with protective devices that work well for lower frequency circuits. The large parasitic capacitance of varistors and avalanche diodes would give an unacceptable mismatch if they were connected to the transmission line for rf circuits. When this section was written (1988), protection of rf circuits was largely an unsolved problem. Military Handbook 419 (1982, p. 1-84) states simply that "effective suppression devices/circuits are not currently available for in-line installation on rf lines above 3 MHz, primarily because of high insertion losses."

There are two techniques that are commonly used to provide partial protection for rf circuits: (1) spark gap in a coaxial mounting (see Chapter 7), and (2) avalanche diode with a rectifier diode in series (see Chapter 9). The spark gap has a remnant that has a large amplitude: the peak voltage can be of the order of 1 kV. The duration of the remnant can be of the order of nanoseconds to microseconds depending on the waveshape of the overvoltage. The large remnant downstream from a spark gap can be sufficient to damage fragile semiconductors. Placing a rectifier diode in series with an avalanche diode can increase the response time of the avalanche diode and create a remnant with a large amplitude (Clark and Winters, 1973; Standler, 1984). This is probably unacceptable when protecting semiconductor devices that have small breakdown voltages.

Schlicke (1974) suggested constructing an overvoltage protection circuit in a lumped element delay line that matched the characteristic impedance of the transmission line. The parasitic capacitance of a metal oxide varistor or an avalanche diode could be used in the delay line, as shown in Fig. 17-11. The value of the characteristic impedance of the lumped line, $\sqrt{L/C}$, should be matched to the impedance of the continuous transmission line, Z_0, in order to prevent reflections from the overvoltage protection circuit. In this way, the large parasitic capacitance of the varistor or avalanche diode would be a less severe limit to the bandwidth of the circuit during normal operation. Varistors that have small values of conduction voltage (e.g., less than 20 V) have a relatively small value of α compared to other varistors and so do not provide tight clamping of the voltage during a surge. However, the parasitic capacitance of a metal oxide varistor is essentially independent of voltage and frequency. In contrast, the avalanche diode has tight clamping of voltage but has a parasitic capacitance that is a strong function of the voltage across the diode, so it is not obvious that the diode

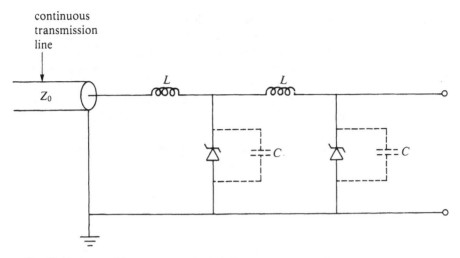

Fig. 17-11. *Lumped-line protection for high-frequency system. The parasitic capacitance of the surge protective devices is shown as a dashed capacitor symbol.*

will be an acceptable substitute for a fixed capacitance. However, this circuit is intriguing and is being investigated in the author's laboratory.

Mitchel and Melançon (1985) described overvoltage protection circuits with a passband from dc to 650 MHz for protection of a wide band amplifier in an oscilloscope. The circuit contains a 100:1 voltage divider with voltage clamping provided by core-switching diodes, as shown in Fig. 17-12. A 50:1 voltage divider is provided by R_1 and R_2, and a 2:1 voltage divider is provided by R_3 and the input resistance of the protected load, an amplifier in an oscilloscope. The overvoltage protection is provided by 1N4151 core-switching diodes. There are a total of 88 diodes, arranged in 22 groups in parallel; one group is shown in Fig. 17-12. The current in the diodes is limited by R_1.

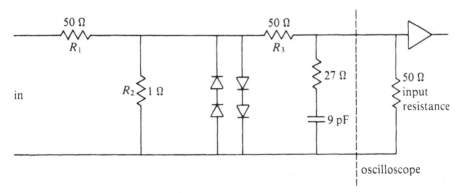

Fig. 17-12. *Protection circuit of Mitchel and Melançon (1985).*

Mitchel and Melançon went to heroic efforts to minimize the parasitic inductance in the resistors in order to achieve a wide bandwidth. R_1 is composed of 15 parallel branches of five 150 Ω carbon composition resistors in series. Each individual resistor has a steady-state power rating of 2 W, so R_1 has a steady-state power rating of 150 W. Similarly, R_2 and R_3 are composed of many carbon composition resistors in parallel. They describe construction techniques to minimize magnetic coupling between the circuit elements.

This input of this circuit could withstand 85 V continuously, and 5 kV for 1 µs. The output voltage is limited to about 3 V.

18

Applications in dc Power Supply

A. INTRODUCTION

It is essential that dc power supplies for electronic circuits be protected against overvoltages, because electronic circuits can be damaged by excessive power supply voltages. U.S. Department of Defense Military Handbook 419 (1982, p. 1-84) states, "Power supplies (5 to 48 V) operating from ac inputs and supplying operating power for solid-state equipment *always require internal transient protection*" [emphasis added].

There are four different situations where the loads are vulnerable to damage, which are illustrated in Fig. 18-1.

1. Transient overvoltages can enter the system through the mains, propagate through the dc power supply module, and affect all of the loads.
2. Transient overvoltages can enter the system on a data line, propagate through an amplifier onto the power supply line, and then affect other loads.
3. Failure of the voltage regulator in the dc power supply module can produce a sustained overvoltage condition that can destroy the loads.
4. A load that requires a dc power supply current with a large time rate of change (owing to both high frequency and large amplitude) will produce a voltage fluctuation on a dc power bus. This voltage fluctuation is caused by parasitic inductance in the dc power supply bus and large values of dI/dt, where I is the power supply current to the load, as explained in Chapter 15. Notice that a large value of

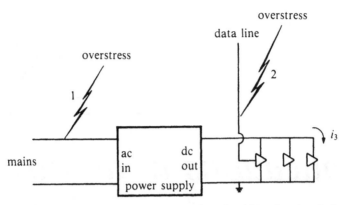

Fig. 18-1. Overstresses on the ac supply mains and on signal lines threaten devices operated from a dc power supply.

di_3/dt in Fig. 18-1 will also affect the power supply voltages at the other two loads.

The effect of item 4 is usually not associated with damage to the loads; however, it can produce excessive noise and upset. It is mentioned here because it is part of the general electromagnetic compatibility problem and because techniques for dealing with it also attenuate destructive overvoltages on the dc supply line. Voltage fluctuations owing to parasitic inductance in the power supply conductors are usually reduced by installing bypass capacitors near each load that has a large time rate of change of dc supply current and by distributing bypass capacitors along the dc power supply conductors. The basic physics is discussed in Chapter 15; some applications information is given in this chapter.

B. SIMPLE LINEAR POWER SUPPLY

1. General

An unprotected linear dc power supply circuit is shown in Fig. 18-2. This circuit is called a *linear* power supply, because the voltage across the filter capacitor, C, is directly proportional to the amplitude of the mains voltage. Later in this chapter, switching power supplies are discussed, which are nonlinear.

For convenience, this discussion considers *only* a circuit with a *positive* output voltage. A circuit with a negative output voltage appears identical, except that the polarity of all diodes and electrolytic capacitors is reversed, and a different model of integrated circuit voltage regulator is used.

The transformer reduces the mains voltage to a lower value in an

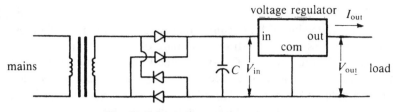

Fig. 18-2. Basic linear dc power supply.

energy-efficient way. The four rectifiers convert the sinusoidal secondary voltage into a waveform that always has the same polarity. The filter capacitor, C, provides a quasi-dc voltage by storing charge when the magnitude of the voltage across the secondary is greater than the voltage across the filter capacitor, and by releasing charge when the magnitude of the voltage across the secondary is less than the voltage across the capacitor.

The voltage regulator module provides a constant output voltage by varying the effective series resistance between the input and output terminals. The regulator module forms a voltage divider with the load resistance. By using negative feedback, the regulator maintains a constant output voltage for a wide range of values of load current and input voltage. Many linear dc power supplies use an integrated circuit three-terminal series-pass voltage regulator. Such circuits can easily provide an output current of a few amperes when an adequate heat sink is attached to the regulator chip. For larger output currents, an integrated circuit is usually used with discrete power transistors. However, for purposes of designing overvoltage protection, all series-pass regulators can be treated in the same general way.

2. Choice of Value of Filter Capacitance

The exact determination of the value of the filter capacitance involves the solution of a nonlinear equation. Although this can be done, the large tolerances on electrolytic filter capacitors (often -10%, $+50\%$ or even wider) make an exact solution unnecessary. A good estimate can be made by assuming that the capacitor discharges for half the period of the mains voltage, Δt, and is instantly recharged when the mains voltage is at a positive or negative peak, and then discharges again. The current in the common terminal of the voltage regulator is negligible compared to the current in the output terminal. The current drawn from the capacitor when it discharges is essentially the same as I_{out}. To simplify matters, assume that I_{out} is constant. The peak-to-peak voltage across the capacitor (called the *ripple* voltage), V_r, is then given by

$$V_r = I_{out}\,\Delta t/C$$

where C is the capacitance of the filter capacitor. This approximation overestimates the value of V_r, because the capacitor actually discharges for a time somewhat less than Δt. An overestimate of V_r results in a conservative determination of the value of C, so one need not be concerned about the error. The maximum value of V_r occurs when I_{out} is a maximum value.

For example, if $V_r = 1$ V, $I_{out} = 1$ A, and $\Delta t = 8.3$ ms (half-cycle of 60 Hz mains), then C is $8300\,\mu\mathrm{F}$. One should choose the next-largest standard capacitance value.

Although this calculation of filter capacitance is commonly made, it does not provide adequate protection against upset due to momentary severe sags of the rms main voltage. Some protection against upset can be obtained by using a value of Δt equal to the duration of several cycles of the normal mains voltage—for example, $\Delta t \gtrsim 30$ ms for 60 Hz mains. Of course, this increases the value of C and therefore increases the cost of the filter capacitor. However, this is the *least expensive way* of getting protection against upset by short-duration sags and momentary outages on the mains.

C. PROTECTION OF LINEAR POWER SUPPLY FROM OVERVOLTAGES

There are four different groups of components that need protection.

1. The transformer needs protection from excessive primary voltage, both common-mode and normal-mode.
2. Rectifiers need protection from excessive current and excessive reverse voltage.
3. The input port of the voltage regulator and filter capacitor needs protection from excessive voltage.
4. The output port of the voltage regulator and the loads that are connected to it require protection from overvoltage and bypassing to reduce noise.

We now discuss how to protect each of these groups of components.

1. Transformer Specifications

A transformer is quite robust when compared to the other components in a dc power supply. Nevertheless, the transformer must be protected, because its failure will be catastrophic for the power supply. If the transformer fails with a short circuit between primary and secondary, the full mains voltage can be applied to the rectifiers, filter capacitor, and voltage regulator with probable destruction of all of them. If the transformer fails with an open circuit in either the primary or secondary winding, the dc power supply will cease to function.

In order to obtain good isolation from the mains, the transformer should have either (1) a properly grounded electrostatic shield between the primary and secondary coils, or (2) the primary and secondary coils of the transformer should be wound nonconcentrically. These specifications reduce the coupling capacitance between the primary and secondary coils. In addition, the transformer should be specified to survive at least 1500 V rms continuously between primary and secondary windings with no degradation in insulation. These requirements ensure that the transformer attenuates common-mode voltages on the mains.

2. Use of Varistors on the Mains

A metal oxide varistor, V_3 in Fig. 18-3, should always be connected across the primary winding of the transformer to provide protection from excessive normal-mode voltage across the mains. This varistor protects not only the transformer but also the rectifiers, filter capacitor, and voltage regulator shown in Fig. 18-2. For example, consider the situation in which the primary is connected to 120 V rms and the secondary voltage is 28 V rms, a ratio of $1:0.23$. If varistor V_3 limits the peak normal-mode primary voltage to 300 V, then the peak secondary voltage might be *expected* to be about 70 V. It was assumed that transient overvoltages are coupled through the transformer in the same way as the normal mains voltage, an assumption that may not be valid. The parasitic capacitance in the transformer, leakage inductance in series with each coil, and different behavior of magnetic core material at high frequencies and large magnetic fields are important in coupling transient overvoltages through transformers. Since these details depend on the construction of a particular transformer and are not given in manufacturer's specifications, it is not possible to predict the coupling of

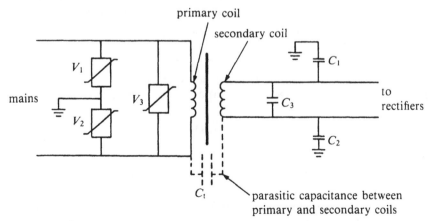

Fig. 18-3. *Redundant protection of dc power supply from overvoltages and noise on the ac supply mains.*

transient overvoltages through transformers until a specific transformer has been characterized.

Two additional metal oxide varistors, V_1 and V_2 in Fig. 18-3, should be connected to provide protection from excessive common-mode voltage on the mains. If the transformer has a small parasitic capacitance between primary and secondary coils so that coupling of common-mode voltages is not a significant problem, varistors V_1 and V_2 may not be as critical as varistor V_3. Specifications for the varistors are described in detail in Chapter 19.

In years past, it was considered good form to use capacitors in the locations shown by V_1, V_2, and V_3 in Fig. 18-3. Common components were ceramic disks with a capacitance between about 5 and 20 nF and a dc voltage rating of at least 1.4 kV. The metal oxide varistors are clearly superior, owing to their capability to clamp at a small voltage even when a large surge current (with a large charge transfer) is present. Because metal oxide varistors with a diameter of 20 mm have a parasitic capacitance of about 2 nF, the use of the metal oxide varistors automatically includes the capacitive shunt for high-frequency noise.

3. Use of Bypass Capacitors at the Secondary Coil

Capacitors C_1 and C_2 in Fig. 18-3 are included to attenuate common-mode noise. Typical values of C_1 and C_2 are between 1 and 10 nF; these two capacitors have the same nominal value in order to avoid conversion of common-mode noise to differential-mode noise. The capacitor C_3, which is shown in Fig. 18-3, provides a shunt path for normal-mode transient currents. A typical value of C_3 is between 0.01 and 0.1 μF. Lasitter and Clark (1970, p. 76) recommended making C_3 a 1 μF nonelectrolytic capacitor. This relatively large capacitance, together with the leakage inductance of the transformer, forms a low-pass filter. However, the parasitic inductance of the 1 μF capacitor may limit the performance of this filter at high frequencies, so an additional 0.01 μF capacitor in parallel would be desirable.

The two capacitors C_1 and C_2 constitute a voltage divider with the parasitic capacitance between primary and secondary coils, C_t. This voltage divider attenuates common-mode transients, in the following ratio

$$V_{\text{out}}/V_{\text{in}} = C_t/(C_1 + C_t)$$

Typical values of C_t are less than 1 pF if a grounded electrostatic shield is used between the primary and secondary and about 30 pF if the two coils are nonconcentrically wound.

Consider an example where C_t is 30 pF, and C_1 and C_2 are each 10 nF. There will be 50 dB of attenuation of common-mode voltages by the transformer and C_1/C_2. If varistors V_1 and V_2 limit the maximum value of the common-mode voltage across the primary coil to 500 V, the common-mode output voltage will be 1.5 V. This places negligible stress on the dc power supply, provided the transformer can withstand a common-mode potential difference of 500 V between the primary and secondary. By requiring an insulation rating of at least 1500 V rms, a safety factor of at least a factor of 3 is obtained.

The voltage rating of capacitors C_1, C_2, and C_3 should be at least three times the rms secondary voltage or 100 V, whichever is greater.

Capacitors C_1, C_2, and C_3, along with varistors V_1, V_2, and V_3, should be installed with minimal lead length to reduce parasitic inductance in series with the shunt path. Techniques for doing this are discussed in Chapter 15. Capacitors C_1, C_2, and C_3 must be mounted near the transformer secondary terminals to minimize electromagnetic radiation.

The use of V_1, V_2, V_3, C_1, C_2, C_3, and a transformer with small parasitic capacitance between primary and secondary helps to reduce the magnitude of transient overvoltages and high-frequency noise that appears at the remaining elements of the dc supply (i.e., rectifiers, filter capacitor, and voltage regulator). There are two general approaches to hardening these remaining elements: (1) increase the voltage ratings of these components, and (2) install additional overvoltage protection.

4. Protection of Rectifiers, Filter Capacitors, and Regulator

Some additional protection for the rectifiers from long-duration overvoltages can be obtained by specifying their maximum reverse voltage rating (also called *peak inverse voltage*) to be four times the normal rms secondary voltage or 400 V, whichever is the greater.

Some hardness for the filter capacitor, C in Fig. 18-2, can be obtained by specifying a dc working voltage that is at least 1.5 times the peak secondary voltage when there is no load on the secondary and the rms mains voltage is at its upper limit. The rated secondary voltage is specified at the maximum rms secondary current. The no-load rms secondary voltage will be greater than the rated full-load secondary voltage. Like many other engineering decisions, the maximum voltage rating of the filter capacitor is a compromise. Larger voltage ratings will increase the probability of survival during an overvoltage, but they also increase the effective series resistance (ESR), volume, and cost of the capacitor. Increased ESR produces extra internal heating in the capacitor, which can reduce the lifetime of the capacitor if ripple currents in the capacitor are large.

A typical value of maximum input voltage, V_{in}, for most integrated circuit regulators is 35 V, although some regulators with $V_{out} = 5$ V have a maximum input voltage as small as 20 V. Some additional hardness for the voltage regulator can be obtained by specifying a model with a large maximum input voltage (e.g., model LM317HV for positive voltages, LM337HV for negative voltages).

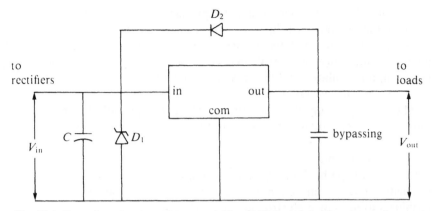

Fig. 18-4. *Protection of series voltage regulator and filter capacitor from overvoltages.*

Simply increasing voltage ratings for the rectifiers, filter capacitor, and voltage regulator is an expensive way of hardening the power supply. Most of the hardening should be provided by varistors on the primary side of the transformer, as shown in Fig. 18-3. If additional hardness is required, an avalanche diode, D_1, can be connected as shown in Fig. 18-4. The avalanche diode is the agent of choice owing to its highly nonlinear relation between current and voltage that provides tight clamping. The breakdown voltage of D_1 should be chosen to be slightly less than the smaller of (1) the maximum tolerable input voltage of the regulator circuit and (2) the voltage rating of the filter capacitor. However, the breakdown voltage of D_1 *must* be greater than the maximum input voltage of the regulator, V_{in}, during normal operation of the dc power supply (i.e., when the mains voltage is high and the load current is small). If avalanche diode D_1 conducts during steady-state operation, then thermal destruction of D_1 is almost certain, because there is no current limiting upstream from D_1. If it is difficult to specify a reverse breakdown voltage that is definitely greater than the maximum value of V_{in} but less than the voltage rating of the filter capacitor or regulator, then it would be reasonable to specify another model of filter capacitor or voltage regulator with a greater voltage rating.

The avalanche diode, D_1 in Fig. 18-4, is definitely recommended if a long cable is present between the source (transformer) and the voltage regulator. Transient overvoltages that enter the system on this cable will not be attenuated by the varistors on the mains or by the capacitive voltage division, as discussed above in connection with Fig. 18-3. Therefore, the avalanche diode is a reasonable choice in this situation.

The voltage regulator can be destroyed by reverse biasing it: the value of $(V_{in} - V_{out})$ must never be more negative than about -0.6 V. Such a condition can arise if there is a short circuit on the input side of the voltage regulator while the output voltage is maintained by bypass capacitors. Such a condition can also arise if a transient overvoltage forces the voltage at the

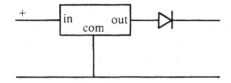

Fig. 18-5. Undesirable protection of series voltage regulator from reverse-bias condition.

output terminal to a large positive value. Protection for the voltage regulator against these hazards can be obtained by connecting diode D_2 as shown in Fig. 18-4. D_2 should be a rectifier that is capable of discharging the bypass and filter capacitors. This diode is normally reverse-biased by only a few volts, so a large reverse breakdown rating is unnecessary for this diode.

The circuit of Fig. 18-5 also protects the positive voltage regulator from positive current passing into the output terminal. However, using a series diode, as shown in Fig. 18-5, has several disadvantages. First, the diode degrades the voltage regulation during normal circuit operation. The voltage across this diode can vary between 0.6 and 1.0 V, depending on the current. Second, the diode may offer little protection from transient overvoltages with a submicrosecond rise time. Since the diode is normally conducting, and an overvoltage causes the diode to be reverse-biased, the diode's response time is on the order of the reverse recovery time. The reverse recovery time is typically on the order of microseconds for most rectifier diodes.

5. Summary

In most applications it will *not* be necessary to use *all* of the overvoltage protection techniques discussed above. The designer should choose appropriate techniques based on the environment, value of the load, and the cost of protection. For many dc power supplies, varistors connected to the mains may provide adequate protection. If the varistors on the mains are inadequate to ensure the survival of the regulator during proof testing, avalanche diode D_1 in Fig. 18-4 can be added. The capacitors C_1, C_2, and C_3 in Fig. 18-3 may be omitted in many applications.

6. Bypassing at Input of Regulator

Sometimes one encounters an unregulated or quasiregulated dc power supply bus that serves one or more distant loads. A voltage regulator should be located near each load. Each voltage regulator should have several bypass capacitors connected in parallel at the input terminals of the regulator. These capacitors should include a large electrolytic capacitor (e.g., 100 μF) as well as a smaller electrolytic capacitor (e.g., 1 to 10 μF) and one or more ceramic capacitors (e.g., 0.01 to 0.1 μF). The parallel combination of capacitors is necessary to counter the effect of the parasitic inductance in the long cable between the dc source and the voltage regulator, as explained in Chapter 15. When environmental specifications permit, the 100 μF capacitor can be an aluminum electrolytic type instead of a tantalum, because the tantalum unit is much more expensive and the superior electrical properties of the tantalum are not important in this application.

A voltage regulator should always be located near the load and within about 30 cm if possible. If the voltage regulator is far from the load, the regulator cannot be expected to maintain a constant load voltage, owing to parasitic inductance and resistance of the wiring.

D. PROTECTION AGAINST TRANSIENTS AT LOADS

There are two general kinds of transient voltage at loads that can cause problems. Interrupting current in an inductive load can cause a severe transient overvoltage that can destroy unprotected semiconductor devices. Changes in current can cause voltage drops across the parasitic inductance in the supply bus, as explained in Chapter 15, which may cause upset of susceptible electronic devices.

1. Suppression at Source

It cannot be too strongly emphasized that transient overvoltages caused by switching inductive loads should be suppressed at their source. A simple and inexpensive way to do this is to connect a rectifier across each inductive load, as shown in Fig. 18-6. This rectifier should be installed with minimal

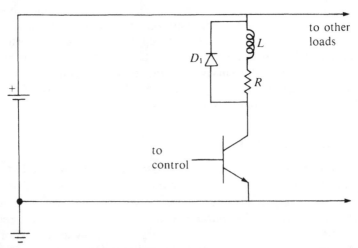

Fig. 18-6. *Clamping of overvoltage caused by interrupting current in an inductive load on a dc supply bus.*

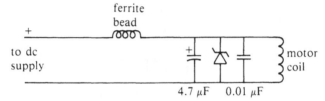

Fig. 18-7. *Suppression of overvoltages and noise at the terminals of a dc motor.*

lead length, directly across the terminals of the coil. When the current in the inductive load is interrupted, the rectifier will be forward-biased and will prevent an overvoltage. This protects the switching transistor shown in Fig. 18-6 and other vulnerable devices that may be connected to the same dc supply bus.

Suppression at the source has two advantages: (1) Suppression at the source minimizes electromagnetic radiation by keeping the dc supply bus clean; otherwise, the bus becomes a transmitting antenna. (2) There are usually more vulnerable devices than inductive loads on a dc bus, so suppressing overvoltages at each inductive load requires fewer protection devices and is therefore less expensive.

Sometimes a noisy load, such as a motor with brushes, must be operated on the same dc bus as sensitive electronic circuits, such as high-gain amplifiers. The circuit shown in Fig. 18-7 may be useful in suppressing noise. The bypass capacitor with the smaller capacitance is installed directly across the motor terminals with the shortest possible length of leads, since this capacitor is only effective at suppressing high-frequency noise. An avalanche diode is also installed directly across the terminals of the motor. The diode clamps noise voltages of either polarity. The reverse breakdown voltage of the diode should be slightly greater than the maximum dc supply voltage. Finally, a tantalum electrolytic capacitor is connected across the terminals of the motor to suppress noise at lower frequencies. Since the electrolytic capacitor is only effective for low-frequency noise, it may have a longer lead length than the other two devices. However, this capacitor should also be installed across the terminals of the motor. The conductors between the dc supply and the motor should be shielded twisted-pair wire or coaxial cable in order to minimize radiation of noise. As a final touch, a ferrite bead can be installed on one of the supply leads to increase the inductance and form a barrier to conduction of noise on the supply leads from the motor.

2. Bypassing

The load can be protected against transients that are introduced on the output side of the voltage regulator by the circuit shown in Fig. 18-8.

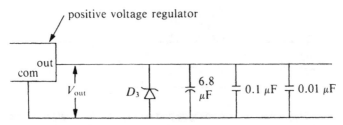

Fig. 18-8. *Overvoltage protection and bypassing on output side of a dc voltage regulator.*

Avalanche diode D_3 shown in Fig. 18-8 protects the loads from positive overvoltages as well as from voltage reversals on the dc power supply bus. The bypass capacitor network shown in Fig. 18-8 is included to provide a stable dc voltage even when the loads are drawing rapidly changing currents from the power supply bus. In addition, they provide limited protection against transient voltages. Provided there are no loads that have large changes in current drawn from the supply (e.g., line drivers, line receivers, one-shots), one clamping diode, D_3, and one 6.8 μF tantalum electrolytic capacitor will serve a single printed circuit board.

A number of 0.1 and 0.01 μF ceramic capacitors should be distributed over the area of the printed circuit board for adequate suppression of transient voltages caused by changing load currents. Any device that has a large time rate of change of power supply current (e.g., a one-shot or line driver), should have network of bypass capacitors located near the device. The combination of a 4.7 or 6.8 μF tantalum capacitor *and* a 0.01 μF ceramic capacitor is commonly used.

The *minimum* value of the breakdown voltage, V_z, of D_3 should be chosen to be about 1.2 times the *maximum* output voltage of the regulator circuit. The words "minimum" and "maximum" should be understood to include effects due to device tolerance, temperature, changes in load current, and so forth. This should preclude the avalanche diode conducting during normal operation of the system. For example, a 6.8 V $\pm$ 5% diode could be used to protect a 5 $\pm$ 0.3 V power supply bus. In this example there would be at least a 1.1 V margin between the maximum supply voltage and the minimum avalanche voltage.

U.S. Department of Defense Military Handbook 419 (1982, p. 1-89) recommends that the minimum avalanche diode voltage at a current of 100 *micro*amperes by 1.05 times the maximum output voltage of the regulator. This is a different way of expressing the same concerns as given in the previous paragraph.

Most severe overvoltages in electronic circuits come from outside the chassis, through interface circuits such as line drivers and receivers. Overvoltage protection should be installed on these interface conductors, as described in Chapter 17.

E. DC CROWBAR

There may be situations for which the avalanche diode of Fig. 18-8 provides inadequate protection of the dc supply bus. De Souza (1967) described a comprehensive crowbar circuit, shown in Fig. 18-9, for protecting loads that were connected to a 28 V power supply. This circuit is called a *crowbar*, because it short-circuits the power bus as if a metal crowbar were dropped across the two conductors. Crowbar circuits have the advantage of being able to conduct large currents for sustained periods of time, unlike many other transient overvoltage suppression circuits. Crowbar circuits have the disadvantage that they interrupt power to critical loads and may cause upset of the system.

The normally closed relay is included to interrupt follow current in either the spark gap or SCR. The fuse, inductor L_1, and resistor R_1 prevent damage to the SCR by large surge currents. If the SCR switches to the conducting state when the gate current is 10 mA, the SCR will conduct at a line voltage of 40.7 V, since

$$(40.7 - 0.7) \text{ V}/4 \text{ k}\Omega = 10 \text{ mA}$$

Rectifier D_1 provides clamping of reverse voltages. This allows the use of an SCR with a low reverse breakdown voltage (e.g., 25 V). The inductor L_1, resistor R_3, and capacitor C_1 form an oscillation suppression network that attenuates transient voltages caused by change of state of the relay, spark gap, SCR, or rectifier D_1.

De Souza claimed that the use of a SCR allowed his circuit to have "a faster response time and a much greater current capacity" than circuits that use a zener diode. There is no doubt that an SCR can conduct larger steady-state currents than an avalanche diode: If we compare an SCR that

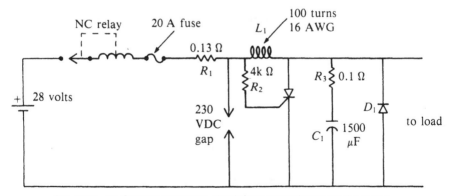

Fig. 18-9. *De Sousa's protection circuit from overvoltages on the mains.*

has a maximum rated power of 10 W with a 40 V avalanche diode that also has a maximum rated power of 10 W, we find that the SCR can carry a steady-state current of 5 A but the avalanche diode can carry only 0.25 A. This difference is due to the difference in conduction voltages: about 2 V for the SCR and 40 V for the avalanche diode. However, De Souza is mistaken about the speed of response. If large currents flow through an SCR before the device is fully turned on, a process that can take a few microseconds, the SCR can be damaged. The avalanche diode turns on much faster.

The circuit in Fig. 18-9 might be improved by including an avalanche diode in series with R_2 so that the cathode of the avalanche diode is connected to the positive line. The breakdown voltage of the avalanche diode could be selected to be slightly greater than the line voltage during normal operation (e.g., 33 V ± 5% for a 28 V line). This could permit the SCR to conduct during small overvoltages that were nearer the normal operating voltage across the line. Alternatively, one could specify an avalanche diode breakdown voltage that was slightly less than the maximum tolerable voltage across the load(s) in order to avoid nuisance crowbar actuations.

Integrated circuits are available to control an SCR for overvoltage protection. These integrated circuits provide a large gate current (more than 100 mA) and fast response (less than 1 μs), and offer a reduction in the number of components in overvoltage protection circuits. The Motorola MC3423 and MC34061A integrated circuits have been commonly used for this application.

F. DC UNINTERRUPTIBLE POWER SUPPLY

If the mains power should be interrupted, the common dc power supply will be unable to continue to supply power to critical loads. A relatively inexpensive uninterruptible power supply is shown in Fig. 18-10.

This circuit uses N rechargeable batteries connected in series to form a reserve power supply. The voltage regulator U_2 regulates the voltage across the batteries to a constant voltage across the critical loads. The voltage

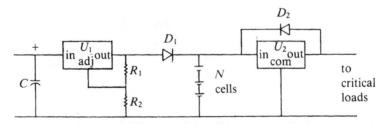

Fig. 18-10. *Simple UPS in a dc supply.*

regulator U_1 regulates the battery-charging voltage. Diode D_1 prevents the batteries from discharging into the regulator U_1. The mains connection, transformer, and rectifiers have been omitted from Fig. 18-10 for simplicity.

Sealed, "maintenance-free," lead-acid cells are particularly desirable for the batteries because they (unlike nickel-cadmium batteries) can be connected indefinitely to a constant voltage source without degrading their life. The number of cells, N, is determined by the load voltage, V_{out}, the minimum desirable value of $(V_{in} - V_{out})$ for regulator U_2, and the fully discharged cell voltage. If we specify

$$(V_{in} - V_{out}) \geq 3 \text{ V}$$

and use lead-acid batteries with a fully discharged potential of 1.6 V/cell, we obtain Eq. 1.

$$N \geq (V_{out} + 3 \text{ V})/(1.6 \text{ V}) \tag{1}$$

The value of N, of course, must be an integer.

The value of the output voltage of regulator U_1, V_1, is determined by the proper charging voltage for N batteries in series and the voltage drop across diode D_1. If the charging voltage per cell is too small, the cells will take a long time to be partially charged and will never be fully charged. If the charging voltage is too large, prolonged connection to this voltage after the cells have been fully charged will damage the cells. A reasonable value for the charging voltage per cell for lead-acid batteries is between 2.25 and 2.40 V. With this information we arrive at Eq. 2.

$$V_1 = (N \times 2.30) + 0.6 \text{ V} \tag{2}$$

The value of output voltage from regulator U_1 is set by the values of R_1 and R_2. The value of R_1 is arbitrarily set to 220 Ω, and the value of R_2 is then calculated from Eq. 3, where

$$V_1 = (V_{a1}/R_1)(R_1 + R_2) \tag{3}$$

V_1 is the potential difference across the series connection of R_1 *and* R_2 and V_{a1} is the potential difference between the output and adjustment terminals of regulator U_1. We can neglect the current that flows in the adjustment terminal of U_1, because it is much smaller than V_{a1}/R_1 when R_1 is 220 Ω. Equations 2 and 3 can be solved for the value of R_2, given V_{a1}, N, and the choice of 220 Ω for R_1.

The circuit of Fig. 18-10 has excellent voltage regulation, owing to the two voltage regulators, U_1 and U_2, in series. The batteries also act like a filter capacitor, since they have an incremental impedance, $\Delta V/\Delta I$, of the order of a few milliohms.

The major disadvantage of the simple circuit in Fig. 18-10 is that the regulator U_1 must be capable of supplying both the current to the loads, through regulator U_2, and the battery-charging current. When the batteries are discharged and the mains are reconnected, the initial battery charging current alone can be between 5 and 10 A for 2.5 A h cells. Such a large current can cause many popular three-terminal integrated circuit voltage regulators to behave as constant-current sources owing to internal "short-circuit protection." Alternatively, the combination of the voltage drop across the input and output terminals of U_1 and the large current can cause internal thermal protection circuit to operate and decrease the output voltage below normal values. Although the internal short-circuit protection and thermal protection circuits save regulator U_1, they may make it difficult to operate the critical loads once the mains power has been restored and the batteries have been discharged.

The disadvantages of the simple circuit in Fig. 18-10 can be avoided by using a separate power supply for the battery charging, as shown in Fig. 18-11. The power transistor Q_1 and avalanche diode D_1 form a voltage regulator for charging the batteries. The battery-charging voltage is the breakdown voltage of the avalanche diode minus the base-emitter voltage drop of Q_1. Resistor R_1 limits the maximum charging current to a safe value for transistor Q_1. Resistor R_2 limits the current in the avalanche diode D_1. Rectifier D_2 prevents the transformer secondary that is shown in the upper half of Fig. 18-11 from supplying charge to the batteries. During normal operation, the voltage at point A in Fig. 18-11 must be greater than the battery voltage so that the batteries are charged. One way to do this is to specify that the transformer have two identical secondary coils. The voltage dropped across the collector-emitter junction of Q_1, as well as across R_1, will make the battery voltage less than the voltage at point A.

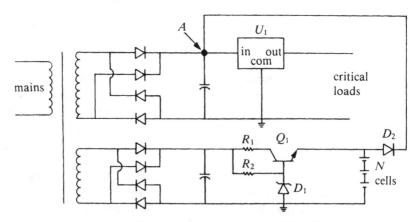

Fig. 18-11. *Better dc UPS than the circuit in Fig. 18-10.*

Building an uninterruptible power supply into a dc power supply is much less expensive than building an uninterruptible power supply for the ac mains. A UPS for the ac mains has all of the components of a dc UPS, in addition to an oscillator and power amplifier to obtain a sinusoidal output voltage. Unfortunately, the chassis of most equipment is already full of components, and there is little space to add a battery pack for an optional dc UPS.

G. SWITCHING POWER SUPPLIES

A different approach to dc power supply design is to use the so-called "switching power supply." The basic circuit is shown in Fig. 18-12. A pair of switches, S_1 and S_2, create a high-frequency rectangular waveform. The switching frequency is usually at least 10 kHz and can be more than 100 kHz. The output voltage of the transformer is rectified and used to charge a filter capacitor, C_2. The control circuit senses the voltage across the load and adjusts the amount of time that each switch is closed. When the value of I_{out} is large or the rms mains voltage is low, the switches will be closed for a longer time. The dc output voltage is regulated by varying the pulse width of the current in the primary of the transformer.

In an implementation of this circuit, the switches S_1 and S_2 are usually either bipolar transistors or power MOSFETs. These transistors are polarity-sensitive, so a bridge rectifier is required upstream. Furthermore, the switching power supply needs a source of current for the primary coil of the transformer when the instantaneous value of the mains voltage is zero, so a bridge rectifier and filter capacitor C_1 are required.

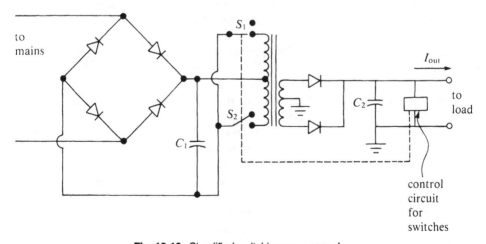

Fig. 18-12. Simplified switching power supply.

There are several striking advantages of this circuit. The switching power supply is much more energy-efficient than an ordinary power supply, because there is no component with a large voltage drop and a large current simultaneously, as in the series-pass voltage regulator. Because the switching frequency is much greater than the frequency of the mains waveform, the transformer core can be much smaller and less massive than in ordinary 50 or 60 Hz circuits. The switching power supply is also insensitive to the frequency and waveshape of the mains voltage.

Switching power supplies in commercial production have been designed to provide the desired dc output voltage and current for any value of mains voltage between 60 and 250 V rms. Such a wide range of mains voltages was previously attainable only by manually changing taps on a transformer for operation at either 120 or 220 V. Furthermore, satisfactory operation of a linear dc power supply at reduced rms mains voltage is possible only by designing the supply so that the secondary voltage is normally relatively large, which makes the power supply inefficient when the rms mains voltage is near the nominal value.

It is easier to design a switching power supply that will provide a satisfactory output voltage during a brief interruption of mains voltage (brief outage). The voltage across C_1 is typically between 200 and 400 V, whereas the voltage across the filter capacitor in a linear supply (C in Fig. 18-2) is typically between 8 and 30 V for dc output voltages between 5 and 20 V. The energy stored in a capacitor is proportional to CV^2, but the volume and cost of a capacitor are approximately proportional to CV. Therefore, storing energy at higher voltage is less expensive and uses a smaller volume than storing the same amount of energy at a lower voltage.

There are several features of a switching power supply that make them difficult to design. Because large current pulses are switched at a high frequency, the internal circuit of the switching power supply radiates electromagnetic noise. The switching power supply must be mounted in a shielded enclosure, and a low-pass filter with large attenuation at the switching frequency must be connected between the switching power supply and the mains. At high switching frequencies it is difficult to find suitable switching transistors and rectifier diodes. However, these difficulties can be overcome, and switching power supplies are now often used to provide more than about 100 W of power. For dc loads that require less than about 50 W, a linear supply is still often used.

Protection of a switching power supply from transient overvoltages on the mains is a difficult job. Rectifier diodes and switching transistors are both directly exposed to the mains. Roehr (1985, 1986) has suggested use of metal oxide varistors, series inductance and resistance, and avalanche diodes. The varistors and series impedance are shown in Fig. 18-13. The varistors clamp the common-mode mains voltage during transient overvoltages. An inductance, L, of the order of $100\,\mu\text{H}$ prevents the filter capacitor, C_1, and switching transistors from being damaged by normal-

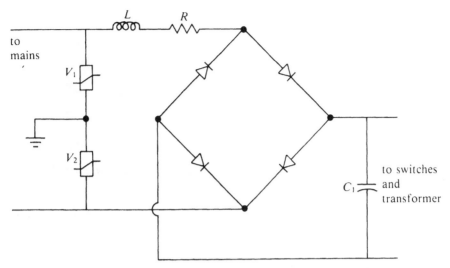

Fig. 18-13. Overvoltage protection for for switching power supply.

mode surges with a brief duration (e.g., an $8/20\,\mu$s waveshape with a peak current of 3 kA). The overvoltage across C_1 has three components: one caused by charging the capacitor with the surge current, one from the passage of the surge current through the equivalent series resistance of the capacitor, and one from the dI/dt of the surge current and the parasitic inductance in the capacitor (Roehr, 1985). The inductance L upstream from the capacitor limits both the surge current and dI/dt in the capacitor.

The series resistance, R, of the order of 1 Ω limits the inrush current drawn from the mains when the switching power supply is first turned on, and C_1 is completely discharged. The resistance also damps any oscillations from the LC network.

Avalanche diodes can be connected across the capacitor C_1 for additional voltage clamping. Although inclusion of avalanche diodes increases the cost of the overvoltage protection circuit, it allows switching transistors with smaller voltage ratings to be used. The increased cost of the avalanche diodes is offset by decreased cost of the switching transistors and an overall increase in reliability.

Roehr (1986) also pointed out that the inductance that was included for surge protection also made a good low-pass filter. He suggested how a common-mode choke and additional inductors and capacitors could attenuate both transient overvoltages and switching noise. This diminished the requirements for the EMI filter module.

19

Applications on the Mains

A. INTRODUCTION

Overvoltage protection devices for the ac supply mains can be divided into four classes.

1. distribution surge arresters (typically 3 kV rms and greater)
2. secondary arresters (175 and 600 V rms)
3. surge protective devices for equipment on a branch circuit
4. surge protective devices inside a chassis

These classes can be distinguished in several ways: location of the protection, values of the specifications, and who is responsible for requesting it.

The local electric utility company installs and maintains the distribution arresters. They are outside the scope of this book. Additional information on such devices can be found in the book by Greenwood (1971) and in proprietary literature from manufacturers of arresters.

The occupants or owners of the building are responsible for requesting the installation of a secondary arrester, which is installed by an electrician at the point of entry or at a distribution panel.

The occupants of the building or users of vulnerable equipment are responsible for requesting installation of SPDs on branch circuits. Some of these SPDs can be easily installed by the user by plugging a box or outlet strip into a wall receptacle. Other SPDS are inside a wall socket and an electrician must remove the old socket and install the protected socket.

SPDs inside a chassis are usually installed by the manufacturer. They may also be installed as a retrofit by repair personnel; however, retrofits may void the manufacturer's warranty and safety agency approvals. SPDs inside the chassis provide redundant protection from external transients and prevent a particular system from polluting the branch circuit with transients that originate inside the chassis.

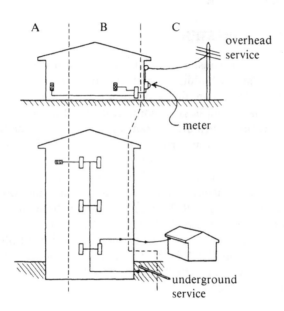

Fig. 19-1. *Location categories in ANSI C62.41-1980.*

Category A. **Long Branch Circuits:**
All receptacles at more than 10 m from Category B and all receptacles at more than 20 m from Category C.

Category B. **Major Feeders and Short Branch Circuits:**
Distribution panel devices, bus and feeder systems in industrial plants, heavy appliance receptacles with short connections to the service entrance, and lighting systems in commercial buildings.

Category C. **Outside and Service Entrance:**
Service drop from overhead power lines to building entrance, conductors between electric meter and distribution panel, overhead line to detached buildings, and underground lines to electric pump motors in water wells.

(Reprinted from ANSI/IEEE C62.41-1980 copyright 1980 by The Institute of Electrical and Electronics Engineers, Inc., by permission of the IEEE Standards Department.)

A somewhat different classification method is given in ANSI C62.41-1980, which uses three categories: A, B, and C, as shown in Fig. 19-1. Category C is the outdoor environment, which consists of the service drop from the utility pole to the point of entry and the conductors between the electric meter and circuit breaker panel. Category B consists of major feeders inside industrial plants and all points on branch circuits less than 20 m from category C. Category A consists of outlets on "long branch circuits" that are more than 10 m from a major feeder and more than 20 m from the point of entry.

Secondary arresters are in category C. Circuit breaker panels are in category B. Most electronic equipment is used in category A, but some may be connected in category B.

B. SECONDARY ARRESTER

The secondary arrester is installed by an electrician or the utility at the point of entry of power into a building, or at the circuit breaker panel inside the building. If the building contains a distribution transformer, the secondary arrester (as the name implies) is connected on the secondary side (low-voltage side) of the distribution transformer. If the building does not contain a distribution transformer, the secondary arrester is usually mounted on the meter box, at the service entrance, or on the distribution panel. A circuit diagram is shown in Fig. 19-2.

Some European countries require that an arrester be connected to the mains at each building. In 1987 there was some discussion about adding such a requirement to the U.S. National Electrical Code for 1990, but such a requirement was defeated. U.S. Department of Defense Military Standard 188-124 (1978, Section 5.1.1.3.12) for long-haul/tactical communication systems *requires* a lightning arrester to be installed at the point of entrance of the mains into the facility.

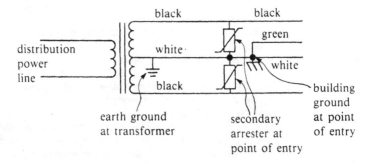

Fig. 19-2. *Connection of secondary arrester at point of entry of the mains into the building for single-phase 120/240 V service.*

One terminal of the arrester is connected to each phase and the neutral of the mains; the common terminal of the arrester must be connected to the electrical ground of the building. Two, three, or four varistors are available in a single container for protecting single-phase, two-phase, or three-phase secondary circuits.

There are two common types of secondary arresters. The older type has a silicon carbide varistor in series with a spark gap; the newer type contains a metal oxide varistor alone for each line.

Common models of silicon carbide secondary arresters are typically rated for either 175 or 600 V rms service so that two models are suitable for use on mains with a nominal voltage of 120, 240, or 480 V rms. These silicon carbide arresters are designed to clamp the voltage at about 2.5 kV at 1.5 kA and at about 4 kV at 10 kA for an 8/20 μs waveform. The spark gap acts as a switch to prevent conduction in the silicon carbide varistor during normal operation of the mains, as explained in Chapter 8. Silicon carbide secondary arresters are suitable for protecting insulation from damage by transient overvoltages. They should offer adequate protection for motors and transformers and prevent fires caused by arcs in electrical circuits. However, the large clamping voltage, as well as the remnant that passes downstream from the spark gap, probably gives inadequate protection for most vulnerable electronic equipment.

A secondary arrester that contains metal oxide varistors (MOVs) can be designed to offer substantial protection for vulnerable electronic equipment inside the building. Specifications for the varistors are given later in this chapter. Arresters with metal oxide varistors have three advantages over silicon carbide-spark gap arresters. The metal oxide varistor does not usually have a series spark gap; deletion of the spark gap decreases the peak voltage of the remnant that travels downstream from the arrester. The clamping voltage of the MOV arrester is less than that for a silicon carbide varistor, because the MOV is more nonlinear. The large parasitic capacitance of the MOV, which is typically about 2 to 5 nF, can form a low-pass filter with the inductance of the distribution wiring upstream from the arrester. This low-pass filter may offer appreciable attenuation of bursts of noise from switching reactive loads in nearby buildings.

C. BRANCH CIRCUIT PROTECTION

Smith (1973) reviewed various types of overvoltage protection devices for use on the mains and concluded that, overall, metal oxide varistors were the best. The author of this book believes this assessment is still true in 1988, when this book was written.

It is not necessary to protect *every* electrical outlet in a typical building. To protect every outlet would be expensive and would probably offer little advantage compared with protecting one or two outlets on each branch.

Criteria for deciding where to install surge protective devices on the mains is discussed in Chapter 20. Briefly, the author recommends (1) a secondary arrester at the point of entry and (2) appropriate protection at each outlet that serves expensive equipment (e.g., computers, electronic instrumentation, videotape recorders, stereo systems, etc.) that is likely to be vulnerable or susceptible to overvoltages. Protection located at outlets in various places throughout the building will offer substantial, but less complete, protection to other equipment that is connected to "unprotected" outlets.

1. Three Varistors

Loads that are connected to a branch circuit can be protected from transient overvoltages by varistors in several different ways. The basic circuit diagram for using varistors on single-phase mains is shown in Fig. 19-3. Specifications for the varistors are given later in this chapter. This circuit is commonly packaged in three different ways: inside an outlet strip, in a box, or inside a special wall socket. The outlet strip format is convenient for computer systems and cabinets that contain various electronic instruments, because a number of outlets are required for the various chassis. The box format commonly has a male connector in the rear of the box that plugs into a wall outlet and several female connectors in the front of the box to provide power to protected equipment.

Varistors V_1 and V_2 are included to clamp the common-mode (CM) voltage to an acceptably low level. These varistors also clamp the differential-mode (DM) voltage, but to twice the value of the CM clamping voltage, because V_1 and V_2 are in series for DM voltages. In order to achieve a smaller DM clamping voltage, varistor V_3 should be included.

Martzloff (1983) discussed the different ways to connect one, two, or three varistor(s) to a single-phase mains with a ground wire. The arrange-

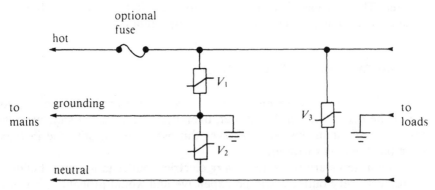

Fig. 19-3. *Connection of metal oxide varistors to the mains.*

ment of three varistors shown in Fig. 19-3 produced the lowest clamping voltages for V_{HN}, V_{HG}, and V_{NG} for an 8/20 μs waveform.

Some low-cost protection modules contain only a single varistor, which is usually connected between the hot and neutral conductors. Although a single varistor is better than none at all, this is a marginal practice. Martzloff and Gauper (1986) showed that use of only V_3 in Fig. 19-3 could produce a large value of voltage between the neutral and grounding conductors when surge currents passed through the inductance of the neutral wire. They measured a peak value of V_{NG} of 2.3 kV at the end of a 75 m length of cable for a 100 kHz ring wave with a peak current of 40 A. If only two varistors are to be used, V_2 and V_3 in Fig. 19-3 are suggested.

Each varistor should, of course, have minimal length of leads to avoid extra parasitic inductance, as explained in Chapter 15.

The fuse shown in Fig. 19-3 is not required for overvoltage protection. If a fuse or circuit breaker is to be included in a surge protection module, it should be placed as shown in Fig. 19-3. Alternative locations for a fuse or circuit breaker and their disadvantages were discussed in Chapter 12.

These varistors can be installed by the user inside a common outlet strip to provide protection against transient overvoltages. Installation of the varistor is more convenient in outlet strips that have solid noninsulated copper bus wire inside for the two mains conductors and the grounding conductor than in models with insulated wire inside. However, installing varistors inside the outlet strip will void the safety agency approval of the outlet strip, since the modified product now needs to be retested.

2. Noise Reduction

A low-pass filter can be installed as part of the overvoltage protection circuit or installed downstream from the overvoltage protection circuit. The low-pass filter will probably attenuate the remnant that travels downstream from the surge protection module. The word "probably" is used because an inductor-capacitor filter has insertion gain near its resonance frequency. Design of a nonresonant low-pass filter that includes overvoltage protection is described later in this chapter.

Another device that can remove disturbances from branch circuits is a line conditioner. Line conditioners are designed to maintain the rms value of line voltage at a nearly constant value and to provide isolation and remove noise. Line conditioners are discussed in Chapter 20.

D. VARISTORS INSIDE THE CHASSIS

The fourth and last class of devices to protect equipment from transient overvoltages on the mains is those that are included inside the chassis of electronic equipment by the manufacturer. Such devices include one or more metal oxide varistors connected to the mains. Such varistors are usually "hidden" inside the chassis with no warning to the user that they are inside. If varistors with relatively small values of V_N are hidden inside the chassis, coordination with other varistors in the environment will be impossible, since the person responsible for the coordination is unaware of them. This is discussed later in this chapter, in the section on coordination. It is recommended that when varistors are connected to the mains inside a chassis, a label be affixed to the outside of the chassis to notify the user.

E. PARAMETERS FOR METAL OXIDE VARISTORS

In Chapter 8 the general criteria for selection of a model of metal oxide varistor were discussed. The specific situation for specifying varistors for protection of equipment connected to the mains is discussed here. There are three major concerns about the performance characteristics: (1) the varistors should not conduct appreciable steady-state current; (2) the varistors should survive expected overstresses; and (3) the clamping voltage should be as small as possible without violating the first concern.

To discuss the first concern, we need to specify the voltage at which the varistor begins to conduct. There is no precise value for this voltage, but it is conventional to use V_N. V_N is the voltage across the varistor when a dc current of 1 mA passes through the varistor. The value of V_N in these specifications is measured at a temperature of 25°C and is the initial value before the varistor has been degraded by passage of surge currents.

We also need to know the peak line-to-ground voltage, V_p, during *nominal* conditions. For example, for mains with a nominal voltage of 120 V rms, $V_p = 170$ V. By referring to variable V_p, rather than a specific numerical value, the specifications are kept in a general form that can be applied to any value of the nominal mains voltage.

The minimum value of V_N, which includes the effects of tolerances, is denoted $\min(V_N)$. To prevent conduction during steady-state conditions, the following requirement must be met:

$$(1 + \beta)V_p < \min(V_N)$$

The factor β is a positive number that accounts for steady-state mains voltages that can be greater than the nominal value.

ANSI C84.1-1982 states that the maximum acceptable utilization voltage (i.e., the voltage at the wall outlet) is 1.05 times the nominal system voltage. The electric utility is required to take corrective measures to reduce the rms voltage if it is greater than 1.05 times the nominal system voltage. It might appear that the value of β should be 0.05. However, this value of β is definitely too small: there are disturbances, called swells, where the mains voltage exceeds the maximum acceptable value according to this standard.

Swells may persist for minutes or even hours before the utility is notified and corrective action is completed. The effect of swells is *not* included in tolerances on mains voltages. Although swells are *not common,* they can destroy varistors with a small value of V_N. Increasing the size of β slightly, for example to 0.10, will be sufficient for many sites, but it is still too small for a few sites.

Another consideration in specifying a value of β is the degradation of the metal oxide varistor after exposure to large surge currents. The arbitrary, but widely used, criterion for end of life of a varistor is a change in the value of V_N by more than $\pm 10\%$. After exposure to surge currents, the value of V_N usually decreases. If the value of V_N decreases to a sufficiently small value, a varistor connected to the mains will begin to conduct appreciable steady-state currents. The heating of the varistor from the steady-state current will further decrease the value of V_N, because that parameter has a negative temperature coefficient. The varistor will then go into thermal runaway and be destroyed. A generous safety margin should be included in the value of β to give a large margin for degradation before the varistor will be damaged. The author's personal preference is for

$$\beta = 0.25$$

for varistors connected to long branch circuits. However, some engineers would recommend even larger values of β, such as 0.5 or even 1.0, but these large values of β lead to rather large clamping voltages during surges.

Most sites will *not* experience a severe swell or large surge currents during a given year that would damage a varistor with the minimum voltage rating (i.e., $\beta = 0.10$). The engineer who designs protective circuits has two choices:

1. A varistor with the minimum voltage rating could be specified. This gives a small clamping voltage, which benefits all sites. However, the varistor fails prematurely at a few sites. This choice trades a small advantage at most sites against a serious disadvantage at a few sites.

2. A varistor with a voltage rating that is appreciably larger than the minimum value could be specified. The protection circuit then has a long lifetime at all sites and a better endurance of large surge currents. However, the clamping voltage is somewhat greater. This choice trades a large advantage at a few sites against a small disadvantage at most sites.

These two choices involve different concerns, and, in the author's opinion, there is no general resolution.

The following numerical values for the minimum initial value of V_N are the author's personal opinions:

$$1.5 \ V_p \leq \min(V_N) \quad \text{for secondary arresters (category C)}$$

$$1.25 V_p \leq \min(V_N) \quad \text{at end of long branch (category A)}$$

For the category A environment, a minimum value of V_N of 210 V is suitable for mains with a nominal voltage of 120 V rms. This corresponds to what is often called a 150 V rms rated varistor.

There is one situation in which there would be no objection to using varistors with the minimum conduction voltage (i.e., $\beta = 0.10$). There can be no abnormally large rms voltages downstream from a ferroresonant line conditioner or "constant voltage transformer." (Line conditioners are discussed in Chapter 20.) If a line conditioner is *always* connected upstream from the equipment in question, the specifications for metal oxide varistors given above might be relaxed, since the equipment will not be exposed to swells. Varistors with a value of V_N as small as 185 V (so-called 130 V rms rating) might be suitable for connection to nominal 120 V rms mains downstream from a line conditioner or an ac voltage regulator. However, there is a real danger in relaxing the specifications on V_N. After varistors with a small value of V_N have been installed, the equipment might later be operated in a different environment, where these varistors would be either marginal or unsuitable. Such events could occur innocently. For example, the line conditioner might fail. It would be removed for service or replacement, and the system might be operated on a temporary basis without the line conditioner. A swell could then reach the varistors and damage them.

1. Neutral to Ground Varistor

These values of V_N are derived from concerns about the voltage that appears between either the hot-grounding or the hot-neutral conductors. Varistors connected between the neutral and grounding conductors, such as V_2 in Fig. 19-3, are exposed to no more than a few volts rms during normal system operation. There are three reasons why it is undesirable to use a smaller value of V_N for the varistor between the neutral and grounding conductors.

1. There is no *a priori* reason to expect V_2 be subjected to less energetic transients than varistor V_1, which is between the hot and grounding conductors. Therefore, V_1 and V_2 should have the same rating for maximum energy that can be absorbed in a surge. For varistors of a given cross-sectional area, devices with a smaller value of V_N have a smaller maximum surge current and energy rating.

2. If V_1 and V_2 have appreciably different values of V_N, then common-mode transient overvoltages will be converted to differential-mode

overvoltages by the different clamping voltages. Common-mode voltages are easily attenuated by isolation transformers, line conditioners, or low-pass filters. It is more difficult to attenuate differential-mode voltages, particularly for transients with durations of tens of microseconds or more. This consideration does not require matched devices for V_1 and V_2, but their nominal values of V_N should be the same.

3. The hot and neutral conductor may be reversed by a wiring error, which would cause V_2 to see the normal mains voltage. Such situations actually exist and are not just theoretical possibilities.

2. Varistor Diameter

Varistors with a larger diameter, but the same value of V_N, can absorb more energy or tolerate larger surge currents without appreciable degradation than varistors with a smaller diameter. However, varistors with a larger diameter are more expensive than varistors with a smaller diameter. Therefore, it is important to specify a large enough diameter to survive most threats but not to overspecify and increase costs. The following discussion considers MOVs in secondary arresters first, then MOVs on branch circuits.

To have a good chance at survival, each varistor in a secondary arrester ought to be rated by the manufacturer to withstand an 8/20 μs waveform with a peak current of at least 30 kA (Martzloff, 1980). Such a current is typical of a cloud-to-ground lightning return stroke; a reasonable worst-case peak current in the return stroke might be 100 kA. However, in most lighting strokes to overhead power lines, there are multiple shunt paths, so that only a fraction of the lightning current passes through a secondary arrester in a building. The other shunt paths to ground include flashover of the power lines to overhead ground wires, power poles, and adjacent trees, as well as through distribution arresters. With varistor technology available in 1988, when this book was written, the author believes that metal oxide varistors for secondary arresters should have a diameter of at least 32 mm.

The peak surge current in a branch circuit inside a building is likely to be much smaller than the peak surge current at the point of entry. ANSI C62.41-1980 gives a maximum representative surge current of 3 kA with an 8/20 μs waveform for category B. With the varistor technology available in 1988, when this book was written, metal oxide varistors suitable for use in category B have a diameter of at least 14 mm.

The author's personal practice is to use varistors with a diameter of at least 20 mm on the mains, even on long branch circuits (category A). There are two advantages to using a larger-diameter varistor: smaller clamping voltage and longer lifetime. Consider two varistors with the same value of V_N, about 240 V, but different diameters—one 14 mm, the other 20 mm. The varistors in this example have a maximum clamping voltage of 400 V at

50 A and 300 A for the smaller and larger diameter, respectively. For 8/20 μs surge currents, the 14 mm diameter varistor has a rated lifetime of 10 surges with a peak current of 1 kA, and the 20 mm diameter varistor has a rated lifetime of 10 surges with a peak current of 2 kA. The advantage of larger-diameter varistors is simply that their area is larger, so that the current density is smaller.

F. COORDINATION OF PROTECTION

In most hybrid protection circuits, shunt devices with smaller clamping voltages are installed downstream, nearer the equipment to be protected. Some series impedance is installed between each pair of shunt protective devices to provide proper coordination. In signal circuits with small currents during normal operation, the series impedance might be a resistance of 10 Ω or more. Such a large impedance would be unacceptable in most mains applications, where normal currents are usually at least 1 A rms. For overvoltage protection on the mains, the series impedance is usually an inductance: either a choke or a few tens of meters of building wiring. For transient overvoltages with short durations (e.g., less than 50 μs), a choke with an inductance of the order of 30 μH or more could provide a suitable series impedance. But for overvoltages with durations of a few milliseconds, it will be difficult to insert an adequate series impedance between shunt protective devices and still maintain normal operation of the load.

Consider an example of two different varistors that are connected in parallel on mains with a nominal voltage of 120 V rms. In this example there is no impedance between the varistors. Perhaps they were connected near each other on the mains, or perhaps we consider long-duration overvoltages so that whatever series inductance that may be between the varistors has negligible effect. The two varistors have the following parameters:

$$V_1: \quad V_N = 190 \text{ V} \qquad \alpha = 45 \qquad R_{bulk} = 0.06 \ \Omega$$
$$V_2: \quad V_N = 285 \text{ V} \qquad \alpha = 45 \qquad R_{bulk} = 0.01 \ \Omega$$

In this simulation V_1 represents a varistor that is often used inside a chassis: it has a small value of V_N (probably too small) and a diameter of about 14 mm. V_2 represents a varistor in the secondary arrester with a diameter of about 32 mm. The bulk resistances of V_1 and V_2 are proportional to their cross-sectional area. The characteristic V-I curve is shown in Fig. 19-4. The two solid lines are the V-I curves for each varistor separately; the dashed line is the V-I curve for the parallel combination. Essentially all of the current passes through V_1, until the total current in the pair of varistors exceeds about 1 kA. The manufacturer's rated lifetime for V_1 is 10 surges with an 8/20 μs waveshape and a peak current of 1 kA. In an environment that has peak surge currents of less than 1 kA, the inclusion of the

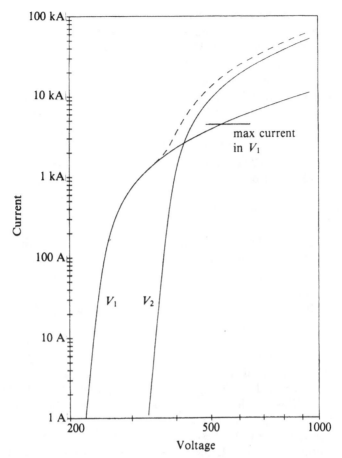

Fig. 19-4. *V-I curve of two varistors, V_1 and V_2. The dashed line shows the V-I curve for the parallel connection of these two varistors.*

secondary arrester, V_2, has little effect on either clamping voltage or lifetime of V_1 (see discussion in Chapter 8). However, during a surge with a peak current between about 5 and 30 kA, inclusion of V_2 could prevent V_1 from being destroyed. Although the parallel connection of V_1 and V_2 are not well coordinated, perhaps it is not a disaster either.

One could insert some inductance downstream from V_2 and upstream from V_1 that would present an appreciable series impedance during the large peak currents in surges with an 8/20 μs waveshape. This would greatly improve the coordination of V_1 and V_2. However, for surges with a duration of 1 ms or more, it will be difficult to insert an appreciable series impedance between V_1 and V_2. One can only hope that the peak current in surges with a duration of more than 1 ms is small.

One solution to the problem of coordinating multiple varistors on the mains is to reverse the normal order of devices. Place the varistor with the *smallest* value of V_N at the secondary arrester. Varistors downstream would have slightly *larger* values of V_N. For example, consider mains with a nominal voltage of 120 V rms: the secondary arrester might have $V_N = 270$ V, the branch circuit varistors might have $V_N = 360$ V, and varistors in a chassis might have $V_N = 390$ V. This approach is also economical, because smaller-diameter varistors can be used in the more common applications. The largest surge current in the network of varistors travels between the source of the overvoltage and the secondary arrester. This reduces radiation from large values of dI/dt in varistors in branch circuits and chassis that otherwise would have occurred if a varistor with a relatively small value of V_N were located on various branch circuits or inside many different chassis. The main disadvantage of this method is that engineers who design protective circuits inside a chassis or for a branch circuit will not want to use varistors with relatively large clamping voltages in their products. And if varistors with large clamping voltages are used in an environment that does not have a secondary arrester with a small clamping voltage, the result could be damaged equipment.

It is sometimes difficult to coordinate multiple surge protective devices when you are aware of them. However, you can't even try to coordinate varistors that are hidden inside a chassis unless you know about them! Poor coordination of multiple varistors makes some varistors superfluous and might lead to damaged equipment. Therefore, it is suggested that if varistors are connected inside equipment, the equipment should be prominently labeled. This would make it easier for conscientious engineers to properly coordinate protective devices in a system. If the varistors have a small diameter, then the label should warn the user to connect additional surge protection upstream. If the varistors have a small value of V_N, then the label should warn the user to connect a line conditioner or voltage regulator upstream. These suggestions for additional labeling will be difficult to achieve in an unregulated consumer market.

G. TRANSIENT SUPPRESSION AT SOURCE

It cannot be too strongly emphasized that transient overvoltages caused by switching reactive loads should be suppressed at their source rather than allowed to pollute the mains. Common sources of transient overvoltages are switching reactive loads (e.g., fluorescent lamps, neon signs, photocopiers, motors in appliances such as vacuum cleaners, and solenoids in vending machines). Inductive loads should have a metal oxide varistor installed across the coil, as shown in Fig. 19-5, to suppress overvoltages. A system that is protected in this way will be a "good neighbor" and is less likely to cause problems when it is installed near a sensitive system. If suppression of transient overvoltages at their source is not feasible, then a system that

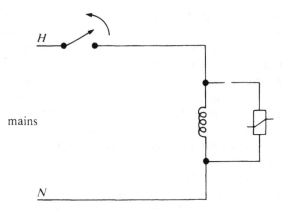

Fig. 19-5. *Clamping of overvoltage caused by interrupting current in an inductive load on the ac supply mains.*

produces transients should be isolated from the mains by a surge protection module, as shown in Fig. 19-3.

H. APPLICATIONS OF SPARK GAP AND MOV ON MAINS

In Chapter 8 it was remarked that a series combination of a properly chosen spark gap and varistor, as shown in Fig. 19-6, could be useful for two different purposes: (1) reduce leakage current, and (2) reduce clamping voltages. The same choice of spark gap will not satisfy both goals.

1. Reduced Leakage Current

Varistors have large parasitic capacitance that can have unacceptably large currents when connected across ac supply mains in some applications. For

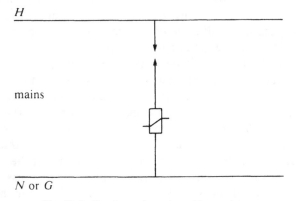

Fig. 19-6. *Spark gap in series with a varistor.*

example, UL 544 specifies that the maximum line-to-ground current is to be less than 0.1 mA rms for patient care equipment used in medical or dental offices. A varistor with a parasitic capacitance of 2 nF has an rms current of 0.09 mA when installed between line and ground on mains with a nominal voltage of 120 V rms and a frequency of 60 Hz. Such a varistor has an inadequate safety margin to meet the specifications in this application. Connecting a spark gap in series with the varistor can decrease the shunt capacitance of the combination to a few pF, which is a trivial value.

Criteria and methods for preventing follow current in spark gaps were presented in Chapter 7.

It is imperative in this application that there be no steady-state operation of the spark gap in the glow region. To achieve this, the spark gap should have a dc firing voltage, V_f, that is greater than the peak value of the mains voltage, V_p. The tolerances on both V_f and V_p must be considered. If the spark gap is not to conduct during normal operation of the mains, the following relation must be satisfied:

$$\max(V_p) < \min(V_f)$$

For good engineering design, a safety factor of 20% might be included so that the relation becomes

$$1.2 \max(V_p) \le \min(V_f)$$

Consider an example of mains with a nominal voltage of 120 V rms. If we allow a +10% tolerance for the normal mains voltage, $\max(V_p)$ is about 190 V. With the 20% safety factor, the value of $\min(V_f)$ in this example is 225 V. If there is a ±20% tolerance on V_f, the smallest *nominal* value of V_f would be 280 V.

The varistor is included in this example to extinguish the arc follow current in the spark gap. (If the series varistor can force the spark gap into the glow regime after the transient overvoltage, the gap will extinguish during a periodic zero crossing of the mains voltage. The gap cannot conduct during the next half-cycle of the mains voltage, because its dc firing voltage was selected to be greater than the peak of the mains voltage.)

To ensure that the arc will be interrupted, the current in the varistor must be less than the arc extinguishing current, I_e, in the gap when the normal mains voltage is across the series combination of the varistor and spark gap. The arc will *probably* extinguish if the current is less than I_e for a few milliseconds. A conservative design will have the magnitude of the current continuously be less than I_e. During operation in the arc regime, there is about 20 V across the spark gap; the remainder of the mains voltage is across the varistor. The current in the varistor must be less than I_e when the voltage across the varistor is $V_p - 20$ V. The 20 V can be neglected, because it is much smaller than V_p; this is a conservative approximation, since the varistor current will be required to be less than I_c at a larger voltage.

We continue the example begun above of an application on mains with a nominal voltage of 120 V rms. A conservative value for I_e is 50 mA; the actual value of I_e for most spark gaps is at least 100 mA. The value of max(V_p) given above was 190 V. If the value of the parameter α for the varistor is known, the value of V_N can be calculated. Given α of 50, V_N is 175 V. At smaller values of α, which are more common, V_N will be smaller than 175 V.

2. Lower Clamping Voltage

The series combination of a gap and an MOV might be used to achieve smaller clamping voltages during surges than is otherwise possible with a varistor. The spark gap acts as a switch to prevent steady-state power dissipation in the varistor. This would allow a varistor with a relatively small value of V_N to be connected to the mains without concern about thermal runaway of the varistor. As discussed earlier in this chapter, a varistor *alone* across mains with a nominal voltage of 120 V rms should have a *minimum* value of V_N of about 210 V. By including the series spark gap, the minimum value of V_N can be decreased to 175 V.

If small conduction currents during normal operation are permissible, the spark gap can be allowed to operate in the glow regime during part of each half-cycle of the mains voltage. It is possible that brief pulses of glow current when the magnitude of the system voltage is near a peak could contribute a large population of ions that could reduce the response time of the spark gap. This is a very old idea, which can be traced back to Wynn-Williams (1926). The continuous current maintains a large population of electrons and positive ions. Smith et al. (1986) showed that with a 0.1 mA keep-alive current, the time to breakdown could be made as small as 0.2 ns, which is one of the fastest spark gaps described in the literature. To obtain the same rate of charge separation as this glow current with radioactive additives would require a dangerously large amount of radioactive material.

This technique would require a spark gap with a dc firing voltage that is slightly less than the peak system voltage (e.g., a 150 V dc gap on mains with a nominal voltage of 120 V rms). In order to reduce the possibility of damage to the spark gap from heat and sputtering, the peak current should be limited to *less* than 0.5 mA, or the duration of conduction should be a small fraction of the period of the normal mains waveform. The line-to-ground current in appliances is limited for safety reasons, in most environments, to less than 0.5 mA rms (UL 1283). Although this concern does not affect protective devices that are connected between the hot and neutral conductors, the same limit of 0.5 mA may also be reasonable to prevent damage to the spark gap by sputtering.

The tolerances for the rms mains voltage, the dc firing voltage of the spark gap, and the value of V_N of the varistor are all of the order of $\pm 10\%$ or more. Hence, precise control of the duration of conduction seems out of reach. However, the current in the glow region could be restricted to less than 0.5 mA by appropriate specifications for the varistor.

The varistors must be specified to prevent follow current in the arc regime immediately after a transient overvoltage. The calculations to ensure this are described above, in the section on the use of a series spark gap and varistor to achieve small leakage currents.

I. OVERVOLTAGE-PROTECTED LOW-PASS FILTERS

Transient overvoltages with a peak value greater than about 500 V are uncommon events in most environments. Bursts of high-frequency noise with a peak-to-peak value between 40 and 100 V occurred at one residential site at an average rate of 0.5 per hour (Standler, 1989). Maximum values of $|dV/dt|$ of the order of 1 kV/μs were common. Because the peak voltage values in these bursts of noise are less than the amplitude of the steady-state mains voltage, overvoltage protection (such as metal oxide varistors) will be ineffective in attenuating these disturbances. However, a low-pass filter could be effective in attenuating these bursts of noise.

There are three basic concerns about the use of filters *alone* as a surge protection device, which are discussed in Chapter 13: (1) it is difficult to obtain adequate attenuation with a filter alone; (2) the filter may be degraded or damaged or may misbehave during overstress; and (3) the filter may resonate, particularly with a reactive source or load impedance. The first two concerns can easily be addressed by using metal oxide varistors together with a filter. The third concern requires compromises in the design of the filter to damp any resonances.

Specific designs are discussed below. Two different environments are considered, which are distinguished on the amount of line-to-ground conduction current that is acceptable: (1) a normal environment where a current of a few milliampers is acceptable, and (2) an environment where line-to-ground currents must be minimized, such as medical and dental patient care areas. The limits on line-to-ground currents are discussed in Chapter 13.

1. Basic Design for Normal Environments

First consider a filter for use in environments where a line-to-ground leakage current of the order of 1 mA is acceptable. The circuit shown in Fig. 19-7 has two stages. The first stage uses nominally identical metal oxide varistors, V_1 and V_2, to provide voltage limiting for common-mode transients. Both the hot and neutral conductors are each clamped at some relatively small potential (e.g., 300 to 500 V) from ground. Inductor L_1 and

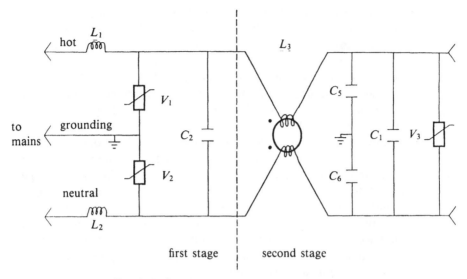

Fig. 19-7. *Basic overvoltage-protected low-pass filter.*

varistor V_1 form a nonlinear voltage divider for transient overvoltages with short durations. L_2 and V_2 act in the same way for transients on the neutral conductor. For high-frequency noise, the inductances L_1 and L_2 form a low-pass filter with the parasitic capacitance of varistors V_1 and V_2 and with capacitor C_2.

The second stage of the circuit shown in Fig. 19-7 contains a common-mode choke, L_3, and bypass capacitors, C_1, C_5, and C_6. The second stage of the low-pass filter shown in Fig. 19-7 attenuates the remnant of any transient overvoltage that passes downstream from varistors V_1 and V_2 and also provides additional attenuation of high-frequency noise. Components L_3, C_5, and C_6 form a low-pass filter for common-mode signals. Owing to imperfections in the symmetry of the two coils in L_3, this choke presents a relatively small inductance to differential-mode signals. This inductance and C_1 form a low-pass filter for differential-mode signals. Additional protection from transient overvoltages is obtained by connecting varistor V_3 as shown in Fig. 19-7.

There are many considerations involved in selecting component values for the circuit shown in Fig. 19-7. The input inductors should have a single-layer construction to prevent insulation breakdown between layers and be rated to withstand at least a 5 kV impulse between terminals, as suggested by Hays and Bodle (1958). The single-layer construction also increases the self-resonant frequency of the choke by decreasing the parasitic capacitances between the terminals. Chokes L_1 and L_2 are connected upstream from L_3, because regular chokes can be more easily designed to withstand high-voltage transients that can common-mode choke L_3.

The inductance values for L_1 and L_2 are limited by two considerations. These inductors are in series with the load current and have a voltage across them during the normal operation of the filter. If the filter is rated for 10 A rms, 60 Hz service, and if a drop of 1 V rms across the filter can be tolerated, the maximum inductance is about 270 μH. This inductance limit applies to the sum of inductances L_1 and L_2 and the inductance of the choke L_3 for differential-mode currents. Subject to this limit, larger values of inductance are more desirable.

Size is the other consideration for inductance value. The input inductors, L_1 and L_2, should have single-layer construction. These inductors should be able to withstand at least 5 kV across the terminals. The high-voltage construction of these inductors will make them relatively large: a length between 7 and 12 cm is typical of single-layer chokes with ferrite cores, an inductance between 10 and 80 μH, and a 20 kV rating. Since devices with larger inductance values have a larger size and there are size constraints on the completed filter module, the size of the module may determine the maximum practical inductance value for L_1 and L_2.

Typical values for the capacitors are 0.5 μF for each of C_1 and C_2, and between about 2 and 5 nF for each of C_5 and C_6. The small values of C_5 and C_6 are chosen to limit the line-to-ground leakage current during normal operation. If larger leakage currents can be tolerated, then additional line-to-ground capacitors can be added, as discussed below.

All lumped-element capacitors have a small parasitic inductance, which may be as small as 5 nH, that makes a capacitor useless at high frequencies, as described in Chapter 15. The parasitic capacitance of the varistor V_3 and capacitor C_1 together provide a small magnitude of impedance over a wider range of frequencies than either device alone.

The specifications for the varistors were discussed earlier in this chapter. The clamping voltage for differential-mode transients at the input of L_3 is twice the common-mode value, since V_1 and V_2 are in series for differential-mode voltages. This peak voltage value is unlikely to damage the dielectric of the capacitors C_1, C_2, C_5, or C_6 or the insulation of the choke, L_3. Although it is no longer critical that the capacitors in this filter be able to withstand high-voltage transients, it is recommended that only capacitors with self-healing dielectrics that are rated for operation at 600 V dc (or more) be used. This provides a greater reliability for the filter.

2. Suppression of Resonances

The filter shown in Fig. 19-7 will have voltage gain at various frequencies owing to resonances of various LC pairs. There is a resonance for common-mode signals owing to L_3, C_5, C_6, and the parasitic capacitances of V_1 and V_2. Another resonance for common-mode signals is owing to L_1, L_2, and the parasitic capacitances of V_1 and V_2. There is a resonance for differential-mode signals owing to L_1, L_2, C_1, C_2, and the nonideal inductance of L_3 to differential-mode signals.

As mentioned earlier, the possibility exists that the filter-load system will have resonances when a reactive load is connected to a low-pass filter. There are two possibilities for reactive loads: inductive and capacitive. Pure inductive loads are unusual. Most inductive loads across the mains (e.g., motors, relay coils) have substantial series resistance that will act to help damp any resonances. However, high-quality capacitive loads do exist: many pieces of electronic equipment have capacitors connected to the mains for suppression of electromagnetic noise. Many products intended for office and industrial applications have conventional low-pass filters connected to the mains, which have a capacitive input. Many electronic products for consumer applications (e.g., television receivers, videotape recorders, stereo receivers, etc.) have simple capacitors across the mains.

Resonances can be avoided by adding resistive dampening, as shown in Fig. 19-8. The circuit shown in Fig. 19-8 includes all of the components shown in Fig. 19-7, plus seven resistors and two capacitors.

Given the types of reactive loads in the world, one cannot suppress resonances in a low-pass filter by merely connecting resistors in series with shunt capacitors inside the filter. Although this would damp internal resonances, it does nothing to solve the resonance owing to an external capacitive load. Instead, resistors, such as R_5 and R_6 in Fig. 19-8, are shunted across the inductor nearest the output port. A crude initial approximation for the value of these resistances can be obtained by calculating the reactance of one coil in L_3 at the resonance frequency.

It is undesirable to connect resistors in parallel with L_1 and L_2, because these inductors may be exposed to high-voltage transients that might damage inexpensive resistors.

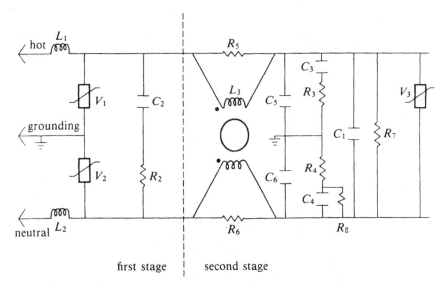

Fig. 19-8. *Overvoltage-protected low-pass filter for commercial environment.*

The series RC networks shown in Fig. 19-8 provides damping of resonances. Such networks are common in switching applications with relays, SCRs, or triacs, where they are known as *snubber networks*. Resistor R_2 damps the resonance to differential-mode signals formed by L_1, L_2, C_2, C_1, and the differential-mode inductance of L_3. The two optional networks R_3, C_3 and R_4, C_4 provide damping of resonances of common-mode signals owing to L_3, C_5, and C_6. Further damping for both differential and common modes may be provided by the resistive component of the impedance of L_1, L_2, and L_3 owing to the losses in the cores of these chokes.

There is no reason to include a damping resistor in series with C_1, since a capacitive load would remove the effect of this resistance. The use of damping resistor R_2 can adequately suppress differential-mode resonances.

Resistor R_7 is included to discharge the capacitors if the filter is disconnected from the mains with no load present. The value of R_7 may be chosen to decrease the potential across capacitors C_1 and C_2 to some small value (e.g., less than 30 V) 1 second after the mains are disconnected at a positive peak of the sinusoidal mains voltage. There is no need to discharge capacitors C_5 and C_6, because their capacitance is so small. If C_3 and C_4 have large capacitances, another discharge resistor, R_8, should be included. The resistances of R_7 and R_8 are too large to have an effect on damping of resonances: these resistors are included only for safety considerations.

3. Design of Filter with Minimum Leakage Currents

A filter can be designed for medical equipment in patient care applications using the principles discussed above. The only unusual criterion is that of very low line-to-ground leakage current. If a generous safety margin is to be provided, the line-to-ground capacitance must be limited to about 500 pF, which effectively prohibits the use of either metal oxide varistors or large capacitances in line-to-ground connections. Some of these difficulties can be overcome with the circuit shown in Fig. 19-9.

The relatively large parasitic capacitance of the line-to-ground varistors, V_1 and V_2, is reduced to only a few picofarads by connecting a spark gap in series. However, the spark gap has a delay time between the application of an overvoltage and the onset of an arc inside the gas. By placing the spark gaps and varistors upstream from chokes L_1 and L_2, we obtain two advantages. First, the inductance in L_1 and L_2 helps create a large voltage across the spark gaps during transient overvoltages with a short rise time, which will reduce the delay time in the spark gaps. Second, the low-pass filters formed by the various inductors and capacitors attenuate the remnant that passes downstream from the spark gaps, V_1, and V_2. Prior to the onset of the arc inside the spark gaps, this remnant will be the entire transient overvoltage across the input port. This idea is not new: Hays and Bodle

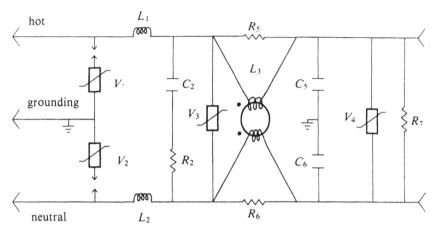

Fig. 19-9. *Overvoltage-protected low-pass filter for medical environment.*

(1958) advocated it 30 years ago. However, the merits of this circuit appear to have been ignored.

Attenuation of differential-mode overvoltages is provided by chokes L_1 and L_2 together with varistors V_3 and V_4. A differential-mode low-pass filter is formed by L_1, L_2, and C_2. Resonance is suppressed by including resistor R_2 in the middle of the filter. Additional attenuation of high-frequency differential-mode voltages is provided by the parasitic capacitances of V_3 and V_4.

A common-mode low-pass filter is formed by common-mode choke, L_3, and capacitors C_5 and C_6. For safety reasons, the capacitance of C_5 and C_6 is limited to about 500 pF each.

The purposes of R_5, R_6, and R_7 in Fig. 19-9 are identical to the components of the same designations in Fig. 19-8, which were explained above.

4. Performance of Filters

Standler (1988b) discussed the performance of the filter circuits shown in Figs 19-8 and 19-9 with both steady-state frequency response and time domain plots of input and output voltage during transient overvoltages. The discussion here is limited to the performance during differential-mode overvoltages. The test circuit is shown in Fig. 19-10. The laboratory surge generator is connected between the hot and neutral conductors at the input of the filter. The filter was energized by the normal 120 V rms mains during the test. Because the filter is designed to be connected to the mains during normal use, it is important to test it when energized by the mains voltage, as discussed in Chapter 22. A device called a *back filter* was used to keep the surge current from passing into the building mains: back filters are discussed

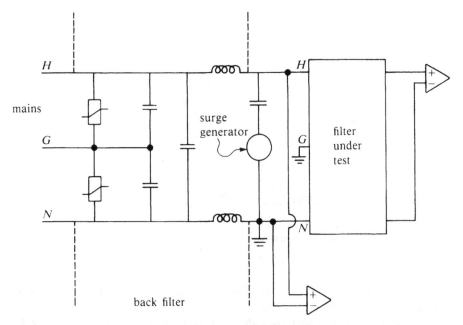

Fig. 19-10. Schematic for testing surge performance of filters.

in Chapter 23. Differential voltage probes and amplifiers in a high-speed digital oscilloscope were used to measure the voltage across the hot and neutral conductors at the input and output of the filter under test. The surge waveform was a $0.5\,\mu s$–$100\,kHz$ ring wave with an effective source impedance of $30\,\Omega$ that was described in Chapter 5.

First we consider the performance of a low-cost commercial low-pass filter module during a differential-mode overvoltage: the voltage across both the input and output ports is plotted versus time in Fig. 19-11. The voltages across the input and output port are essentially identical, because this small filter has little attenuation at frequencies below about $200\,kHz$. When common-mode overvoltages were applied across the input port of this filter, there was evidence of internal dielectric breakdown in the filter. However, this internal breakdown did not affect the output voltage, probably because of the shielding inside the filter.

The performance of the overvoltage-protected filter (schematic shown in Fig. 19-8) is shown in Fig. 19-12. The surge generator was adjusted to give a peak voltage of $6\,kV$ across an open circuit during this test. The voltage across the output port is clamped to less than $370\,V$ during this ring wave with a peak current of $129\,A$. Notice that the output voltage also has a much smaller value of dV/dt than the input voltage. Varistors alone would clamp the peak voltage. However, the combination of varistors and low-pass filter is needed to reduce the value of dV/dt to very small values.

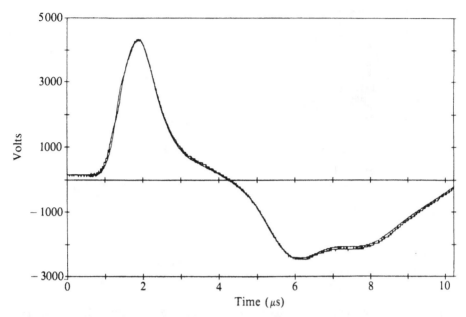

Fig. 19-11. *Plot of voltages across input and output ports of a commercial low-pass filter module that contains no surge protective devices. The input and output voltages are essentially identical. The surge was applied and the voltage measured, between the H and N conductors.*

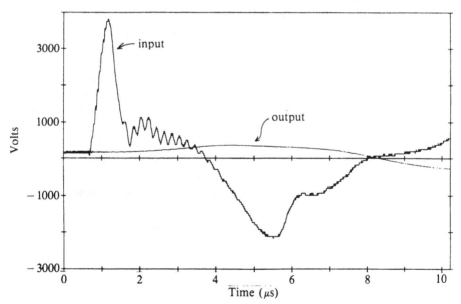

Fig. 19-12. *Plot of voltages across input and output ports of the filter shown in Fig. 19-8. The surge was applied, and the voltage measured, between the H and N conductors.*

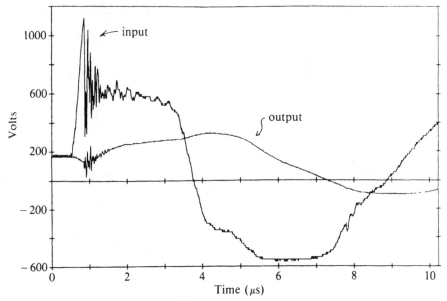

Fig. 19-13. Plot of voltages across input and output ports of the filter shown in Fig. 19-9. The surge was applied, and the voltage measured, between the H and N conductors.

The performance of the overvoltage-protected filter with low leakage currents (schematic shown in Fig. 19-9) is shown in Fig. 19-13. The worst-case overstress for this circuit is a waveform with a moderately short rise time and relatively small peak voltage so that the spark gaps just barely conduct. After some experimentation, a 100 kHz ring wave with a peak voltage of 1.5 kV across an open circuit was selected for use in this test. The voltage across the input port reaches a large value at about 0.9 μs before the spark gaps conduct. Between about 1.2 and 3 μs (and again between 5.5 and 7 μs), the magnitude of the voltage across the input port is clamped to slightly less than 600 V by varistors V_1 and V_2 in series with conducting spark gaps. The voltage across the output port is clamped at less than 340 V during this test. Except for the burst of noise around 1 μs, the slope of the output voltage is much smoother than the slope of the input voltage. The burst of noise at 1 μs is probably radiation from the large value of dI/dt in the spark gaps that was picked up by the loop area of the voltage probes at the output port.

The use of varistors together with low-pass filters can reduce the magnitude of both the peak voltage and the slope dV/dt. This reduction may be helpful in avoiding both damage and upset of electronic systems.

20

Mitigation of Disturbances on Mains

A. INTRODUCTION

Computers are vulnerable to damage and susceptible to upset (temporary malfunction) from disturbances on the low-voltage ac supply mains. Such disturbances include transient overvoltages, high-frequency noise, reductions in rms voltage (sags, brownouts), and outages (Key, 1979; Kania et al., 1980; Duell and Roland, 1981; Speranza, 1982; Nash and Wells, 1985). Koval et al. (1986) analyzed 11 years of operation of a mainframe computer and found that a mains disturbance had a 36% probability of damaging equipment and a 31% probability of destroying files.

A variety of devices are manufactured to provide disturbance-free power to computers and other sensitive equipment (Martzloff, 1985; Nash and Wells, 1985; Standler and Canike, 1986). This chapter reviews some appliances and techniques that can mitigate disturbances on the ac supply mains and gives some simple design guidelines for protecting computers and other sensitive electronic equipment from damage and upset by disturbances on the mains.

There are several ways that disturbances on the mains can affect operations. Damage to equipment increases costs in three ways: (1) the equipment must be repaired or replaced; (2) until the equipment's operation is restored, employees are idle or must use less efficient means to accomplish work; and (3) after the equipment is restored to service, it may be necessary to pay overtime wages to reduce the backlog of work that accumulated when the equipment was out of service. Temporary malfunction of electronic equipment also increases costs. If the malfunction is obvious, the job can be repeated (with some wasted time). If the malfunction is not detected, decisions can be made with faulty information.

An example of an obvious malfunction is the failure of a word processor when the power is interrupted momentarily; the information stored in the computer's semiconductor memory is lost, and some typing will need to be repeated. Such a malfunction is usually an annoyance in a word processor. However, the same type of malfunction that occurs in a computer that analyzes clinical laboratory data in a hospital can adversely affect the care of patients who contributed samples on that day. The cost of a malfunction is strongly dependent on the situation and may be impossible to determine accurately (Koval et al., 1986).

The scope of this chapter is limited to protection of small electronic systems that operate from single-phase mains with a nominal amplitude of 120 V rms and a frequency of 60 Hz. "Small" electronic equipment is defined as a system that consumes less than 800 W of power. Probably the most common application is protection of desktop computer systems, but protection of electronic instrumentation and other critical loads follows the same principles. The general principles are probably valid for operation from single-phase 250 V rms, 50 Hz, but the author has had no experience with this. Operation of systems on three-phase power or on single-phase power at 400 Hz in ships and aircraft is an entirely different matter, which is excluded from the scope of this book.

The chapter begins with a discussion of several wiring practices and operating techniques that may help to reduce the severity of disturbances to which equipment is exposed. Various appliances that provide regulation of the rms voltage, isolation, and power to loads during interruptions of the mains are described. This chapter concentrates on line conditioners and uninterruptible power supplies, which can provide comprehensive protection from disturbances. The chapter concludes with a list of simple rules for protecting small electronic systems from disturbances on the mains.

B. WIRING AND OPERATING PRACTICES

1. Use of Steel Conduit

It is often recommended that the mains wiring be enclosed in rigid steel conduit for shielding (Lasitter and Clark, 1970, p. 4). There are two advantages to shielding the mains conductors. First, transient overvoltages and noise conducted on the mains will not be coupled onto other cables, such as those for computer data and telephones. This is desirable where showering arcs are generated inside the building and conducted on the mains. Second, electromagnetic radiation in the interior of the building will not be coupled onto the mains. The steel conduit may also give appreciable attenuation to high-frequency noise and transient overvoltages with sub-microsecond durations, as discussed in Chapter 4.

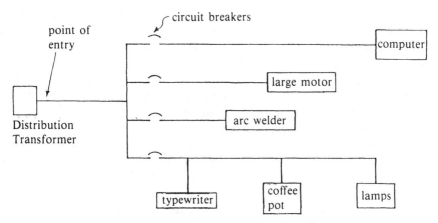

Fig. 20-1. *Use of a dedicated line for a computer.*

2. Dedicated Line

Dedicated lines are often recommended for computers and other sensitive loads. Nash and Wells (1985) dispelled the notion that a dedicated line will remove disturbances from the mains. The major reason to use a dedicated line for a computer, as shown in Fig. 20-1, is to avoid interruption of power to the computer owing to excessive current in another load. For example, in Fig. 20-1, excessive current drawn by the coffee pot, which trips the circuit breaker, will interrupt power to the typewriter, because they are on the same branch. When this circuit breaker trips, power is not interrupted to the computer, because the computer is on a dedicated line.

There is another, less important reason to use a dedicated line. The rise time of disturbances with a submicrosecond duration will increase as the distance from the source increases (see Chapter 4). Use of a dedicated line forces these short rise time disturbances to travel from the source to the circuit breaker panel and then down the dedicated line to the computer. This is a longer path than, for example, from the coffee pot to the typewriter in Fig. 20-1. The extra length of line between a source of transients and the critical load is useful in another way than described in Chapter 4. The parasitic inductance of the line forms a low-pass filter with both (1) the parasitic capacitance of metal oxide varistors and (2) the capacitors in EMI filters at or near the critical load. This low-pass filter will decrease the magnitude of dV/dt in bursts of noise.

3. Turn-on Sequence

When an electronic power supply is switched on, a much larger amount of current is drawn from the mains during the first half-cycle of operation than is drawn in the steady state. This unusually large current can cause a sag in the mains voltage. To avoid this disturbance and to ensure proper handshaking between devices in a system, the user should establish an orderly sequence of turning on equipment. The peripherals should be turned on individually, one at a time, in some definite sequence. The computer is usually turned on last. The exact sequence depends on the system and may be found in the computer operator's manual.

The worst thing to do is to leave all of the individual switches on the equipment in the "on" position and energize the entire system simultaneously by flipping one switch (e.g., on an outlet strip, on a UPS, or on a line conditioner). Switching everything simultaneously on may even open a circuit breaker or fuse in the distribution panel, the line conditioner, or the UPS.

There should also be a definite turn-off sequence. Usually external disk drives are switched off first, followed by other peripherals. The computer is usually turned off last.

4. Bunched Power Cords

To avoid electromagnetic coupling, power cords to critical equipment and conditioning appliances should be compactly bundled and secured with a plastic cable tie, as shown in Fig. 20-2. Bundling power cords to SPDs and line conditioners reduces electromagnetic radiation from large surge currents and values of dI/dt in these cables. Bundling power cords to critical

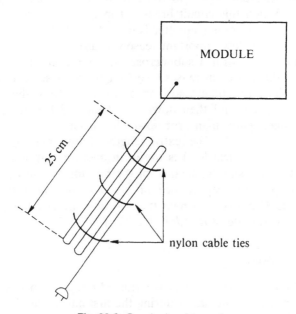

Fig. 20-2. *Bunched mains cord.*

equipment reduces the loop area for pickup of electromagnetic radiation. To further reduce coupling, power cords upstream from SPDs and line conditioners should be kept as far as possible from power cords to critical electronic systems. Careless arrangement of power cords can defeat some of the benefits of expensive power conditioning equipment.

C. APPLIANCES

A number of different types of appliances are sold to remove disturbances from the mains and provide "clean power" to vulnerable or susceptible loads such as computers and electronic instruments. Line conditioners (which provide regulation of the rms voltage, low-pass filtering, and isolation) are discussed first. There are two widely used types of line conditioners for loads that consume less than about 1 kW of power: tap-switching and ferroresonant. Several other voltage-regulating appliances are also discussed briefly: autotransformers, electronic voltage regulators, and motor-generator sets. Finally, uninterruptible power supplies (which supply power to critical loads when the mains have been interrupted) are discussed.

1. Definition of Line Conditioner

It has been suggested that an ac line conditioner is a device that accomplishes *all* of the following three items (Standler, 1984):

1. Provide voltage regulation. When the rms input voltage is between 95 and 130 V rms, the output voltage should be between 110 and 125 V rms for no load to full load.
2. Provide at least 50 dB differential-mode attenuation at frequencies above 100 kHz.
3. Provide isolation: no more than 1 pF capacitance between the input and output terminals. This, when combined with capacitance shunted across the load, attenuates common-mode transients.

2. Tap-Switching Line Conditioners

A tap-switching transformer, shown in Fig. 20-3, can be used to provide regulation of the rms voltage in a line conditioner. A standard two-coil (primary and secondary) transformer can be provided with multiple terminals (called *taps*) on either coil to compensate for variations in the magnitude of the input (primary) voltage. An electronic circuit senses the rms output voltage and selects the appropriate tap. Depending on the number of taps, one can obtain arbitrarily good regulation. A typical specification is for the output voltage to remain constant within ±5% for an

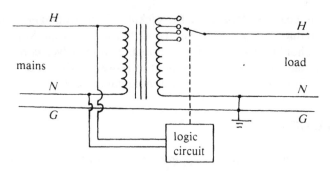

Fig. 20-3. *Tap-switching transformer.*

input voltage change of ±12%. Although this is worse regulation than that of a ferroresonant circuit, it may still be acceptable for most critical applications. The tap switching is usually done with a triac, which is faster and more reliable than a power relay.

The second requirement of a line conditioner is that it provide differential-mode attenuation for high-frequency noise. This can be accomplished by including some inductance in series with the transformer primary. If a relatively large series inductor is inserted in the primary side, the voltage across both primary and secondary coils of the transformer will be reduced. Compensation for this effect can be obtained by increasing the number of turns in the secondary coil. The large series inductance in series with the primary coil gives much better rejection of noise at frequencies below 0.5 MHz than typical inductor-capacitor filter modules for attenuation of EMI. The low-pass filter will probably attenuate the remnant that travels downstream from the large protection module. The word "probably" is used because an inductor-capacitor filter has insertion gain near its resonance frequency.

The third requirement for a line conditioner is isolation. This is achieved with an electrostatic shield between the primary and secondary coils in the transformer, as explained in Chapter 14. Isolation alone offers no protection from differential-mode transient overvoltages.

The bond between neutral and grounding conductors at the secondary of the line conditioner, shown in Fig. 20-3, is required to satisfy the U.S. National Electrical Code.

Standler and Canike (1986, pp. 177–178) were able to destroy the electronic control circuit in tap-switching line conditioners from two well-known manufacturers by applying transient overvoltages of no more than 6 kV to the input port of the line conditioners. One of these line conditioners, after damage by overvoltage, had an output voltage of 150 V rms for an input voltage of 120 V rms. This failure mode could damage electronic power supplies that were connected to the line conditioner. This outcome underscores the importance of connecting metal oxide varistors upstream from the line conditioner to limit the magnitude of an overvoltage to a value that the line conditioner can withstand.

The electronic control circuit in one tap-switching line conditioner that was rated for a 120 V rms, 500 VA load would not drive a 15 µF capacitor that was connected to the output of the line conditioner to simulate a capacitive load (Standler and Canike, 1986, p. 48).

3. Ferroresonant Line Conditioner

The physics of operation of a ferroresonant transformer circuit has been reviewed by Grossner (1983, pp. 160–162). The following description is grossly oversimplified but serves to convey the general idea. The ferroresonant transformer uses a core that is operated in "saturation"; that is, the magnitude of magnetic induction, B, is essentially independent of the magnitude of the magnetic field, H. Since the magnitudes of the input voltage, current in the primary coil, and the magnetic field, H, are all proportional, this makes the value of B essentially independent of the rms input voltage. If a resonant circuit were not present, the output voltage would be a crude square wave with the same frequency as the input voltage. An inductor-capacitor resonant circuit can be designed to convert the output voltage to a qausisinusoidal waveform. If a sinusoidal output waveform is desired, a "harmonic-neutralized" design should be specified, which will typically provide less than 3% harmonic distortion.

A schematic of the basic ferroresonant transformer is shown in Fig. 20-4. The bond between the neutral and grounding conductors at the output port is required by the U.S. National Electrical Code. The author has observed ferroresonant line conditioners sold by a major manufacturer that omitted this bond.

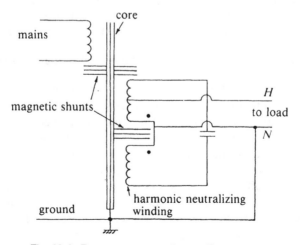

Fig. 20-4. *Ferroresonant transformer (Sola, 1954).*

The ferroresonant circuit has several outstanding advantages compared to tap-switching regulators:

1. The ferroresonant transformer has excellent voltage regulation when the rms load current is constant or slowly changing: output voltage is typically 120 ± 3 V for input voltages between 95 and 138 V (no load to full-rated load conditions).

2. The ferroresonant transformer is inherently short-circuit-proof. If the output terminals are shorted, the magnitude of the output current will be about 1.5 to 2.0 times the maximum rated load current. Under these conditions the transformer will act as a constant current source. This will *not* harm the transformer.

3. The ferroresonant circuit tends to ignore brief losses of input power. There is essentially no change in rms output voltage when the input voltage is zero for durations of 2 to 4 ms, since the resonant circuit continues to oscillate for several cycles without additional energy input. The resonant circuit also gives a larger attenuation for differential-mode noise at frequencies below 10 kHz than typical LC filter modules.

4. The ferroresonant transformer has few components and has a high reliability. The only components in a ferroresonant transformer are one transformer (with multiple windings) and one capacitor. There are no moving parts and no semiconductors.

5. Loads connected to a ferroresonant line conditioner receive substantial protection against transient overvoltages on the mains (Martzloff, 1983; Standler and Canike, 1986, pp. 128–177). Although a ferroresonant transformer can be difficult to damage by transient overvoltages, it is still reasonable to connect varistors upstream from a ferroresonant line conditioner: the smaller the overvoltage at the input port of the ferroresonant line conditioner, the smaller the transient overvoltage at the output port of that line conditioner.

However, the ferroresonant transformer has several major disadvantages:

1. Because the transformer core is driven into saturation, the transformer can be inefficient. If the ferroresonant transformer is operated at half its maximum rated load, the typical efficiency is only about 65% owing to large losses in the core. (If the unit is operated with its rated load and with an input voltage that is approximately the same as the output voltage, typical efficiencies are between 80 and 95%. The latter figures are often cited by vendors as evidence of "good efficiency.") The ferroresonant transformer core operates at 45 to 85°C above ambient temperature owing to power dissipated in the core. The heat burden from the ferroresonant transformer can be a serious consideration in computer rooms that need cooling.

2. Ferroresonant transformers contain an air gap in the core (Wroblewski, 1978) that can radiate a relatively large time-varying magnetic field. This field modulates the electron beam in the CRT of a computer and can make the display unpleasant or impossible to read. The author has used one brand of a ferroresonant transformer that made text on an IBM PC monochrome monitor unpleasant to read even when the transformer was 2 m from the CRT. This particular display interface is quite susceptible to interference from magnetic fields at the frequency of the mains (Baishiki and Deno, 1987). In contrast, another brand of ferroresonant transformer was used within 1 m of four different models of CRT displays with no discernible effect on the display. The radiated magnetic field is *not a general contraindication* for ferroresonant line conditioners; however, it is definitely a problem for some models of ferroresonant line conditioners. Tap-switching transformers should not have an air gap.

3. The mechanical stress in the core of the ferroresonant transformer often produces audible hum, which may annoy personnel near the transformer.

4. Because a resonant circuit with a fixed resonance frequency is used, the device is sensitive to changes in frequency of the input waveform. For a typical ferroresonant transformer, if the input frequency deviates from the design frequency, the output voltage will change by about 1.5% to 2% for each 1% change in input frequency. This effect is inherent in the performance of a resonant circuit that is not driven at the resonance frequency. It is not a serious problem when the input power is obtained from public utilities, which have tightly regulated frequency. However, when the input power is obtained from local generators that are driven by an internal combustion engine, the error in input frequency is often several percent.

5. The ferroresonant transformer is massive. A 500 VA ferroresonant unit has a mass of about 21 kg, about 75% more than for a tap-switching line conditioner of the same VA rating.

The limited output current of a ferroresonant transformer, which can be an advantage when isolating faults, can produce a severely distorted voltage waveform when electronic power supplies are connected to the transformer. Electronic power supplies, both "linear" and switching types, draw large currents from the mains near the positive and negative peaks of the mains voltage and nearly zero current elsewhere on the mains waveform. When electronic power supplies are switched on, the current drawn from the mains during the first half-cycle can be much larger than usual (Standler and Canike, 1986, pp. 96–109). Some engineers specify that the VA rating of ferroresonant transformers be twice the total nameplate VA rating of electronic loads that will be connected to the transformer in order to avoid distorted voltage waveforms. However, switching power supplies are quite tolerant of distorted waveforms.

When a ferroresonant transformer is switched on, an unusually large current is drawn from the mains to saturate the core of the transformer. All loads connected to a ferroresonant transformer should be off before the transformer is switched on to avoid drawing an even larger current from the mains and creating a disturbance on the mains.

When a ferroresonant transformer is switched off, it can inject a high-voltage transient back into the mains (Standler and Canike, 1986, pp. 82–94). Interruption of magnetizing current in any transformer can produce an overvoltage (see Chapter 2), but the larger magnetizing currents in a ferroresonant transformer may generate more severe overstresses. This is not a serious problem: it can easily be solved by connecting metal oxide varistors upstream from the ferroresonant transformer.

4. Autotransformer

The autotransformer is another type of ac voltage regulator. The auto-transformer (also known as a *variable ratio transformer* or by the trade name Variac) provides no isolation, since the primary and secondary share turns on the same coil. The autotransformer does not meet the requirements for a line conditioner and will not be discussed further.

5. Electronic Voltage Regulator

Another type of line conditioner is the electronic circuit shown in Fig. 20-5. Internal circuits convert the mains voltage to a constant voltage (dc). The regulated dc power is then converted back to ac by an oscillator. An amplifier and output transformer drive the load. An unprotected electronic regulator is vulnerable to transient overvoltages, as are other unprotected dc power supplies. This electronic regulator removes all power disturbances except blackouts. Protection against blackouts can be obtained simply by

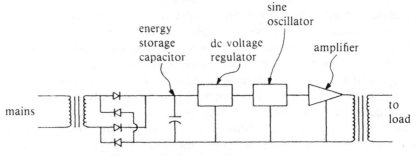

Fig. 20-5. *Electronic ac voltage regulator.*

placing a battery bank across the energy storage capacitor in the dc power supply and increasing the capacity of the dc supply so that it can both charge batteries and operate the loads. When batteries are added to the electronic voltage regulator, we obtain a true uninterruptible power supply, a device that is described later in this chapter.

Electronic voltage regulators are much more expensive than the other types of line conditioners. If one is going to spend the money for an electronic ac regulator, one may as well pay the relatively small additional cost of a true uninterruptible power supply. Electronic voltage regulators are the only common device that will correct for deviations in the frequency of the mains. However, such frequency variations are rarely a problem with power from a commercial utility.

6. Motor-Generator Set

A very different way to obtain the functions of a line conditioner is shown in Fig. 20-6a. Isolation is obtained by having an electric motor turn the shaft of a generator: there is no electrical connection between input and output sides of this circuit. The shaft between the motor and generator can be made of insulating material to provide complete isolation. Attenuation of differential-mode noise is provided by the large moment of inertia of the rotating machinery: the rate of rotation of the generator cannot change rapidly. An additional advantage of this arrangement is that interruptions of

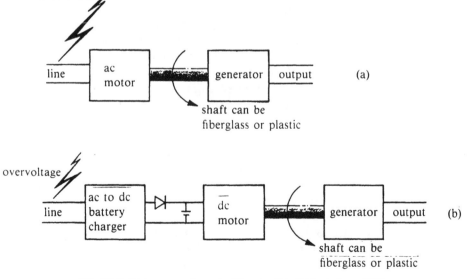

Fig. 20-6. *Motor-generator sets. The circuit in (b) includes a UPS.*

mains voltage, up to several hundred milliseconds in duration, do not significantly affect the output voltage, since the rotating parts of the generator, motor, and shaft act as a flywheel and store mechanical energy.

The motor/generator set can be modified, as shown in Fig. 20-6b, to be an uninterruptible power supply.

Motor-generator sets are widely used with loads with power consumptions greater than about 2 kVA, including mainframe computer systems.

7. Uninterruptible Power Supply (UPS)

The line conditioner is designed to solve most of the problems encountered with poor-quality ac power. However, it does not solve the problem of temporary loss of ac power (outages and flickers). To provide continuous power to critical loads when the mains are interrupted, an *uninterruptible power supply* (UPS) is required. There are two types of UPSs—a true UPS and a standby UPS. The true UPS is always on line supplying power to the critical loads. The standby UPS normally powers the loads from the mains, when the rms mains voltage falls to a low value, the standby UPS switches state and operates the critical loads from batteries and an inverter inside the UPS. Both types of UPS circuits are described below.

A UPS is particularly nice to have for a computer system, so the system does not "crash" during brief ac power outages. If the duration of the outage is longer than the UPS can operate the load, then the computer system can be shut down in an orderly way with all data or information preserved on magnetic disks or magnetic tape. Large computer systems require cooling, usually air conditioning, to prevent thermal damage to the electronics. It is usually uneconomical to purchase a UPS that has adequate capacity to operate the cooling system and computer. Since the computer should be operated without cooling for *less* than 20 minutes, this places a limit on the maximum practical size of the UPS for large computers.

If operation of critical loads during ac power interruptions of more than about 30 minutes is desirable, a diesel engine and a generator may be more economical than a UPS that is operated from lead/acid batteries. This is particularly true for large systems that include air conditioners.

a. True UPS. All disturbances on the mains can be removed by a true uninterruptible power supply (UPS). A block diagram of a true UPS is shown in Fig. 20-7a. An oscillator and power amplifier generate a disturbance-free sinusoidal waveform that delivers power to the critical loads. When the mains voltage is within reasonable limits, power from the mains is converted to dc, which charges the storage batteries and also operates the oscillator and power amplifier. When the rms mains voltage is too low (e.g., during a sag or outage), the batteries operate the oscillator and power amplifier. There is thus no interruption of power to the critical

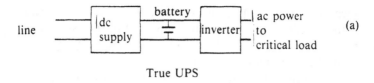

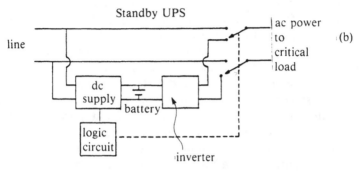

Fig. 20-7. Schematics of two types of UPSs.

loads, provided the battery voltage is adequate. Typical designs of true UPS units include a transformer in the battery charger and another transformer in the power amplifier. These transformers isolate the critical loads from the mains. Negative feedback in the power amplifier can be used to give good regulation of the rms output voltage and to minimize distortion of the output waveform. A typical true UPS can supply its full-rated output current during an outage with a duration of 10 minutes. Longer operating times are available by increasing the capacity of the battery bank (and the size of the battery charger).

Because the power to the critical loads is always generated in the true UPS, which is connected between the mains and the critical loads, disturbances in the power delivered to the building by the utility and disturbances created elsewhere in the building have no effect on the loads connected to the output of the true UPS. This makes the true UPS an ideal way to avoid disturbances on the mains. There is one major disadvantage to a true UPS: it is quite expensive. A true UPS for small computer systems must be designed to deliver between 200 and 800 W of power continuously with a mean time between failures of more than 10^4 hours. This is a difficult engineering assignment, and therefore a true UPS is quite expensive. A true UPS may cost more than the critical loads that are connected to it.

b. Standby UPS. A block diagram of a standby UPS is shown in Fig. 20-7b. The standby UPS has only one function: to supply power from batteries when the input voltage is less than about 105 V rms. Normally the output of the standby UPS is directly connected to the input. During times of low rms input voltage, a relay inside the standby UPS switches. The critical load then receives power from the batteries and the electronic inverter circuit inside the standby UPS. Power to the critical loads is interrupted for a few milliseconds while the relay in the standby UPS switches. This brief interruption is unlikely to cause problems, as discussed below. Typical standby UPS modules can operate their full rated load from internal batteries for an interval between about 10 and 30 minutes. When the input ac power is restored, the batteries are recharged.

By eliminating the requirement for continuous operation of the loads from the UPS, the cost of the standby UPS is much less than for a true UPS of the same maximum power rating and battery capacity.

When the standby UPS switches from the mains to the internal battery-operated inverter, the power at the output port is briefly interrupted. The duration of the interruption is dependent on the design of the particular standby UPS, but typical durations are between 1 and 10 ms. This interruption is equivalent to a loss of no more than one half-cycle of the mains, which should cause few problems, since electronic power supplies have internal capacitors that store energy. Most personal computers will operate for *at least* 10 ms after the mains are disconnected (Rosch, 1986). However, the interruption in ac power may interrupt the operation of certain "hair trigger" power supplies (Rosch, 1986). When this situation arises, it is likely to be less expensive to replace the electronic power supply with one more tolerant of brief outages than to replace the standby UPS with a true UPS.

8. Examples of Laboratory Tests

One model of ferroresonant line conditioner and two models of tap-switching line conditioners were tested for steady-state voltage regulation and dynamic regulation during sudden changes in line voltage or load current (Standler and Canike, 1986). All three line conditioners were rated by their manufacturers for service on single-phase 120 V rms mains with a frequency of 60 Hz and a load of 500 VA. Such line conditioners are suitable for operation of most desktop computer systems. Devices to test were chosen from several well-known brands in April 1985.

The circuit shown in Fig. 20-8 was used to determine the steady-state voltage regulation. A variable autotransformer was used to obtain various rms voltages. The input and output voltages of the line conditioner under test were measured with two digital true rms digital meters, which were interfaced to a computer for data collection and analysis.

Figure 20-9 shows the results when a 35 Ω load was used. At the nominal 120 V rms voltage, this load is 410 VA, or about 80% of the nameplate rating. For comparison, a curve is shown for "no device," which is just a

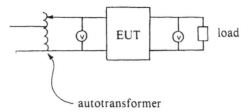

Fig. 20-8. Steady-state regulation tests.

piece of wire between input and output ports. The tap-switching line conditioner has four different output/input ratios, which correspond to four different taps. The ferroresonant line conditioner clearly provided the best regulation of output voltage in this test. The dashed lines at an output voltage of 110 and 125 V rms indicate somewhat arbitrary limits on minimum and maximum mains voltages for sensitive loads (Duell and Roland, 1981).

Figure 20-10 shows the results when a 150 Ω load was used. At the nominal 120 V rms voltage, this load is 100 VA, or about 20% of the nameplate rating. The curve for the tap-switching line conditioner is similar to that for a larger rms output current. However, the ferroresonant line

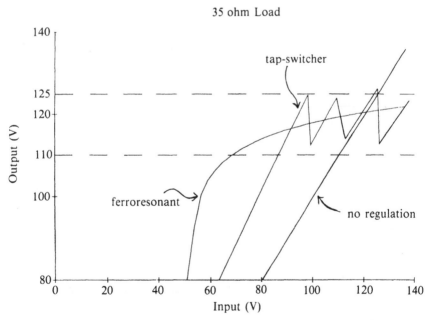

Fig. 20-9. Rms output voltage as a function of rms input voltage for three different devices with a 35 Ω load.

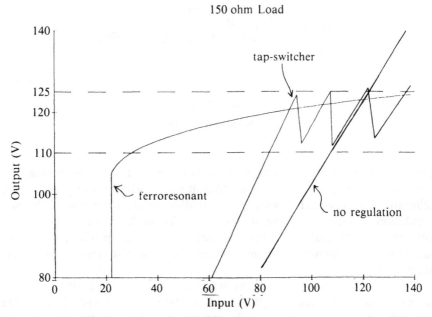

Fig. 20-10. *Rms output voltage as a function of rms input voltage for three different devices with a 150 Ω load.*

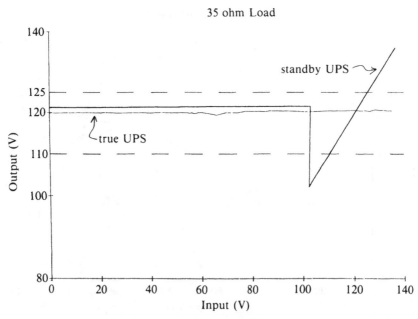

Fig. 20-11. *Rms output voltage as a function of rms input voltage for two types of UPSs with a 35 Ω load.*

conditioner is able to provide an acceptable output voltage for very small input voltages.

The tap-switching line conditioners had discontinuities in steady-state output voltage between 10 and 14 V rms caused by changing a tap. This is not nearly as smooth as the steady-state output voltage of the ferroresonant transformer shown in Figs. 20-9 and 20-10.

Two UPS units were also tested with the circuit of Fig. 20-8 and a resistive load equal to the nameplate rating of the UPS. The results are shown in Fig. 20-11. The standby UPS has an output voltage that is equal to the mains voltage for mains voltages greater than 102 V rms.

To test the response of the line conditioners to a sudden change in rms voltage, the circuit shown in Fig. 20-12 was used. An electromechanical relay was switched, and the voltage across the input port of the EUT was increased or decreased 26 V rms from the initial value. This simulated either the beginning or ending of a sag. The relay often produced noise and a notch for durations between 3 and 10 ms. Such noisy switching may be typical of field conditions. The voltage across the input and output ports of the EUT was monitored with a Tektronix 7612D digitizer with two Tektronix 7A13 differential amplifiers. A pair of probes were used to measure the voltage difference between the hot and neutral conductors.

Figure 20-13 shows the rest results when a tap-switching line conditioner is inserted in the circuit of Fig. 20-12 with a 150 Ω load, about 20% of the nameplate rating. The noise on the input of the tap-switching line conditioner is transferred to the output, because there is little attenuation of differential-mode noise at low frequencies.

Figure 20-14 shows the test results when a tap-switching line conditioner is inserted in the circuit of Fig. 20-12 with a 35 Ω load. Notice that the voltage across the output port is about the same as at the input port during the switching between 43 and 63 ms. This is to be expected, because the tap-switching transformer has fixed gain. The voltage at the input port is

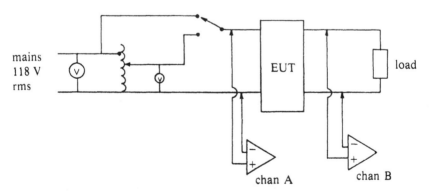

Fig. 20-12. *Dynamic line regulation test.*

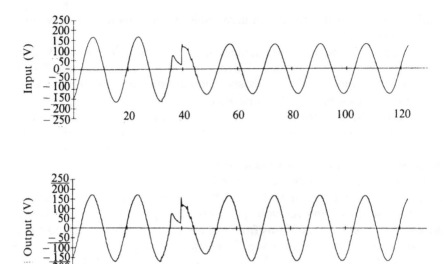

Fig. 20-13. *Dynamic line regulation test of a tap-switching line conditioner with a 150 Ω load. The input voltage was switched from 116 to 90 V rms.*

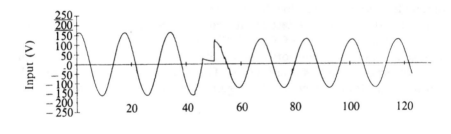

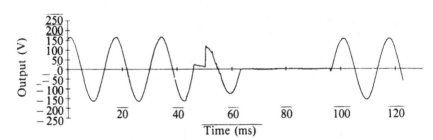

Fig. 20-14. *Dynamic line regulation test of a tap-switching line conditioner with a 35 Ω load. The input voltage was switched from 116 to 90 V rms.*

interrupted for about 4 ms at about $t = 48$ ms in Fig. 20-14. However, the output shows a complete outage for about 33 ms after the input voltage switched. The low rms input voltage during the switching may have caused the dc supply voltage for the control circuit to drop out of regulation, which would upset the control circuit. This is an example of how line-conditioning equipment can exacerbate a disturbance on the mains.

Figure 20-15 shows the test results when the ferroresonant line conditioner is inserted in the circuit of Fig. 20-12 with a 35 Ω load. Notice that the voltage across the output port is much smoother than at the input port during the switching between 35 and 45 ms. The ferroresonant transformer is a tuned circuit that is resonant at 60 Hz, so it easily rejects disturbances at other frequencies. The reduced output voltage between 40 and 60 ms is due to the low rms input voltage.

Figure 20-16 is an expanded view of a similar test with the ferroresonant transformer. Notice that the switching noise at the input side of the ferroresonant transformer is greatly attenuated on the output side.

The line conditioners were also tested with a variety of different overvoltages (Standler and Canike, 1986, pp. 128–179). The ferroresonant line conditioner had better performance during the overstress tests than the tap-switching line conditioners: not only did the ferroresonant unit survive without degradation (unlike the tap-switching models), but the voltage across the output of the ferroresonant was usually smaller than or about the same magnitude as the voltage out of the tap-switching models for a given overstress applied to the input of the line conditioners.

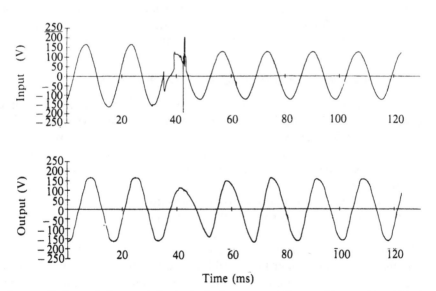

Fig. 20-15. *Dynamic line regulation test of a ferroresonant line conditioner with a 35 Ω load. The input voltage was switched from 116 to 90 V rms.*

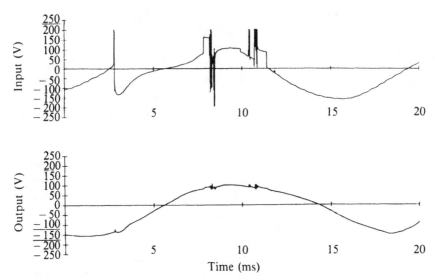

Fig. 20-16. *Dynamic regulation test of a ferroresonant line conditioner with a 35 Ω load. The input voltage was switched from 90 to 116 V rms.*

In the author's opinion, the ferroresonant line conditioner generally performed better than the two tap-switching line conditioners in these tests. In most tests the differences in performance between the two types of line conditioners were small and might not correspond to an appreciable difference in a practical situation. The ferroresonant line conditioner is the device of choice for operation of critical systems during brownouts and severe sags, owing to its performance at low rms input voltages.

D. COORDINATION OF UPS AND LINE CONDITIONER

There are a number of situations where the combination of a line conditioner and UPS is desirable. There are two possible ways to use a UPS with a line conditioner, which are shown in Fig. 20-17. The two arrangements differ in whether the line conditioner is upstream or downstream from the UPS. When the UPS should be upstream from the line conditioner is discussed first.

If the UPS has a nonsinusoidal output waveform (square wave inverters are often used in less expensive UPS models), then a ferroresonant line conditioner could be placed downstream from that UPS, as shown in Fig. 20-17a, to provide a more nearly sinusoidal waveform to the load.

By using a ferroresonant line conditioner downstream from the UPS, as shown in Fig. 20-17a, more nearly continuous power to the load is obtained when the standby UPS is changing state, since the resonant output circuit continues to oscillate for a few milliseconds in the absence of input power.

(a)

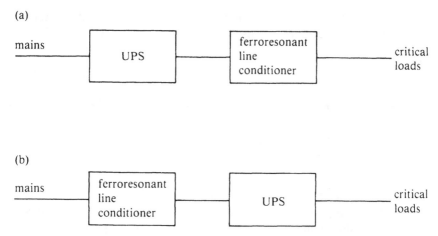

Fig. 20-17. Two ways to use a line conditioner and UPS together.

However, placing a ferroresonant line conditioner downstream from a standby UPS has a serious disadvantage. When a ferroresonant transformer is turned on, an abnormally large current is drawn to saturate the core of the transformer. When an outage occurs, the ferroresonant transformer will draw an abnormally large current from the inverter in the standby UPS when the output of the standby UPS is switched to the inverter. This combination of the large magnetizing current in the ferroresonant transformer and the load current may trip the circuit breaker in the inverter. Similarly, in normal startup the UPS is turned on first, then the ferroresonant line conditioner. The large initial current required by the ferroresonant transformer *may* cause the circuit breaker at the UPS output to trip when the ferroresonant line conditioner is turned on.

The circuit of Fig. 20-17b is preferable when the mains voltage is often less than 110 V rms during brownouts or sags. By including the line conditioner with its voltage regulation upstream from the UPS, the UPS sees essentially normal rms voltage during conditions with low mains voltages. This prevents the UPS batteries from being drained during brownout conditions, which may have a duration of several hours. This also prevents switching of a standby UPS during momentary sags, which reduces the wear and tear on the standby UPS. Since low rms voltages are the most common disturbance in most environments, the circuit of Fig. 20-17b *would be preferred* in most situations.

E. RATIONAL PROTECTION

Nash and Wells (1985) suggest that a selection of power conditioning equipment should be "based on the requirements of the equipment to be protected and the power-line problem that can be expected." There are three approaches that might be used (Nash and Wells, 1985):

1. Treat all installations the same way (based on the assumption that disturbances on the mains are universal).
2. Consider each installation as a special case, examine data on disturbances at the site, examine data from the manufacturers of the equipment to be protected, etc.
3. "If it ain't broke, don't fix it."

The first approach will be too expensive if high standards of reliability are imposed needlessly on noncritical systems. The second approach has the difficulty that the manufacturer of the equipment to be protected may not have (or may not wish to release) data on the vulnerability or susceptibility of the equipment. Since comprehensive testing by the user is too expensive, the second approach may not be realistic, even though it looks good on paper. Nash and Wells (1985) endorse the third approach as the "only logical" alternative when the quality of power at the site is unknown and "the sensitive equipment to be protected has ill-defined power requirements." However, they also say that "the education received this way is fine as long as the cost of the tuition is not too high." It appears that the third approach is analogous to waiting to be injured before deciding to purchase a first-aid kit. It is well known that transient overvoltages do exist and can damage equipment. It is also well known that outages and severe sags exist and cause upset of computers. Effects of disturbances on the mains should be considered *before* sensitive equipment is installed.

A proper selection of power conditioning equipment should use a judicious combination of the first two approaches. There is a large amount of general data in the literature about disturbances on the mains (see Chapter 3). It is well-known that reductions in rms voltage (sags, brownouts), interruptions of mains power (flickers, outages), and transient overvoltages occur. Why wait for them to occur at a particular site before deciding to install SPDs and conditioning appliances?

Some engineers recommend routinely monitoring for disturbances at a site with critical equipment. It is unlikely that the "true" rate of disturbances can be determined during a brief monitoring interval, because disturbances are rare events at most sites. If a survey of disturbances at a site shows no disturbances of a particular type, then it is *not* justified to conclude that type of disturbance will not occur at that site (Martzloff and Gruzs, 1987). Conditions at a site may change in the future, with a resulting increase in the rate or severity of disturbances. Since the typical user is not aware of all of the loads connected to the mains in the vicinity, changes in the environment are unpredictable by the user. Monitoring is a valuable research tool. Monitoring may also be a useful diagnostic tool for users who are experiencing unexplained problems with electronic equipment. However, the author believes that routine monitoring of sites is unnecessary.

Before overreacting to possible threats and automatically specifying protection from all possible disturbances, it is appropriate to consider both the likelihood of a particular type of disturbance and the cost of recovery from damage and upset if the system is not protected from that disturbance. This weighted average of the cost of a partially unprotected system should be balanced against the cost of providing the protection. It is not necessary (or even possible) to do an exact calculation, but even a crude estimate can provide useful guidance.

F. RECOMMENDED PROTECTION METHOD

A cost-effective approach to protection is to use a combination of three different modules, each of which removes some (but not all) disturbances and each of which is relatively inexpensive. A block diagram of the recommended method for protecting electronic equipment is shown in Fig. 20-18. Each of the three modules is discussed in the following paragraphs. Recommendations are also given for when it may be appropriate to include or omit a particular module. It is not justifiable to include all three modules for all electronic systems.

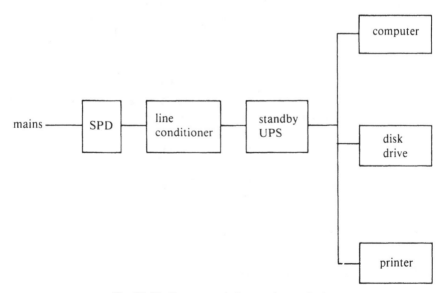

Fig. 20-18. *Recommended protection method.*

1. Overvoltage Protection

It is recommended that all systems be protected from transient overvoltages. Such protection is inexpensive and easy to accomplish. Three metal oxide varistors can be connected as described in Chapter 19. When minimum cost is more important than complete protection, the varistor between the hot and grounding conductors can be omitted.

Adding a low-pass filter to the overvoltage protection module, as described in Chapter 19, may increase the protection from large magnitudes of dV/dt, but it also increases the cost. The low-pass filter is probably unnecessary if a line conditioner is connected downstream from the overvoltage protection.

By placing the overvoltage protection module upstream from the other two modules, electronic circuits in the line conditioner and standby UPS are protected from damage by transient overvoltages.

2. UPS

The most expensive module in the protection system shown in Fig. 20-18 is the standby UPS. If continuous operation of loads during severe sags, flickers, and outages is not required, then the UPS can be omitted.

3. Line Conditioners

Many electronic devices use switching power supplies that are designed to operate properly over a wide range of rms mains voltages (e.g., from 80 to 130 V rms). These switching power supplies receive little benefit from the voltage regulation in the line conditioner. If a UPS is part of the system, then a line conditioner should probably also be included, since the voltage regulation in the line conditioner would still permit operation during brownouts, without draining the batteries in the UPS. However, if the UPS is omitted, one might also omit the line conditioner, provided all of the loads can operate satisfactorily at reduced voltages.

The manufacturer's specifications may be checked to determine the suitability of operation with low mains voltage. One should be careful to check all of the devices in a system. Loads that consume relatively large amounts of power are likely to have a switching power supply because of its greater efficiency. However, loads that consume relatively small amounts of power (e.g., an external modem) are likely to use a regular linear power supply, which may not operate satisfactorily at low rms mains voltage.

4. Does This Scheme of Protection Work?

Standler (1988d) reported that a group of five desktop computers were operated for a total time of over 1.8×10^4 h with neither damage nor upset related to mains disturbances. These systems were operated routinely during local thunderstorms, brownouts, and other adverse conditions. Each of these systems was protected with all three of the following: an overvoltage-protected low-pass filter, either a ferroresonant or tap-switching line conditioner, and a standby UPS.

5. Use of Overvoltage-Protected Low-Pass Filter

Providing critical loads, such as computer systems, with protection from transient overvoltages and noise would be expected to reduce the upset and damage rate for the systems. Many kinds of appliances (e.g., ac voltage regulators, isolation transformers, and line conditioners) have been recommended by engineers to protect vulnerable equipment from disturbances on the mains. These appliances are expensive: a unit that is rated for service at 120 V rms, 500 VA costs at least $400. In many applications regulation of the ac mains voltage is no longer critical. For example, many switching power supplies that are intended for use on nominal 120 V rms mains will operate reliably with any input voltage between 80 and 130 V rms. Typical desktop computer systems use switching power supplies and do not require a regulated ac mains voltage. When ac voltage regulation is not needed, a low-pass filter that includes protection from transient overvoltages may be an acceptable substitute for more expensive devices such as line conditioners. When isolation at dc and at the frequency of the mains is not required, the common-mode choke in a low-pass filter may be an acceptable substitute for noise reduction by line conditioners or isolation transformers. Under these conditions, an overvoltage-protected low-pass filter, as described in Chapter 19, may be a less expensive substitute for other appliances.

G. ZONES OF PROTECTION

It may be decided to provide protection for each critical system against one or more of the following disturbances: transient overvoltages, variations in rms voltage (sags, brownouts, swells), or interruptions of power (flickers, outages). Regardless of the type of power conditioning selected, it is important that *all* of the loads in a particular system be connected to the same conditioned power. A typical small computer system requires ac power for the computer, CRT monitor(s), and printer. Some systems also have a plotter, external disk drive(s), external modem, or laboratory instruments. All of the devices that are connected to the computer should have the same conditioned ac power.

Consider a violation of this rule: an external disk drive is connected to the raw mains, and the computer is connected to conditioned power.

Transient overvoltages on the mains threaten to destroy the disk drive. Furthermore, transient overvoltages could propagate from the disk drive to the computer on the data interface cable. In this way, the computer is vulnerable to damage from transient overvoltages on the mains, although it is operated from conditioned power. Upset is possible when there is an outage. The computer will continue to function (operating from power supplied by the UPS), but the disk drive will not function. With modern disk-based operating systems, the computer's operation will be greatly limited. Whenever the computer attempts to read or write information from or to the disk, the system will "hang," because the disk will not respond. Any data that were being written by the computer to the disk at the time of the outage will probably be lost.

The simplest solution is to have one large line conditioner or UPS that serves all of the computer system. However, if some of the peripherals are to be located more than about 4 m from the computer, separate line conditioners for the remote peripherals may be advisable.

In addition to conditioning the mains that supply power to all of the devices in the system, there must be overvoltage protection on all data lines that go outside the system. Typical examples of such data lines are the telephone line that is connected to the moden in a computer system, local area networks that allow transport of information between workstations, and cables to remote peripherals (e.g., high-speed line printers, plotters). Protection of data lines is discussed in Chapter 17.

Any equipment that is powered from a different mains branch circuit, isolation transformer, motor-generator set, line conditioner, or UPS is declared to be in a different zone. Although equipment in each different zone should be adequately protected from transient overvoltages on the mains, the value of ground potential may vary between zones (owing to different values of surge currents and the different values of resistance and inductance of the different grounding conductors). Differences in ground potential appear as common-mode voltages on data lines and may cause damage or upset of equipment.

Standler and Canike (1986, pp. 183–185) proposed the following general rule for identifying when it is desirable to install overvoltage protective devices. A workstation is a cluster of equipment that is connected by data interfaces and power cords. Any conductors that go outside of the zone surrounding each workstation need protection. Fig. 20-19 shows a distributed system. The bold lines in Fig. 20-19 are the ac power conductors; the dashed lines separate three zones. The data lines that go to the plotter in Fig. 20-19 need protection at both ends of the cable, because the plotter is outside of the zone for the workstation.

It is unnecessary to protect data lines that are only connected between devices that are in the same zone. For example, it is unnecessary to protect the interface ports for the printers shown in Fig. 20-19.

This general rule also applies to the mains connection, since the mains always goes outside of the zone surrounding the workstation.

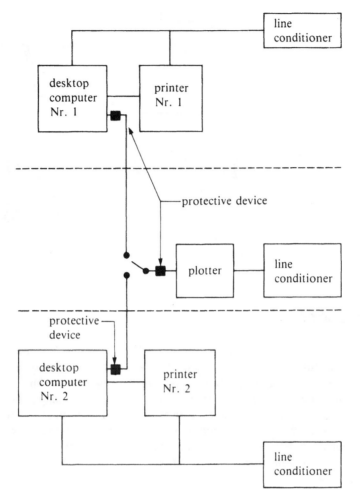

Fig. 20-19. *Zones for protection of a distributed system.*

H. CONCLUSION

It is recommended that all systems be protected from transient overvoltages by metal oxide varistors, since overvoltages can damage electronic equipment. If a system is also susceptible to upset by *commonly occurring* changes in rms voltage, the addition of a line conditioner and possibly a standby UPS is also recommended. If continuous operation of a system is critical, then the combination of varistors, line conditioner, and a standby UPS is recommended.

21

Circuits That Avoid Upset

A. INTRODUCTION

Nearly all overvoltage protection circuits allow a small fraction of the incident transient overvoltage, called the *remnant*, to propagate to the protected devices. In a properly designed protection circuit, the remnant will have insufficient energy, current, or voltage to damage protected devices. However, there is still concern that the remnant could be misinterpreted as valid data. Such misinterpretation is called *upset*. The threshold for upset is often within the normal range of input voltages to the system. Therefore, one cannot discriminate against upset on the basis of voltage levels alone.

B. UPSET AVOIDANCE

Some practical suggestions that can be used to reduce the possibility of upset are discussed below.

Brown et al. (1973) recognized that decreasing the response of the circuit to high-frequency signals would decrease the vulnerability of the circuit to upset by fast transient overvoltages such as EMP. There are four particularly helpful suggestions that can be made:

1. A low-pass filter (with a cutoff frequency of 10 kHz or less) can be inserted in series with the input data line.
2. Analog interface devices with the smallest acceptable gain-bandwidth product should be used.

3. Relatively slow digital logic devices should be used in interface circuits.
4. Edge-triggered logic should be avoided in digital interface circuits.

Brown et al. (1973) also suggested the use of error detection codes in digital transmissions to decrease the vulnerability to upset. This is a special case of bandwidth reduction, since transmitting redundant information in an error detection code reduces the amount of information that can be transmitted per unit time.

Use of balanced, differential line and input amplifier with both large common-mode rejection ratio (CMRR) at frequencies above 1 MHz *and* a large range of permissible common-mode input voltage will decrease vulnerability to common-mode transient overvoltages. Although this is a good suggestion, it does nothing to avoid upset from differential-mode transient overvoltages.

C. DIGITAL CIRCUITS

Digital circuits appear to be more easily protected against upset than analog circuits. There are two reasons why this is so:

1. The voltage in a digital circuit has only two valid states, high and low. Values of voltage between these two states can be recognized as inappropriate. This is in contrast to an analog circuit for which a proper signal voltage is allowed to vary continuously over some range. In practice, many analog signals when converted to a digital format require 12 bits (which is equivalent to 4096 different states). If the analog signal has a range of 20 V (e.g., -10 to $+10$ V), the voltage increment per state is about 0.005 V. This is a much smaller value per state than for digital signals, where a margin of 1 V (or more) separates the two states.
2. Digital data can be easily stored in a memory and transmitted redundantly with a large time delay between transmissions. If both transmissions have identical content, there is a negligible chance that a transient overvoltage corrupted both transmissions in the same way. There is no convenient analog memory that would permit a corresponding operation.

Noise on digital lines can be rejected by using interface devices with Schmitt-trigger inputs. A typical 7414 or 74LS14 Schmitt-trigger digital input requires that the input signal have a change in voltage of at least 0.8 V before the output voltage will change. This allows the system to reject noise that has a magnitude less than this value.

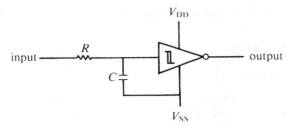

Fig. 21-1. *Use of a low-pass filter on the input of a Schmitt trigger gate.*

The use of a low-pass filter with digital logic is particularly simple if logic circuits of the complementary metal oxide semiconductor (CMOS) family are used. The large input impedance of the CMOS gate, which can be modeled as a $10^{12}\,\Omega$ resistance in parallel with a $5\,\text{pF}$ capacitance, constitutes a negligible load for the output of a simple RC low-pass filter as shown in Fig. 21-1. The value of R can be as large as $1\,\text{M}\Omega$ without introducing complications in the circuit design. If electrolytic capacitors are used for C, solid tantalum units with a capacitance value of less than about $10\,\mu\text{F}$ would be preferred, owing to concerns about dc leakage current. The two power supply connections to the CMOS logic are labeled V_{DD} and V_{SS} in Fig. 21-1, where V_{DD} is the more positive supply. Typical values are

$$V_{\text{DD}} = 5 \text{ to } 10 \text{ V}$$

and

$$V_{\text{SS}} = 0 \text{ (ground)}$$

The output from the low-pass filter no longer has the sharp edges that are characteristic of a good digital signal. Therefore, the output of the low-pass filter must be connected to the input of a Schmitt-trigger gate to properly interpret the degraded digital signal. Incidentally, a large value of R will provide substantial overvoltage protection for the CMOS input, as discussed in Chapter 17.

Brown et al. (1973) suggested the use of a delay line and an AND gate to verify that digital input data are stable, as shown in Fig. 21-2. A Schmitt

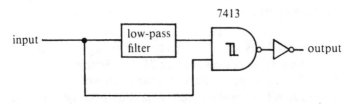

Fig. 21-2. *Simple circuit to detect a stable logic signal and avoid upset.*

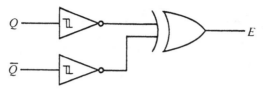

Fig. 21-3. *Simple circuit to detect complimentary data: E is high if Q and Q̄ are identical (Buurma, 1978).*

trigger input circuit must be used for the logic device, owing to the degradation in rise and fall times of the signal by the low-pass filter. The Schmitt-trigger input also provides noise immunity, as discussed above. The output of the circuit in Fig. 21-2 is high only if the input has been stable in the high state. When this circuit's output is in the low state, one cannot tell if the input signal is supposed to be in the low state or whether a high state has been corrupted by a negative-going transient. Thus this circuit is appropriate for use in systems where a high state indicates a critical operation (e.g., launch weapons) and the low state indicates a routine or benign condition. This circuit does not guarantee immunity to upset: a long-duration transient could still corrupt the data at the output.

Valid data on a balanced digital line require that the two lines have complementary states. Buurma (1978) described the use of an exclusive-or gate with Schmitt-trigger inputs to detect invalid conditions on a balanced digital data line, as shown in Fig. 21-3. Whenever the output of the exclusive-or gate is in the low state, the data are invalid. Conversely, when the output of the exclusive-or gate is in the high state, the data are valid. The circuit in Fig. 21-3 is not available in a single integrated circuit package, but it certainly could be fabricated in a single package if the market were to demand it. Again, this circuit does not guarantee immunity to upset: common-mode transients on Q and $\bar{Q}$ could still corrupt the data at the output E in Fig. 21-3.

D. COORDINATION OF OVERVOLTAGE PROTECTION AND UPSET AVOIDANCE

Since overvoltage protection is desirable on all interface circuits, it is reasonable to consider including a transient overvoltage detection circuit with the protection circuit(s). The output of the overvoltage detection circuit would be used to inhibit acceptance of data during or immediately after a transient.

Knight (1972) suggested a novel transient protection circuit similar to that shown in Fig. 21-4. The avalanche diode and light-emitting diode (LED) serve as a transient overvoltage detection circuit. The light from the LED causes the switch on the right-hand side of Fig. 21-4 to close, thus

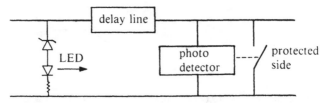

Fig. 21-4. Optical transmission path circumvents finite response time of switch (Knight, 1972).

protecting the load. The delay line compensates for the response time of the photodetector and switch. Although Knight (1972) advocated this circuit for protection applications, it is a simple modification to remove the shunt switch and use the output of the photodetector to indicate the presence of overvoltages. This output could be used to inhibit the acceptance of data.

The circuit shown in Fig. 21-4 responds only to positive overvoltages; however, bipolar circuits are a simple extension, as shown in Fig. 21-5. In the circuit of Fig. 21-5, avalanche diodes D_1 and D_2 provide the final stage in a circuit for protection from overvoltages (for simplicity, the earlier stage(s), which may include a spark gap, are not shown). Diodes D_1 and D_2 are essential, because diodes D_3 and D_4 cannot be relied on for protection. The shunt path through D_3 and D_4 includes a resistance R_1, and probably considerable parasitic inductance, that defeats any protective function. The pair of LEDs, D_5 and D_6, along with phototransistor Q_1, are contained inside a single package, (e.g., model H11AA1, which is available from several manufacturers).

The value of R_1 is determined so that the LEDs are not damaged by overvoltages. The circuit will be designed to operate one LED, D_5 or D_6, at 10 mA and 1.1 V. Let V_U denote the magnitude of the voltage at which the upset detection circuit will respond. Of course, the magnitude of the

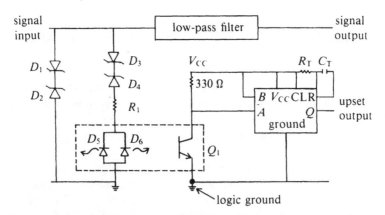

Fig. 21-5. Isolated electronic circuit to inhibit system and prevent upset during overvoltages.

clamping voltage of the series combination of protection diodes D_1 and D_2 must be greater than V_U, or D_3 and D_4 will not conduct. Let V_z denote the breakdown voltage of the series combination of diodes D_3 and D_4. The following relation is obtained from Kirchhoff's voltage law:

$$V_z + (10\,\text{mA} \times R_1) + 1.1\,\text{V} = V_U$$

This equation has two unknowns, V_z and R_1.

The LEDs must not be damaged by a large magnitude of current during a transient overvoltage. Let V_c denote the magnitude of the clamping voltage of the series combination of protection diodes D_1 and D_2. The maximum current in either LED should not exceed about 60 mA; this occurs at about 1.3 V across the LED. Kirchhoff's voltage law can be applied to obtain:

$$V_z + (60\,\text{mA} \times R_1) + 1.3\,\text{V} = V_c$$

This equation also has two unknowns, V_z and R_1. These two equations can be solved simultaneously to complete the design.

When one of the LEDs is illuminated, the phototransistor conducts, and the voltage at the A input terminal of the 74221 monostable multivibrator goes to the low state. This triggers the multivibrator, which provides a positive pulse at its output terminal. The pulse duration is a function of a timing resistance, R_T, and capacitance, C_T. The multivibrator automatically resets itself at the end of the pulse.

Optical isolators with a phototransistor output have a slow response. The low-pass filter or delay line in the circuit shown in Fig. 21-5 should have a delay time of at least several microseconds in order to be certain that the output of the upset detection circuit has responded before the remnant has propagated to the output port of the delay line.

One could substitute an optical isolator with a photodiode output and decrease the response time to less than $0.1\,\mu\text{s}$. Although this greatly increases the cost of the optical isolator, the total circuit cost may decrease owing to the smaller delay line that can be used with a faster optical isolator. Suitable optical isolators with a pair of photodiodes and internal amplifiers include the Hewlett-Packard HCPL-2630.

An alternate overvoltage detection circuit might have a PIN photodiode, D_1, detect luminosity from a spark gap, as shown in Fig. 21-6. To avoid upset of the electronics from the large value of dI/dt in the spark gap, a fiber optic cable could be used to route the light from the spark gap to the photodiode. The signal from the photodiode is detected with a high-speed comparator. Resistors R_2 and R_3 bias the comparator so that the output is normally in the low state. When the photodiode conducts, the voltage across R_1 is greater than the voltage across R_3, and the output of the comparator switches to the high state. Typical values of the resistors are

$$R_1 = 2\,\text{k}\Omega \qquad R_2 = 4.7\,\text{k}\Omega \qquad R_3 = 1\,\text{k}\Omega$$

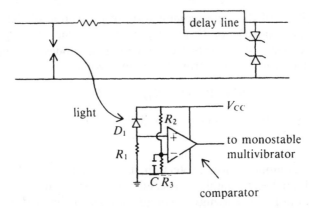

Fig. 21-6. *Isolated circuit to detect light from spark gap and inhibit system to prevent upset.*

The capacitor C is a 0.01 μF ceramic bypass capacitor that is included to ensure a stable voltage at the inverting input of the comparator.

The spark gap in the circuit in Fig. 21-6 must have a transparent glass case (e.g., Siemens button-type gap) in order to flood the photodiode with intense light during a transient overvoltage. Also, the photodiode should be specified to have adequate response to blue light, since this color predominates when the spark gap operates in the arc region.

E. OVERVOLTAGE DETECTION ON MAINS

Overvoltages on the mains can be detected with the circuit shown in Fig. 21-7. The varistor, V_1, is a metal oxide varistor with a clamping voltage of less than 1 kV at the maximum expected surge current. Avalanche diodes

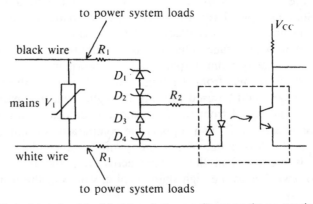

Fig. 21-7. *Isolated circuit to detect transient overvoltages on the ac supply mains.*

D_1 and D_2 may have a nominal breakdown voltage of 200 V and a steady-state power rating of at least 5 W. These diodes provide the discrimination between normal mains voltages and a transient overvoltage. Avalanche diodes D_3 and D_4 and resistor R_2 protect the LEDs in the optical isolator from excessive current. Suitable types of avalanche diodes for D_3 and D_4 have a breakdown voltage of 6.8 V and a steady-state power rating of 0.5 W. Diodes D_3 and D_4 do not need to have large power ratings, since the current in them is less than the current in diodes D_1 and D_2 (owing to the shunt path through the LEDs), and the voltage across D_3 and D_4 is much less than the voltage across D_1 and D_2. The optical isolator is the same type as discussed above in Fig. 21-5, such as model H11AA1.

The value of resistors R_1 is determined by the maximum voltage across the varistor and the maximum tolerable current in D_1 and D_2. Because this circuit is isolated from ground and the energy deposited in the resistance is large, the resistance has been divided into two units of equal value. If the varistor maintains the voltage at less than 1 kV and D_1 and D_2 can tolerate a current of 25 mA (5 W/200 V = 25 mA), then

$$2R_1 \geq 1 \text{ kV}/25 \text{ mA}$$

or

$$20 \text{ k}\Omega \leq R_1$$

Resistor R_2, along with diodes D_3 and D_4, protects the LEDs from excessive current. The maximum allowable steady-state LED current, about 50 mA, is greater than the maximum steady-state current in diodes D_1 and D_2. Therefore, all of the current in D_1 and D_2 can pass through the LEDs. This maximizes the current in the LEDs, and thus makes the optical isolator faster responding. The series combination of diodes D_3 and D_4 will not conduct at a voltage of less than

$$[(6.8 \times 0.95) + 0.6] \text{ V}$$

or about 7.0 V. A $\pm 5\%$ tolerance on, the breakdown potential of 6.8 V is taken into account. The value of R_2 is determined so that when 25 mA passes through D_1 and D_2, there is just 7.0 V across D_3 and D_4. Since the LED has a drop of about 1.1 V when conducting, the value of R_2 is 240 Ω.

The output of the phototransistor in Fig. 21-7 can be connected to a multivibrator as was shown in Fig. 21-5. The presence of low-pass filters downstream from the varistor V_1 is standard in mains overvoltage protection circuits (see Chapters 13 and 19). These low-pass filters will delay the remnant. Tests need to be performed in the laboratory to verify that the delay time of the low-pass filter is at least as long as the response time of the overvoltage detection circuit of Fig. 21-7.

F. APPLICATION OF UPSET AVOIDANCE CIRCUITS

In the circuits shown in Fig. 21-5 through 21-7, a delay line or low-pass filter is necessary between the overvoltage sensor and the connection of the input line to the system, in order to inhibit the system *before* the peak of the overvoltage reaches the system. In addition to inhibiting the system, one might also activate an electromechanical latching mechanism that would signal maintenance crews to check for possible damage caused by the overvoltage.

Since the overvoltage sensors in Figs. 21-5 through 21-7 depend upon voltage discrimination, there will be a "window of susceptibility" for magnitudes of input voltage that exceed the upset threshold but are less than the level needed to activate the overvoltage sensor. This window of susceptibility makes the upset problem very difficult to solve.

It is important to recognize that *not all* input conductors to a system need to be connected to an overvoltage sensor. Input lines should be connected to an overvoltage sensor if they either (1) are connected to transducers that are illuminated by exterior electromagnetic fields or exposed to direct injection of current by lightning, (2) carry data of a particularly critical nature, or (3) have relatively lengthy or poorly shielded transmission lines. The outputs of multiple overvoltage sensors should be connected to an OR gate so that a single digital signal is obtained to inhibit the system.

Hardening a circuit to prevent upset needs to be done during the design of the equipment to be protected. Effective upset hardening is unlikely to be accomplished by later connecting a box to each input. There is general agreement that hardening a system against upset is much more difficult than protecting it from destructive overvoltages.

G. UPSET DUE TO INTERRUPTION OF MAINS

An electronic system can also be upset by transients on the ac supply mains. If spark gaps are used for protection against overvoltages on the mains, the line current must be interrupted for several cycles to stop the power follow in the spark gap (see Chapter 7). This is why the lights in buildings flicker or blink during thunderstorms: lightning hits an overhead power line and drives spark gap(s) into the arc mode (which puts an effective short circuit across the line), and the arc is interrupted by automatically reclosing circuit breakers.

To avoid these brief losses of ac power, one could prohibit the use of spark gaps (and other devices that inherently have a follow current problem) as SPDs on power lines. Metal oxide varistors offet an effective substitute for small and moderate surges. If the power line is buried underground, the very large surges that are associated with direct lightning strikes are largely avoided. However, for overhead power lines, spark gaps appear to be necessary for economical protection of the distribution system. Moreover, there are natural spark gaps such as insulator flashover to ground and arcs between the line and adjacent trees.

Uman and Standler (1981) described a circuit to detect either electric field changes from lightning or brief interruptions of power on the ac supply mains. When either event was detected, a relay changed state. This relay could be used to sound an alarm or to disconnect vulnerable equipment (e.g., short-circuit vulnerable transducers at the end of a long cable).

Digital computers are particularly susceptible to upset from unanticipated interruption of power. When the mains voltage is interrupted, the filter capacitors in the dc power supplies inside the computer will be the only source of power to keep the computer's memory operational. Significant upset protection can be obtained by making these filter capacitors large enough to operate the entire computer for at least 1 second. Alternately, an "uninterruptible power supply" (UPS) can be connected between the power line and the computer. The operation of a UPS is described in Chapters 18 and 20 for dc buses and ac mains circuits, respectively. A typical commercial-grade UPS for the ac mains that can supply 3 A at 120 V rms for 20 minutes costs about $1500 in small quantities. Use of a UPS is an economical way to avoid upset during outages with a duration of a few minutes. Protection against outages with a duration of no more than 1 or 2 seconds may be obtained by less expensive means.

Another technique to avoid upset due to temporary loss of ac power is to use a voltage-responsive upset detection circuit to switch the load into a standby state. For example, computers with CMOS static random access memory (RAM) can *retain* their data for weeks or months when powered from small Ni-Cd or Li batteries that can be mounted on the printed circuit board that contains the memories. This technique is used in the "continuous memory" feature of some hand-held calculators. The quartz-crystal oscillator and counter that determines the date and time in common desktop computers is also operated from batteries when the computer is off. This saves the operator the bother of entering the date and time whenever the computer is turned on. Double-layer electrolytic capacitors, with capacitance values of nearly 1 F, can also be used to provide operation of CMOS static RAM when power is off for a few days.

Part 4

Validating Protective Measures

22

Testing

A. INTRODUCTION

1. Types of Tests

When different types of objects are tested, the goals and procedures are naturally different. The object being tested may be equipment that is being tested for vulnerability or susceptibility to overvoltages. The object being tested may be a surge protection module that contains one or more surge protective devices (SPD). It may be of interest to measure the surge current (input) and voltage across the protected port (output). Or the object being tested may be a component such as a spark gap, varistor, or avalanche diode. Tests of components usually measure current and voltage to determine the V-I characteristic curve, and from analysis of these data parameters in the model of the device are determined, such as α and V_N for a varistor. It is conventional to call the object being tested *equipment under test* (EUT) or *device under test* (DUT).

Equipment should be tested for both damage and upset conditions. Surge protective devices or modules to be used to protect unspecified loads only need to be tested for their performance during overvoltages. Without having a particular load attached, one cannot test an SPD for ability to avoid either damage or upset.

There are four types of tests, each of which has a different purpose: (1) design tests, (2) production tests, (3) qualification tests, and (4) diagnostic tests.

When a prototype device or equipment has been developed, it needs to be tested to characterize its performance and to find the stresses required for failure. This information from such *design tests* is often used to improve the design, so weaknesses and defects should be characterized. The final design (production model) should be tested thoroughly. The results of design tests can be compared with design goals, but the result of these tests is usually information for design engineers rather than a pass/fail verdict.

A representative sample of the product being manufactured in large quantities is usually tested to assure that the production process is adequate. These are called *production tests.* It is common to use the results of production tests to declare a batch of products as acceptable and ready for shipping to customers. One might also use the results of these tests to detect trends in parameter shifts that might indicate future production problem.

Qualification tests are done to verify that a product meets its specifications. The result of a qualification test is usually evaluated on a pass/fail basis against requirements of the specifications, including tolerances.

Diagnostic tests are done to determine the cause of unexpected events, such as failures of equipment in the field. The goal of diagnostic tests is to answer the question, "Why did it happen?" The equipment or device cannot pass or fail this test: it has already failed.

2. Test Plan

Before attempting high-voltage testing, the engineer should formulate a test plan. Such a plan may as well be written since it can also be used in the final report, which is nearly always written. There are several key parts to a test plan:

1. define purpose of test
2. choice of waveforms, peak values
3. which conductors to test?
4. EUT operating during test?
5. statistical considerations
6. criteria for pass/fail
7. laboratory technique
8. safety

The purpose of the test was discussed above. Parts 2 to 6 are discussed in the following sections. Laboratory technique is discussed in Chapter 23, and safety is discussed in Chapter 24.

B. TEST WAVEFORMS

1. Transient Control Levels

There is a wide variety of overstress waveforms to which a device or system may be exposed in service. It is difficult, if not impossible, to test a product with a collection of overstress waveforms that faithfully represent all possible overstresses in the field. To make overstress testing a tractable task, Fisher and Martzloff (1976) proposed the concept of *transient control levels* (TCL).

The basic concept in establishing TCL is to select a few test waveforms, possibly just one. The waveshapes are somewhat arbitrary but contain features that are representative of real events. This avoids the problem of duplicating an unknown environment.

Determining the TCL is simple. Test are made with a series of values of open-circuit voltage levels starting at 100 V, increasing in a series of 1, 2, or 5 times an integer power of 10, until either damage or upset occurs or 10 kV is reached. The final value of open-circuit voltage for which neither damage nor upset occurs is the TCL value for that equipment: operation of equipment at the TCL or at smaller magnitudes of peak voltage should not cause problems. This gives a simple rating for the vulnerability and susceptibility of electronic equipment and systems.

Surge protective devices and modules would also have a TCL rating. This rating would be the largest magnitude of voltage to appear across the "protected" port of the protector. The protector should be tested with the same series of peak voltages as described for testing equipment, except that it is rarely necessary to expose the SPD to an open-circuit overvoltage as large as 10 kV. The maximum magnitude of open-circuit overvoltage for an SPD must be agreed between the consumer and manufacturer before the TCL can be determined. Some guidance on reasonable worst-case magnitudes can be obtained from relevant standards, such as ANSI C62.41.

If a particular electronic system had a TCL of 500 V, one would need to use a protector that would clamp the voltage across this system to less than 500 V. This is a simple and effective way of determining economical protection from transient overvoltages.

There is a balance between (1) hardening equipment so that it will withstand large overvoltages and (2) installing surge protective devices so that equipment is not exposed to large overvoltages. Martzloff (1977) states that a key concept of TCL is that: "Equipment is designed to fit the protective level that practical suppressors can guarantee not to be exceeded, rather than to retrofit suppressors after the equipment is found to have transient problems." The TCL concept was advocated by Fisher and Martzloff for use in purchase specifications and regulatory specifications. The TCL concept should be applied not only for equipment operated from the ac supply mains but also for other systems, including telephone and aerospace equipment.

There are a number of possible criticisms that could be made against the TCL concept. For example, peak voltage is not the only criterion of failure: energy transfer, value of dV/dt, and charge transfer may also be important. To some extent, the effects of these other variables are provided by specifications for the TCL test waveform, including the output impedance of the surge generator. Fisher and Martzloff (1976, p. 128) state that protection from overvoltages "is more likely to be achieved through the successful passing of even an imperfect test than it is in the avoidance of all but perfect tests."

Fisher and Martzloff (1976) proposed the 0.5 μs–100 kHz ring wave that was later adopted in ANSI C62.41-1980 as the basic TCL test waveform. Additional test waveforms should be added to the TCL specifications when there is evidence that systems that are properly protected according to TCL tests with the 100 kHz ring wave are failing in the field because of exposure to markedly different kinds of overstress waveforms. Martzloff (1983) has specifically stated:

> Test waves should *not,* however, be misconstrued as representing natural phenomena. They are "realistic" (which is not the same thing as "representing reality") only to the extent that the conclusion drawn from surviving the test wave is validated by better survival in the field than for those devices that do not survive the test wave.

2. Choice of Test Waveforms

Whenever possible, the test waveform should be chosen from the various standard overstress waveforms, some of which were described in Chapter 5. This allows other laboratories to more easily reproduce or verify measurements and minimizes the inventory of surge testing equipment. The standard test waveforms form a common language that can be spoken by manufacturers of SPDs, manufacturers of vulnerable equipment, users, and regulatory agencies.

For products that are connected to the low-voltage mains, the standard document ANSI C62.41-1980 describes a simple selection of reasonable worst-case test waveforms. This standard is being revised while this book is being written, and it is not possible to predict the contents of the next edition of ANSI C62.41. However, it seems likely that there will be a broad selection of representative waveforms and severity levels that recognizes the diversity of existing environments. As indicated in Chapter 3, more information has been published about overvoltages on the mains than about all other environments together.

For products that are connected to telephone lines, various regulatory agencies have established standard overvoltage test waveforms, some of which are described in Chapter 5. More information on the telephone environment is given in Chapter 3.

Overvoltages in environments other than the mains and telephone communications have not been described extensively in the archival literature. Overvoltage testing of products for these relatively unknown environments is therefore largely uncharted territory. An engineer faced with testing a product for these environments might be able to make an analogy with a better-understood environment.

There is a definite hazard in automatically deciding to protect and test against the "worst-case" overvoltage. Such protection is likely to be expensive both in engineering design and testing and in cost of components. Once this expense is recognized, management may decide to omit the overvoltage protection to cut costs. A more realistic and reasonable goal is to recognize that the majority of environments will never experience a "worst-case" event during the expected lifetime of the product.

It is not necessarily a good idea to test protective *components* with a waveform that resembles a worst-case transient, or even a typical transient. Laboratory characterization of protective devices should be done with a waveform that is appropriate to elicit the desired information. For example, inductance in the package can be determined from measurements taken at two different values of dI/dt that have similar values of I and similar total energy, $\int VI\, dt,$ at the two data points. Neither of these two waveforms is required to have dI/dt values that are similar to those of expected transients. As another example, one might use steady-state currents to determine the effects due to device heating, even though continuous overvoltages may not be anticipated in a specific application of the protective device.

However, proof-testing of transient protection circuits *must* be done with a waveform whose voltage, current, and charge transfer are all similar to the anticipated overstress, as described earlier in this chapter in the discussion of TCL philosophy. Any effort that falls short of this requirement provides little or no assurance that the transient protection circuit is adequate.

Unless there is a specific contractual requirement, it is probably not possible to justify precise amplitudes and waveshapes for overvoltage testing. It is important to recognize that the choice is often somewhat arbitrary. Certainly it is possible to select gentle or harsh test specifications that are ludicrous. For example, requiring an SPD that is intended only for use on a long branch circuit inside a building to survive an $8/20\,\mu s$ current waveform with a peak current of 20 kA is too harsh. On the other hand, testing an SPD for use on the mains with a *maximum* stress of an $8/20\,\mu s$ waveform with a peak current of 5 A is too gentle, although such a test might be representative of some environments. The test specifications should be plausible, and there need to be some tests with large amplitudes that appear adequate to colleagues and customers.

3. Test Sequence

After one or more test waveforms have been chosen, the engineer must choose the peak value of current or voltage for that waveform. Usually several different peak values are chosen for each waveform. It is customary to begin overvoltage testing with the waveform and peak value that has the smallest energy transfer. If the SPD or EUT has adequate performance during this waveform, then the peak value is increased according to a predetermined plan, and the test is repeated. ("Adequate performance" means that the device under test met its design goals or specifications, including being intact after the test. Criteria are discussed later in this chapter.) Eventually, the SPD or EUT will fail, or the maximum output of the surge generator will be reached. If the SPD or EUT fails, one knows that the failure level is somewhere between the previous test that was survived and the latest test. The difference in peak voltage or current between successive tests determines the resolution in the failure level, whenever that level is reached. Smaller increments in the peak volage or current give better resolution but also require more time and labor during testing. A compromise between conflicting requirements depends on the requirements and resources for each particular job.

If the maximum output of the generator was reached without failure, then one may wish to use another waveform with greater energy transfer owing to either longer duration or larger surge currents. Perhaps the SPD or EUT will survive greater blasts than the design goals or specifications required. Then one might halt the tests with the knowledge that the SPD or EUT exceeded all expectations. In the process of obtaining these data on failure threshold, one can collect much data that are useful in characterizing the performance of the SPD or EUT.

An alternative method is to initially blast the SPD or EUT with a "worst-case" waveform and peak value. If the SPD or EUT has adequate performance during this test, it is not always justifiable to conclude that everything is fine, as will be explained in the next section. And if the SPD or EUT fails, nothing was learned about its performance under lesser stresses, which are probably more typical of most environments. However, *if* there are no blind spots, then using a single worst-case waveform may be an acceptable part of a qualification test.

4. Blind Spots

A blind spot is a region where an SPD or EUT will exhibit poor performance but will have significantly better performance with larger overstresses (i.e., greater peak voltage or current, longer duration, etc.). The classic example of a blind spot is a hybrid SPD where the first element is a crowbar device, such as a spark gap or thyristor. A schematic of this type of circuit is shown in Fig. 22-1. The spark gap and avalanche diode are shown because they are commonly used, but they are illustrative examples of a hazard that occurs with a variety of different devices.

If the overstress waveform has a short rise time and a large peak open-circuit voltage, the spark gap will conduct quickly and shunt the surge current away from the resistor, avalanche diode, and protected load. For example, consider the situation in which the diode conducts at 6.8 V. If the overstress has a value of dV/dt of $10 \, \text{kV}/\mu\text{s}$, the spark gap may conduct about $0.1 \, \mu\text{s}$ after the overstress begins. At that time the voltage across the

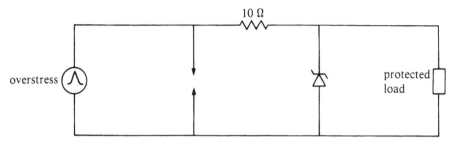

Fig. 22-1. Example of blind spot.

spark gap is 1 kV. The energy deposited in the avalanche diode before the spark gap conducts is about 70 μJ, a trivial amount. After the spark gap operates in the arc region, the voltage across the spark gap is about 20 V, so the current in the resistor is about 1.3 A. If all of this current passes through the diode, the power dissipated in the diode will be about 9 W. Even a small avalanche diode that is designed for transient overvoltage applications can probably survive this level of stress for at least several milliseconds.

Now consider the blind spot. If the peak value of the overvoltage is less than the dc conduction voltage of the spark gap, the gap will not shunt current away from the avalanche diode. For a spark gap with a dc firing voltage of 220 V, the worst case is a 200 V dc overstress. The power dissipated in the avalanche diode will be about 130 W, and the diode will almost certainly fail quickly. Connection of this circuit to the mains with a voltage of 120 V rms will produce similar results.

To uncover blind spots, one should include some tests with a combination of relatively small peak voltages, long rise times, and long durations. Knowledge of the conduction voltages of spark gaps and thyristors in the SPD will provide useful guidance in designing appropriate tests of blind spots.

Another example of a blind spot might be a circuit that was satisfactorily protected by inadvertent flashover of insulation when the test waveform has a large open-circuit voltage and a short rise time but that had no protection for voltages that did not cause flashover. Inadvertent flashover can occur between two closely spaced traces on a printed circuit board, or in wall outlets on the mains.

5. Tests with Both Polarities

ANSI C62.45-1987 states explicitly that "in all cases, surge testing should be performed with both polarities." Many SPDs, such as avalanche diodes, rectifier diodes, and thyristors, have very different properties depending on the polarity of the current and voltage. Likewise, when active devices are used in vulnerable loads, they have markedly different properties depending on polarity. The only way to discover whether a specific situation is sensitive to polarity is to test with both polarities.

C. TEST CONDITIONS

As discussed in the section on a test plan in the introduction to this chapter, there are a several important issues that must be decided before one goes into a high-voltage laboratory to do testing. Each of these is discussed below.

1. Which Conductors to Test

The determination of which conductors to test in cables going to or from the EUT is difficult if there are many conductors. Since a voltage is applied or measured between two nodes, each firing of the overvoltage generator will involve one specific combination of conductors. If the device under test has many conductors, a complete test that involves at least one firing of the surge generator for each specific combination of conductors will take a long time. To appreciate this, consider the relation for the number of combinations, N_2, of a system of m conductors taken two at a time:

$$N_2 = m!/[2(m-2)!]$$

If there are 50 conductors, then there are 1225 pairs of conductors. This large number does not include cases where two or more conductors are connected to one node.

If two or more conductors are allowed to be connected to one node, then the total number of combinations, N, is

$$N = \sum_{i=2}^{m} \frac{m!}{i!(m-i)!}$$

If there are just 4 conductors, then N is 11. If there are 5 conductors, then N is 26. Many cables in military systems, aircraft, and computers have more than 20 conductors. If there are 20 conductors, then N is about 10^6. If each single high-voltage surge requires 1 minute, then about 2 years of continuous work will be required to apply one surge to each of the combinations of a 20 conductor system. How can one find time to test all of the possible combinations in these multiconductor systems? Fortunately, there is a simplification that can be made: test all of the differential modes and the one common mode, which are called the *basic tests*. Although the set of basic tests excludes many combinations, it is a practical simplification.

If the equipment under test were a linear system, then any overvoltage could be expressed as the superposition of various differential and common modes. This would form a scientific basis for justifying using only the few tests in the basic set. However, when the equipment under test contains active electronic devices, which is commonly the situation, the system is not linear, and superposition is not valid. Nevertheless, when one is faced with an overwhelmingly large number of combinations, the basic set is a welcome simplification.

Consider an example of a system of four conductors, which can be labeled A, B, C, and G. There are three differential modes: AB, BC, AC, as shown in Fig. 22-1. There is always one common mode: connect one terminal of the surge generator to G, and couple the other terminal of the generator to all of the other conductors, as shown in Fig. 22-2. This simplification reduced the number of configurations to test from 26 to 4, which is an 85% reduction in the number of combinations to test. The reduction will be more dramatic as the total number of conductors increases.

Of course, one cannot always ignore all of the configurations outside the set of differential and common modes. During design or diagnostic testing, if the equipment under test does not behave as well as hoped in any of the basic tests, then tests of additional combinations of conductors may be useful in finding the Achilles' heel of the EUT. Additional tests may also be desirable to test specific combinations that are expected to experience unusual overvoltages.

Coupling elements, shown as capacitors in Fig. 22-2, are required when the transient overvoltage is superposed on normal system operating voltages. Coupling elements are discussed in Chapter 23. It is generally desirable to do overvoltage testing while the system is operating, as will be discussed in the following section.

When testing a subset of the conductors that go to the EUT, there is a question of what to do with the remaining conductors that are not connected to the surge generator. One choice is not to connect them to anything. Passage of surge currents down a cable may then induce voltages and currents on the unconnected conductors and couple a small stress to these conductors. Although this may be realistic of field operating conditions and therefore desirable for some types of testing, this may not be desirable for laboratory experiments whose purpose is to stress only certain SPDs. Another choice is to connect all unused conductors to ground at the chassis of the EUT. This short-circuits induced currents on the unused conductors, but is not typical of how the EUT will be used in the field. A third choice is to terminate unused conductors in a resistive load at the end of the cable farthest from the EUT. The value of the resistive load should simulate the impedance of longer cables or the output or input impedance of the circuit to which the EUT is normally connected.

Since the testing of equipment that is connected to single-phase mains is a common laboratory exercise, a specific example of the set of basic tests for single-phase mains is shown in Fig. 22-3.

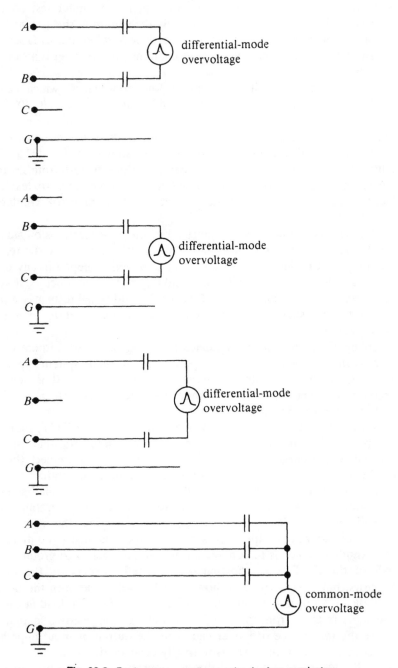

Fig. 22-2. Basic surge coupling modes for four conductors.

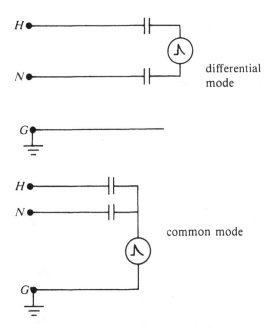

Fig. 22-3. Basic surge coupling modes for single-phase mains.

A practical example of testing equipment connected to the three-phase mains was shown in Fig. 22-2: *A, B,* and *C* are the three phases in the mains, and *G* is the earth ground. The neutral conductor is not shown in Fig. 22-2. In routine, *basic* tests of three-phase mains, it is not necessary to apply any differential-mode surges to the neutral conductor (see ANSI C62.45-1987). However, the common-mode surge should be coupled to the neutral conductor in addition to each of the phases.

Some equipment has both power and data lines connected to it. During tests of upset susceptibility, it may be desirable to apply one type of common-mode surge to the data conductors simultaneously with a different common-mode surge on the power conductors. For example, one might apply a 0.5 μs–100 kHz ring wave to the data conductors and a combination of 1.2/50 μs voltage, 8/20 μs current waveform to the power conductors. Simultaneously applying surges to both data and mains conductors simulates the effects of a lightning strike. A representative peak current might be 10 A per data conductor and 500 A for the mains connection of the EUT. For example, if there are four data conductors, plus a ground connection, then the short-circuit current from the surge generator that is connected to the common-mode data conductors would be 40 A. There is no guarantee that the current will divide evenly among the different conductors; however, this appears to be a reasonable way of determining surge currents. Problems in synchronizing two different surge generators might be avoided by using a single generator, perhaps a 0.5 μs–100 kHz ring wave, and inserting a resistor in series between the generator and each data line to reduce the surge currents in the data lines.

2. EUT Operating During Test

If one wants to test only the performance of insulation (dielectric withstand tests), then it is not necessary to have the EUT operating during the overvoltage tests. This greatly simplifies the testing, since one terminal of the surge generator can be connected to ground and also directly connected to one node of the EUT. The other terminal of the surge generator can be connected directly to another node of the EUT. No coupling devices are required between the surge generator and the EUT.

However, as discussed in Chapter 1, the failure modes of semiconductor devices depend on currents and voltages across the terminals of the semiconductor device prior to the application of the overvoltage. Testing of nonconducting semiconductors covers up failure modes such as second breakdown and suddenly reverse-biasing a conducting PN junction. If the vulnerability of semiconductors is being tested, then the semiconductors should be tested when operating in different states. If an SPD is being tested for adequate protection of semiconductor devices, then some of the tests should be done with the semiconductor devices operating. If a protection circuit under test could have a power-follow current (because it contains a spark gap or thyristor), then it is essential that some powered testing be performed to evaluate the adequacy of measures to cope with the follow current. Of course, if the susceptibility of an EUT to upset is being tested, then the EUT must be operating when the overvoltage is applied.

If the device to be protected is a logic gate, then it would be adequate to test the SPD with the gate (1) initially in the high state, (2) initially in the low state, and (3) with the gate unenergized. The gate should also be tested with a surge during a transition. It is not so easy to determine all of the relevant cases for an analog system.

If the EUT is connected to the mains, then surges should be applied at various phase angles of the mains. It is common to choose the positive peak, zero crossing, and negative peak of the normal mains waveform. One should also consider additional points, such as odd-integer multiples of 45° points of the sinusoidal mains voltage.

3. Statistical Considerations

The purpose of production and qualification tests is often to evaluate the ability of a component or a system to not be damaged or upset by a transient overvoltage. A significant part of the test plan is to decide how many samples to test. If the specifications being tested are near the maximum limits of the component or system, then one can expect to observe some failures when testing many samples. For example, consider a group of devices that has a surge current failure threshold with an *average* value of I_m and a standard deviation of σ. If the devices are tested with a surge current of $I_m - \sigma$, then one would expect to see some failures due to statistical variations in actual failure thresholds. However, if the devices are tested with a surge current of $I_m - 3\sigma$, then one might observe no failures even when testing a group of 100 devices.

The number of samples to test should be increased when the specifications are near the actual failure threshold of the device. Selecting a large safety margin between the typical failure threshold and the specified limits will increase the reliability of the device (when it is used within rated specifications) and decrease the labor during testing.

In addition to deciding how many samples to test, one must also decide what percentage can fail and still have an acceptable product. One might decide that protective devices and modules must have zero defects. To guarantee zero defects, it is necessary to test *every* single device or module before it is shipped to a customer. Although testing every module is costly, the costs can be reduced by designing an automatic test fixture.

It is important to verify that qualification tests do not contribute substantially to reducing the lifetime of a product by blasting it with a severe stress. Components that are blasted to determine their ability to withstand "worst-case" stresses or repeatedly stressed to determine their end of life should *not* be used in equipment that is sold to customers.

During tests of susceptibility of equipment to upset, it is necessary to apply many pulses in order to have confidence that at least one pulse occurred during briefly occurring states where the equipment was more susceptible. To minimize the time required for testing, one might want to generate repetitive pulses at a rate of at least 20 pulses per half-cycle of the normal mains waveform. To properly uncover states where the equipment is most susceptible, the pulses should be distributed along the normal mains waveform rather than clustered at a few particular values of phase of the mains waveform. Therefore the ratio between the pulse repetition frequency and the frequency of the normal mains waveform must not be an integer and should preferably change slightly with time.

Another consideration in doing overvoltage protection with repetitive pulses is the need to verify that the average power dissipation ratings of SPDs will not be exceeded. For example, the EFT waveform specified in IEC 801-4 can deliver a maximum energy of 4 mJ into a 50 Ω load. If the pulse repetition rate is 2.5×10^3 per second, then the maximum time-averaged power into a 50 Ω load is 10 W. This is more than most varistors and avalanche diodes can tolerate for prolonged periods, although these SPDs will not have the same power as a 50 Ω load. Before testing with repetitive overvoltages, the engineer should estimate the time-averaged power in the SPDs and verify that the ratings will not be exceeded.

4. Criteria for Pass/Fail

There are four key criteria for a pass/fail judgment:

1. The peak voltage across protected port during overstress must be less than the maximum acceptable value.
2. The energy in the remnant that appears across the protected port must be less than maximum acceptable value. A 50 Ω load might be connected across the protected port during measurement of the remnant. This forms a reproducible requirement for qualification tests. However, during design and diagnostic tests, one might connect typical semiconductor devices across the protected port.
3. All components must be intact and functional after test.
4. The leakage current should be measured before and after prescribed overvoltage tests. The change in leakage current must be less than the maximum acceptable value.

The peak voltage across the protected port is easy to determine and is often used as a substitute for a more complicated (and often unknown) criterion for acceptable performance.

Restrictions on the amount of energy in the remnant help assure the survival of a resistive device connected to the protected port. To minimize the energy in the remnant, one would want both a fast response time and a low clamping voltage. It is more common to protect semiconductor devices than a 50 Ω resistor. However, since it is very difficult to identify all of the ways that semiconductors can fail, it is simpler to connect a representative semiconductor and see if it survives.

If the protective circuit is damaged during the test, declaration of failure is obvious.

The leakage current in the devices is monitored so that degradation of devices is held to an acceptable level. The leakage current in varistors and avalanche diodes is often measured with a dc bias voltage or current. This leakage current should be measured both before and after testing to detect degradation of the SPD.

A "worst-case" overstress should be applied to at least one sample during a design test. It would be nice if the sample could survive and have acceptable performance during at least one worst-case overstress. But, as discussed in Chapter 16, survival of worst-case overstresses should not be required automatically.

Design tests should include a determination of the lifetime of protective devices or modules. The end of life can be defined as either failure of unit (e.g., fuse opens, catastrophic failure) or a change in the device conduction voltage by more than ±10% from the virgin value. Surge protective devices may reach the end of their life by (1) degradation or destruction by transient overvoltages, (2) storage or operation at elevated temperatures, or (3) time (e.g., decay of radioactive additives in spark gaps).

23

High-Voltage Laboratory Techniques

A. INTRODUCTION

This chapter contains a discussion of several different topics on high-voltage laboratory technique:

1. how to couple the surge generator to the equipment under test
2. choice of oscilloscope (analog vs. digital) and pitfalls of sampling
3. probes for measurement of high voltages, differential amplifiers, and probe compensation
4. measurement of surge currents
5. production of surges in the laboratory

There are many other sources of relevant information: textbooks on laboratory practice and analysis of experimental data (statistics), application notes from manufacturers of laboratory equipment, and tutorial standards such as ANSI C62.45. Safety is discussed in Chapter 24. The techniques discussed in this chapter are not generally applicable to electrostatic discharge.

B. COUPLING METHODS

When the surge generator is connected to an *un*energized device or equipment, the connections are simple, as shown in Fig. 23-1.

When a surge is applied to an energized circuit, one must use some type of coupling device. There are two broad types of coupling mechanisms: series and shunt, which are illustrated in Fig. 23-2. Series coupling uses a

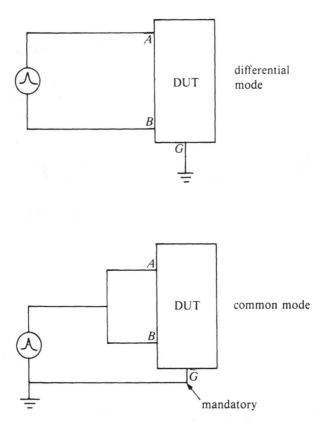

Fig. 23-1. *Application of overstress test waveform to unenergized conductors.*

transformer in series with one conductor. Shunt coupling of short-duration waveforms uses coupling capacitors.

Transformer coupling has several disadvantages. The secondary voltage of an air core transformer can have short rise time, but air core transformers cannot transfer large amounts of energy. Iron core transformers can transfer large amounts of energy but have a limited $\int V\, dt$ value owing to saturation of the core that limits the low-frequency performance (for long-duration non-oscillatory waveforms). Iron core transformers with good low-frequency performance tend to have a long rise time owing to leakage inductance.

By placing the coupled overvoltage in series with one conductor, the SPD or EUT is being surged on that one conductor with respect to all other conductors. This makes it impossible to do differential-mode surges with series coupling. To do a common-mode surge with series coupling would require the coupling transformer to be inserted in series with the grounding conductor of the EUT, a practice that is dangerous.

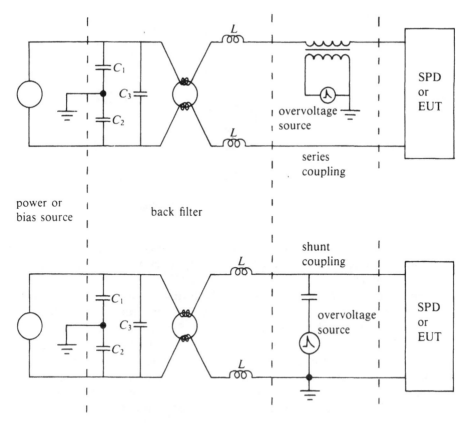

Fig. 23-2. *Series and shunt coupling and back filter for application of surges to equipment powered from the mains.*

1. Shunt Coupling

Shunt coupling is more versatile in that multiple shunt paths can be introduced between the EUT and the surge generator. Uninvolved conductors can be removed from the test simply by not coupling them to the surge generator. It is simple to do normal-mode surges with shunt coupling as shown in Fig. 23-2. It is also simple to do common-mode surges with shunt coupling, as shown in Fig. 23-3.

The coupling capacitor must be rated for operation with the expected short-circuit current of the surge generator. The capacitor dielectric must be rated to withstand at least twice the maximum open-circuit voltage of the surge generator. The factor of 2 is included to cover possible oscillations that may occur when the surge is coupled to a circuit. The capacitance value must be large enough so that there is a small voltage drop across the capacitor at the end of the transient.

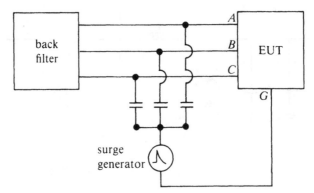

Fig. 23-3. *Common-mode shunt coupling.*

For example, the electrical fast transient specified in IEC 801-4 uses a $0.033\,\mu$F coupling capacitor between the output terminal of the generator and each conductor being surged.

Shunt coupling of long-duration waveforms would require a large coupling capacitor, which would present an unacceptably small load impedance to the mains. Long-duration waveforms can be coupled to the mains with a metal oxide varistor. For coupling to the mains, the varistors should at least meet the guidelines given in Chapter 19. The author uses varistors rated for use on 250 V rms mains and with a diameter of 32 mm to couple surges (up to 500 A, 8/20 μs waveshape) to the nominal 120 V rms mains. The large voltage drop across these varistors is a disadvantage of this method.

The use of a parallel combination of a large capacitance, such as 3.3 μF, and a suitable varistor would make a good coupler for a wide variety of waveforms. The varistor described in the preceding paragraph has a parasitic capacitance of about 3 nF. The use of the parasitic capacitance of the varistor to bypass the inductance inherent in the larger capacitor provides a low-impedance path for a wider frequency range.

2. Back Filter

A back filter is inserted between the power or bias source and the coupled surge, as shown in Fig. 23-2. The back filter is nearly always a simple inductor-capacitor network for waveforms with a duration of less than about 100 μs. The back filter has two purposes: (1) to prevent damage to the power or bias source, and (2) to provide a large series impedance so that most of the surge current goes to the DUT or EUT. In general, the back filter should provide attenuation of both common-mode and differential-mode voltages.

Typical values of the series inductance, L, in the back filter are between 100 and 300 μH. This inductance should be a single-layer choke with insulation that can withstand high voltages, as described in Chapter 12. The line-to-neutral capacitor, C_3, usually has a value of several microfarads.

The line-to-ground capacitors, C_1 and C_2, have capacitance values limited to several nanofarads, as discussed in Chapter 13. To provide adequate filtering of common-mode overvoltages, a common-mode choke is included in the back filter. Common-mode chokes were described in Chapter 12.

The author of this book strongly recommends adding metal oxide varistors in parallel with each line-to-ground capacitor, C_1 and C_2, in the back filter shown in Fig. 23-2. A third varistor may be connected in parallel with C_3 to provide a lower clamping voltage for differential-mode surges. Varistors with a diameter of *at least* 20 mm should be considered for this application. If the varistors interfere with the testing, then the series inductance in the back filter should be increased or varistors with a larger value of V_N should be used. If the varistors have a value of V_N that is more than twice the peak of the normal mains voltage, then the varistors will give rather little protection to vulnerable equipment on the mains.

The back filter and the mains modify the output waveform of the surge generator. As discussed in Chapter 5, standard overstress test waveforms have appreciable energy at low frequencies. However, the back filter cannot have large attenuations at these low frequencies because of practical limits on the values of inductance and capacitance in the filter. Therefore, some of the surge current from the generator will pass through the inductors in the back filter. To avoid unacceptable alterations in the surge waveshape at the DUT or EUT, one must consider the back filter as part of the surge generator (Richman, 1985). The back filter and either the bias source or mains must be connected when measuring the open-circuit voltage waveforms.

For long-duration waveforms, such as the $10/1000\,\mu s$ wave, a simple inductor-capacitor filter is difficult to design for an ac power circuit. In this situation a motor-generator set or isolation transformer may be useful in keeping common-mode long waves out of the mains. Such an isolation transformer blocks power follow current that results from flashover to ground inside the EUT. The differential-mode output impedance of the generator or transformer is too small to be useful as a back filter, so there is no suggestion about how to test energized equipment with differential-mode long waves.

C. INSTRUMENTATION

1. Time Domain or Frequency Domain

There was a brief discussion in Chapter 5 on the virtues of time domain versus frequency domain. Nearly all measurements of single-shot phenomena, such as transient overvoltages, are made with either an oscilloscope or a digitizing voltmeter, which provide time domain data. A frequency spectrum analyzer that requires a steady-state signal* is an inappropriate tool for measurements of transient overvoltages, because transient overvoltages are most definitely not steady-state phenomena!

The instruments necessary to measure currents and voltages in a high-voltage laboratory are both similar to and different from instruments in a conventional electronics laboratory. One of the most important instruments in both laboratories will be an oscilloscope. However, the criteria for the selection of the oscilloscope, and especially the probes, will be different in the two laboratories. The unusual features of high-voltage instrumentation will be discussed in the following paragraphs.

2. DC Voltmeters

High-voltage probes can be purchased for common digital and analog voltmeters to measure constant voltages up to about 40 kV. The high-voltage probe contains a series resistor that forms a voltage divider with the input impedance of the voltmeter. If the voltmeter has a 10 MΩ resistance (common with electronic meters), a 10^{10} Ω series resistance will allow kilovolts to be measured when the meter indicates volts. The high-voltage probe also contains adequate insulation to help protect people from electrical shock. If the large resistance in a high-voltage probe is made from a carbon film, it will have appreciable changes in resistance with temperature and age. Even when these probes are used with voltmeters that have an error of ±0.1%, it is difficult to measure high voltages with uncertainties of less than 5%.

Higher-accuracy measurements of steady-state waveforms can be made with an electrostatic voltmeter. An electrostatic voltmeter is fabricated from a parallel plate capacitor, similar to a tuning capacitor in a radio. One plate is rigidly mounted to the instrument case, and the other plate is free to rotate on jeweled, low-friction bearings. A spiral spring provides a restoring torque to balance against the torque provided by the electrostatic force from charge on the plates of the capacitor. Electrostatic voltmeters are unusual in that they are not sensitive to polarity and they respond to the rms value of the input voltage. An electrostatic voltmeter with a full-scale reading between 3 and 100 kV has a capacitance of about 10 to 20 pF. Electrostatic voltmeters in commercial production can have a maximum error of less than 0.5% of the full-scale value over a range from about 30% to 100% of the full-scale reading.

---

* A steady-state signal is one whose frequency spectrum is essentially constant in time. If the spectrum of the signal changes slowly, so that changes are negligible during the time required for the spectrum analyzer to complete its sweep, then the signal is approximately steady state.

3. Relationship Between Bandwidth and Rise Time

Measurements in the time and frequency domain can be related by mathematical analysis. For example, it is common to use rectangular waveforms to determine the bandwidth of electronic circuits. The input waveform has essentially an instantaneous rise time; the degradation in the rise time of the output waveform gives information about the high-frequency response of the circuit being tested.

For a circuit or system whose high-frequency behavior is governed by a single time constant (e.g., a single resistor-capacitor network), the rise time, t_r, and the upper 3 dB point in the frequency response, f_c, is given by

$$t_r = 0.35/f_c$$

This relation is often used to characterize systems that are not governed by a single time constant, although there is some error in this misuse of the relation.

When there are two or more systems arranged in series, each with a single time constant, the effects of these time constants must be correctly considered. For example, consider a system composed of a phenomenon being measured, a voltage probe, and an oscilloscope amplifier. If t_x = rise time of the phenomenon (probably unknown), t_p = rise time of probe alone, t_a = rise time of oscilloscope amplifier alone, and t_0 = rise time observed on display of oscilloscope, then it can be shown that

$$t_0 = \sqrt{t_x^2 + t_p^2 + t_a^2}$$

and the rise time of the phenomenon, t_x, is given by

$$t_x = \sqrt{t_0^2 - (t_p^2 + t_a^2)}$$

The observed and actual rise times are approximately equal (with less than 5% error) when

$$t_p^2 + t_a^2 < t_0^2/10$$

For example, if the phenomenon has a rise time of 5 ms, then the bandwidth of the probe and amplifier together must be greater than 220 MHz in order to neglect the response of the probe and amplifier. This relationship is useful when specifying equipment for a surge testing laboratory.

4. Oscilloscope for High-Voltage Transients

In an ordinary electronics laboratory an oscilloscope is commonly used to give a stable display of a repetitive signal, which is usually a sine or rectangular waveform with constant amplitude. In a high-voltage laboratory, the waveform to be measured is a single event that may occur with a repetition rate of one per minute. To study the events in a high-voltage laboratory, the oscilloscope must have some kind of storage capability.

There are three ways to obtain storage capability in an oscilloscope: (1) use an analog oscilloscope with a storage CRT, (2) use an analog oscilloscope with a photographic camera, or (3) use a digitizing oscilloscope. Each of these methods has both advantages and disadvantages. The maximum speed at which the electron beam in the CRT can produce a visible trace on the phosphor is called the *writing rate*. The writing rate limits the utility of analog oscilloscopes with fast single-shot events.

The storage CRT is expensive—at least 40% more than a conventional oscilloscope of the same bandwidth in 1986. The fastest storage oscilloscope in commercial production in 1986 had a writing speed of 0.27 cm/ns. If used with a sweep rate of 5 ns/div (one division = 9 mm), the fastest vertical speed would be only 0.22 div/ns. This could be greatly improved by using "reduced scan," which limits the electron beam to 36 mm vertically and 45 mm horizontally, but then the resolution is poorer. The author believes that an analog storage oscilloscope is not a good choice for recording fast events.

The fastest conventional oscilloscope in commercial production in 1986 had a writing rate of 20 cm/ns when used with photographic film with an exposure index of 3000 and an f/1.9 lens. Faster film and lightly fogging the film prior to exposure would produce an even greater effective writing rate.

Digitizers are not limited by writing rate of the display, since information is stored in a memory and not on a phosphor screen of a CRT. The measure of speed of a digitizer is a combination of the bandwidth of the amplifier and the sampling rate. Although high-speed digitizers are much more expensive than conventional oscilloscopes of the same bandwidth, digitizers have several advantages that may outweigh their high initial cost.

In many situations it is desirable to do mathematical computations with the recorded data. For example, one might want to find the energy in a pulse, W, from computation of

$$W = \int V(t)I(t)\, dt$$

or find the rms value, V_{rms}, of a high-frequency signal

$$V_{rms} = \sqrt{\frac{1}{T}\int_0^T V(t)^2\, dt}$$

Mathematical analysis of analog oscilloscope records requires (1) a photograph of the CRT display and (2) manual digitization of the waveform in the photograph. Using instant print photographic film is expensive, and using spools of 35 mm negative film requires time-consuming processing. Manual digitization of the prints is tedious: a human operator moves cross hairs of a digitizing sight to each point and presses a button that causes the machine to measure the coordinates at the location of the cursor. As the operator moves the cursor and digitizes successive points, a piecewise linear representation of the waveform is obtained.

In the mid-1970s, high-speed analog-to-digital converters became available. These could be used to record data in digital form, store it temporarily in semiconductor memory, and transfer it to a computer for processing and storage on a magnetic disk. Although electronic digitizers are more expensive than an analog oscilloscope and a photographic camera, the elimination of the manual processing of photographic records make the digitizer cost-effective for scientists and engineers who want to use mathematical algorithms to process the data.

In addition to the considerable advantage in processing the records, electronic digitizers can have much better resolution than analog oscilloscope displays. Consider an oscilloscope display that has a size of 80 × 100 mm and a trace that is 1 mm in width. The CRT can be represented by an array of digital picture elements: 80 elements vertically and 100 elements horizontally. An analog-to-digital converter with 8-bit dynamic range has 2^8 or 256 distinct states, which correspond to vertical picture elements on a conventional display. The number of samples in the digital record is commonly either 512, 1024, 2048, or 4096; which correspond to horizontal picture elements on a conventional display.

An 8-bit digitizer with 1024 samples per record has about three times better vertical resolution and about ten times better horizontal resolution than an analog CRT. The additional vertical resolution is useful for capturing rare large-amplitude events while preserving detail in routine events with smaller amplitudes. The additional horizontal resolution is useful for observing both small rise times and long durations in the same waveform. With an analog oscilloscope, one would need to take one photograph with a fast sweep rate and then repeat the experiment with a slower sweep rate. Since high-voltage transients in a laboratory are often not exactly reproducible, the ability to observe all of the relevant features in a single waveform is an advantage of digitizers. Naturally occurring transients (e.g., lightning) are definitely not reproducible events, so the user of an analog oscilloscope must either own two sets of oscilloscopes or make the unhappy choice between measuring the rise time or the duration of the event.

One often wishes to record signals that occur immediately prior to the event that triggers an oscilloscope. The signal can be fed into the external trigger input and a coil of coaxial cable that acts as a delay line, as shown in Fig. 23-4. The amount of delay is typically about 1 µs per 200 m of delay line. Aside from the use of delay line, there is no way that an analog

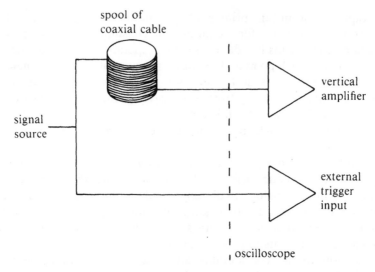

spool of
coaxial cable

vertical
amplifier

signal
source

external
trigger
input

oscilloscope

Fig. 23-4. *Use of delay line to record pretrigger signal with an analog oscilloscope.*

oscilloscope can record events that occur prior to the trigger. In contrast, digitizers can record any fraction of their memory prior to the trigger event. The method is simple. The digitizer samples the data continuously and stores it in a memory that is organized on a first-in, first-out basis. A complete record has N samples, of which M are to be reserved for pretrigger events. When the trigger events occurs, another $N - M$ samples are digitized, then the machine stops. Then the contents of its memory are M pretrigger samples and $N - M$ post-trigger samples.

5. Pitfalls with Sampling

Some manufacturers of digitizers specify a bandwidth for their digitizers that assumes a repetitive signal can be sampled on multiple occurrences and digitized in equivalent time. This technique is used in sampling units for conventional oscilloscopes (e.g., Tektronix 7S12 or 1S1 modules). However, this technique is *not* appropriate for most overvoltage measurements, because the signal is not exactly repeatable. When specifying a digitizer, one should be certain that the maximum "real-time" or "single-shot" digitizing rate is sufficiently fast.

It is common to require a digitizing rate that is 10 times the maximum frequency that one expects to observe in the data. This requirement gives at least three points between the 10% and 90% points of the rising or falling edge of a rectangular pulse in a single-pole system.

If the signal consists of a single sinusoid, it is possible (Nyquist's theorem) to make measurements with a sampling rate that is twice the frequency of the sinusoid. If the sampling rate is smaller than Nyquist's limit

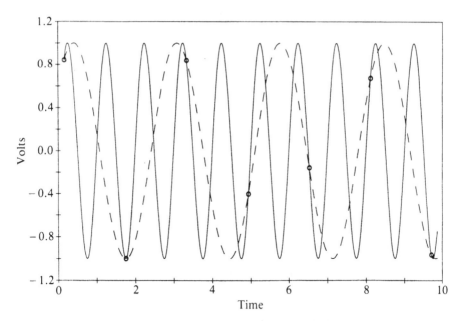

Fig. 23-5. *Classical example of undersampled waveform. Solid line is the true sinusoidal waveform, samples are at the dots, and dashed line is the apparent (aliased) sinusoidal waveform.*

and the signal is a pure sinusoid, passing a smooth curve through the data will produce an apparent waveform that is completely erroneous. This phenomenon, known as aliasing, is illustrated in Fig. 23-5. A signal of frequency f is shown as a solid line in Fig. 23-5. Samples of the true signal, shown as seven circles in Fig. 23-5, are obtained at a rate of 0.63f. The dashed line in Fig. 23-5 is the alias—the erroneous sinusoid that fits the sampled data. In this example the alias has a frequency of 0.37f. The alias always has a lower frequency than the true signal.

Since transient overvoltages do not resemble sinusoidal waveforms, discussion of the alias may not be helpful when using a digitizer in a high-voltage laboratory. Waveforms in a high-voltage laboratory often closely resemble a step function with an exponential decay. A simulation of such a waveform is shown in Fig. 23-6. The solid line is the "true" waveform that is a concoction of a decaying exponential that begins abruptly at $t = 48$ ns and the superposition of two sinusoids with frequencies of 63 and 182 MHz. This waveform is sampled every 20 ns; the 10 samples are shown as circles in Fig. 23-6. The dashed line is the piecewise linear waveform that connects adjacent samples. The amplitude of the discontinuity at $t = 48$ ns is underestimated with an error of about 25%, because the waveform was inadequately sampled. Another interesting feature of Fig. 23-6 is that the four samples after the discontinuity at $t = 48$ ns lie nearly on

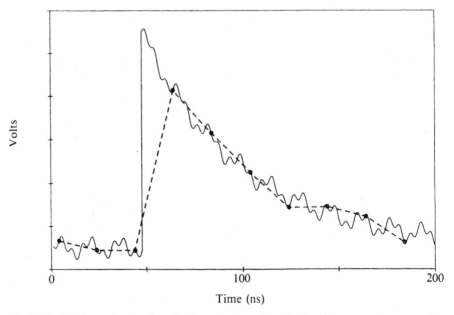

Fig. 23-6. *Effect of undersampling of pulse waveform: Solid line is the true waveform, samples are at the dots, and dashed line is a piecewise linear reconstruction of the sampled signal.*

a straight line. Inspection of the sampled data would lead to the conclusion that the decay of the voltage is linear with time, although the truth of the matter is that the decay of the actual waveform is exponential with time and modulated by two sinusoids. The last four sampled points in Fig. 23-6 are arranged concave downward, although the actual waveform is not. Undersampling can clearly lead to a fallacious picture of reality.

If a plot of a digitized waveform resembles a succession of trapezoids or triangles, then the sampling rate was probably inadequate. Naturally, an event that has a pulse width that is less than the interval between samples may be completely missed.

When using a digitizer, it is good advice to occasionally double the sample rate. If the plot of data taken at the faster sample rate appears similar to the plot of data at the usual sample rate, the usual rate is probably all right.

While the pitfall of inadequate sampling is unique to digitizers, there is a similar pitfall that occurs. with an analog oscilloscope. Consider an analog oscilloscope and voltage probe that have a combined bandwidth of 100 MHz. The oscilloscope will give a fallacious picture of complex high-frequency waveforms such as a square wave with a repetition rate of 100 MHz.

Another pitfall of using a digitizer is the discrete nature of the data. If the least significant bit (LSB) of vertical scale is ΔV and the sample interval is Δt, then inspection of the raw data will always show a "slope" that is an integer multiple of $\Delta V/\Delta t$, possibly zero. When there is a sudden change that has an amplitude of many LSBs, computation of the slope from the raw data may be a satisfactory estimate of the true slope. Otherwise, the curve must be smoothed by an appropriate numerical technique before the slope can be determined.

6. High-Voltage Probes

When a voltage measurement with a bandwidth greater than about 1 MHz is desired, a properly compensated probe is required. Common probes in an electronics laboratory are known as 10X, because they attenuate the input signal by a factor of 10. A simplified schematic diagram of a 10X probe is shown in Fig. 23-7. The probe is designed for use with an amplifier that has an input impedance of 1 MΩ shunted by a parasitic capacitance, C_{in}, which is usually between 10 and 50 pF. At low frequencies, the two resistances give a factor of 10 attenuation. By adjusting the value of C_x, the voltage attenuation at high frequencies can also be set to a factor of 10. Because C_{in} is a parasitic element whose value cannot be predicted, it is not possible to properly adjust the value of C_x unless the probe is connected to the amplifier with which it will be used. The process of adjusting the value of the capacitor in the probe is known as "compensation," because one compensates for the unknown value of C_{in}.

The easiest way to understand the circuit analysis of probe compensation is to consider the low-frequency and high-frequency behavior independently, the asymptotic solution. A more sophisticated approach, which requires an understanding of differential equations to be convincing, is that

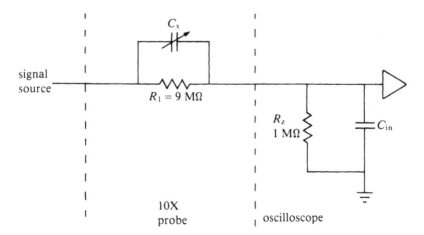

Fig. 23-7. *10X voltage probe for oscilloscope.*

the network is frequency-independent when the two time constants $R_1 C_x$ and $R_2 C_{in}$ have an identical value.

Common probes in an electronics laboratory are suitable for measurements with a peak voltage of less than 500 V. Use of these probes with greater voltages may cause errors due to heating of the 9 MΩ resistor inside the probe, or worse—damage to the resistor due to excessive voltage. If the series resistor or capacitor C_x in the probe flashes over, the input circuit of the oscilloscope or digitizer will almost certainly be destroyed. Using these low-voltage probes to measure the voltage across the protected port of an SPD is risky.

Two probes are commonly used for measurements of high-voltage: (1) the Tektronix model P6009 has a bandwidth of 120 MHz and a maximum dc voltage of 1.5 kV, and (2) the Tektronix model P6015 has a bandwidth of 75 MHz and a maximum dc voltage of 20 kV. For continuous signals at frequencies above 20 MHz, the maximum voltage for these probes is less than their dc voltage rating.

When one needs wider bandwidth than can be obtained with the voltage probes in the preceding paragraphs, one uses coaxial attenuators in a 50 Ω system.

7. Differential Amplifiers

In the high-voltage laboratory it is difficult to find a good ground reference for measurements. When large transient currents pass through real conductors, voltage drops occur owing to the parasitic inductance and resistance of the conductors, as shown in Fig. 23-8. It is therefore desirable to measure the voltage difference between two points by using two probes and either two single-ended amplifiers, as shown in Fig. 23-9a, or one true differential amplifier, as shown in Fig. 23-9b.

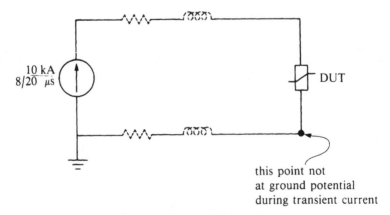

Fig. 23-8. Need for differential measurements.

(a)

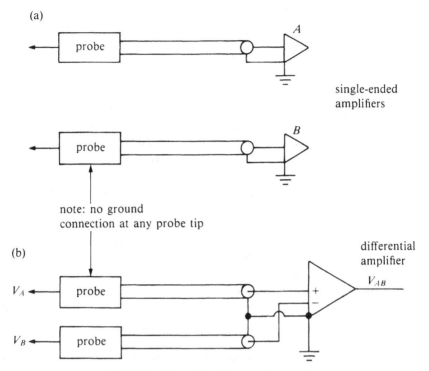

Fig. 23-9. Differential voltage measurement.

Radiated magnetic fields can induce voltages in conducting loops. By avoiding the use of a ground lead on oscilloscope probes, one avoids a loop area that can introduce large errors in measurements. The shield of the coaxial cable is connected to chassis ground of the oscilloscope or digitizer, but no connection is made to the shield near the probe tip (see Fig. 23-9).

Many engineers are accustomed to using two single-ended amplifiers, denoted A and B, to make a pseudodifferential measurement. They use the $A + B$ mode with B inverted, which produces a display of $A - B$. However, this method gives poor rejection of high-frequency common-mode signals. Recall from Chapter 4 that common-mode signals appear simultaneously on both A and B. To use this method, one must adjust the gain of either the A or B amplifier while observing a repetitive rectangular waveform with a fast rise time applied to the A and B probes simultaneously. The gain of one channel is adjusted for a minimum common-mode signal. If the nominal gain of the channels is changed, one must repeat the fine-tuning of the gain of one channel.

A second technique is to use two single-ended amplifiers to make independent measurements. For example, one probe and amplifier measures the voltage between point A and chassis ground at the amplifier, V_{AG}. The other probe and amplifier measures the voltage between point B

and chassis ground at the amplifier, V_{BG}. During analysis of the data, one measurement is subtracted from the other to obtain the voltage difference between points A and B:

$$V_{AB} = V_{AG} - V_{BG}$$

This method is particularly convenient when using high-speed digitizers.

The third technique is to use a true differential amplifier. There are few true differential amplifiers with both wide bandwidth and large common-mode rejection ratio (CMRR). However, if a true differential amplifier with adequate bandwidth is available, then using the differential amplifier generally gives more trustworthy results than using two single-ended amplifiers. Manufacturers of differential amplifiers are listed in Appendix B.

When any of these three differential techniques are used, the probes must be properly compensated. When a differential amplifier is used, there are three steps to this procedure.

1. Properly compensate probe A with its amplifier input, as shown in Fig. 23-10a. Ground the unused (inverting) input of the differential amplifier.

2. Properly compensate probe B with its amplifier input, the inverting input of a differential amplifier. Ground the unused (noninverting) input during this step.

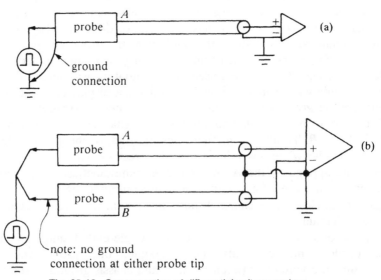

Fig. 23-10. *Compensation of differential voltage probes.*

3a. *If* the probes have an adjustable dc attenuation, obtain the optimum low-frequency CMRR value. Ground both the *A* and *B* inputs. Adjust the vertical zero position to coincide with a convenient graticule line. Then, unground the amplifiers and connect both probes to a large dc signal—e.g., at least 10 V. Adjust the dc attenuation of one or both probes to give the *minimum* indication.

3b. Adjust the compensation of one probe to maximize the high-frequency CMRR of the probe-amplifier system. To do this, connect both probes to the same rectangular waveform with a short rise time, as shown in Fig. 23-10b, and connect the amplifiers for differential operation, and adjust the compensation of probe *B* to produce the *minimum* indication.

It is customary to adjust the compensation of probe *B* if it is used to measure a nominal reference or "ground" voltage, because this adjustment changes the "proper" compensation of the probe.

If one has two single-ended amplifiers *A* and *B* (see Fig. 23-9a), a similar method can be used to maximize the CMRR of the system.

1. Connect probe *A* to amplifier *A*, and compensate this probe in the usual way.

2. Connect probe *B* to amplifier *B*, and compensate this probe in the usual way.

If one wants to optimize the CMRR for a single gain setting, there is a third step. If the nominal gain at the oscilloscope front panel is to be changed during the experiments, then the third step should be skipped.

3a. Obtain the *A−B* mode on the display. Ground both the *A* and *B* inputs. Adjust the vertical zero position to coincide with a convenient graticule line.

3b. Connect probe *A* to amplifier *A*, and connect probe *B* to amplifier *B*, and set both amplifiers to the same nominal gain. Both probes *A* and *B* should be connected to the same signal source. The inputs of both amplifiers should now be *un*grounded.

3c. Obtain the optimum low-frequency CMRR value. Use a large dc signal—e.g., at least 10 V. If the probes do not have adjustable dc attenuation, then adjust the variable gain setting on one channel to give the minimum indication. If the probes have an adjustable dc attenuation, then adjust it to give the minimum indication.

3d. Obtain the optimum high-frequency CMRR value. Connect both probes to the same rectangular waveform with a short rise time, and adjust the compensation of probe *B* to produce the minimun indication.

8. Effects of High Voltage on Compensation

In most cases the calibration and compensation of probes for measurements of surge voltages will be done with small voltages (e.g., less than ±15 V) in an electronics laboratory. Unfortunately, this will not ensure good results at high voltages. Resistors for high voltage are usually fabricated from metal or carbon film. The resistance of these materials is a weak function of voltage: the resistance decreases by a few parts per million per volt. This voltage dependence is negligible when working at usual voltages encountered in electronic circuits. It is not negligible when doing precise high-voltage work. When 10 kV is placed across a resistor, its resistance may decrease to 95% of its resistance value at a few volts. In precise work, calibrations and probe compensation should be checked with high-voltage pulses.

9. Noise Reduction

When two voltage probes are used to measure a voltage difference, the cables for the two probes should be gently twisted and firmly secured with plastic spiral wrap. This minimizes induction of noise on the probe cables by the time rate of change of magnetic field, dB/dt, by minimizing the loop area between the probes and by giving a random orientation to the small area between the two cables.

The large transient currents in the surge testing laboratory can produce much larger values of dB/dt than in the environment of a typical electronics laboratory. The value of dB/dt can be further enhanced when spark gaps quickly change from the insulating to highly conducting state, as explained in Chapters 7 and 15.

High-voltage probes are often connected directly to conductors that contain large surge currents. Because high-voltage probes tend to be long and bulky, there can be a substantial loop area formed between the two probe tips. Since the magnetic field is largest at the surface of current carrying conductors and decreases with distance from the conductors, loop area near the conductors will give more induced voltage than the same loop area farther from the conductors. To avoid this problem, one can connect a short (5 to 15 cm) insulated conductor between the measurement point and the tip of the high-voltage probe. The short conductors for each probe should be twisted together to reject errors due to changing magnetic fields.

D. CURRENT MEASUREMENTS

Since current is often the input variable in overvoltage laboratory experiments, it is important to discuss how to measure current. There are two common ways: use a current transformer or a current-viewing resistor. These two devices are discussed in the following paragraphs.

1. Current Transformers

When a conductor carries a current, a magnetic field surrounds the conductor. This effect can be exploited to measure currents, by using a so-called "current transformer," which is shown in Fig. 23-11. The conductor whose current is to be measured forms the primary coil of the transformer. The secondary coil is wrapped on a torus that is formed of ferrite or transformer steel. The current in the primary wire establishes a magnetic field around the wire, including inside the torus. The magnetic field is approximately perpendicular to the plane of each loop of the secondary coil. By Faraday's law of induction, the voltage across the secondary is proportional to the time derivative of the magnetic flux. This makes the voltage across the secondary proportional to dI/dt, where I is the current in the primary wire. The voltage across the secondary is integrated with a passive RC network that is contained inside the current transformer.

Although Fig. 23-11 is easy to understand, it omits two critical details: the secondary coil is wound as a Rogowski coil, and it has a partial electrostatic shield. The Rogowski winding is shown in Fig. 23-12. By threading the return path for the secondary coil along the core of the transformer, one can reduce the sensitivity of the output to the component of ambient magnetic fields that are normal to the major diameter of the core (Leonard, 1965). A partial electrostatic shield can be wrapped around the windings to reduce the sensitivity of the output to time-varying ambient

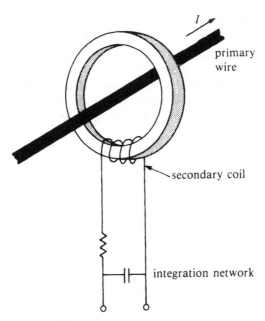

Fig. 23-11. Current transformer.

primary wire

secondary coil

integration network

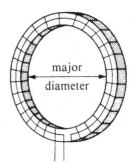

Fig. 23-12. *Rogowski winding of current transformer.*

electric fields. The electrostatic shield must not completely enclose the loop, or there will be a current in the shield owing to the changing magnetic field of the primary current.

The location of the primary wire inside the core of the current transformer does not affect the output voltage, owing to the symmetry of the core of the transformer. However, when the primary wire is at a much higher potential than the measuring circuit, it is essential to use adequate insulation between the primary wire and the current transformer. The primary wire may be threaded through a hollow plastic cylinder, which is inserted inside the current transformer.

For measurements of current at high frequencies or when a long cable must be connected between the current transformer and the oscilloscope or digitizer, one should use a current transformer whose output impedance is matched to the impedance of the long cable. A quantitative criterion for "high frequency" or "long cable" is that the shortest wavelength of interest should be less than 10 times the length of the cable. In practice, the current transformer will probably have an output impedance of 50 Ω, and the transmission line will be coaxial cable with a 50 Ω characteristic impedance. The transmission line should be terminated at the oscilloscope or digitizer with a coaxial load (or the amplifier's input impedance) that has the same impedance as the line.

The current transformer is not suitable for measurement of either low-frequency or very high-frequency currents. The response is approximately constant between the upper and lower 3 dB points. Typical values for the lower frequency limit are between 1 and 500 Hz. Typical values for the upper frequency limit are between 2 and 40 MHz. Models are available with different sensitivities that produce 1 V output for currents between 1 and 100 A. When measurements are to be made of currents of tens of kiloamperes, an insensitive current transformer (e.g., 0.01 V/A) is used with a wide band attenuator network (e.g., 20 or 40 dB) to reduce the output voltage to a level that is safe for the amplifier at the input of the oscilloscope or digitizer.

The output voltage of a current transformer is not proportional to the primary current when the core is near saturation. The core may be saturated in either of two ways. A pulse current may exceed the value of $\int I\,dt$ for saturation, owing to a large magnitude of current or a long duration. A constant (dc) primary current will bias the core and, if excessive, will cause the core to saturate. The values of the limits for a particular current transformer are available from the manufacturer.

One of the advantages of using a current transformer is that no ground connection is made at the transformer. Therefore there are no ground loops for error voltages in the cable to the amplifier, and the current can be measured in conductors that have a potential difference of many kilovolts from ground. Because of the symmetrical design of the current transformer, it is possible to reject errors due to stray magnetic fields.

When common- and differential-mode currents need to be measured, the easiest way is with a current transformer, as shown in Fig. 23-13. The common-mode current could also be measured by placing just the ground lead through the current transformer. However, this is not recommended in

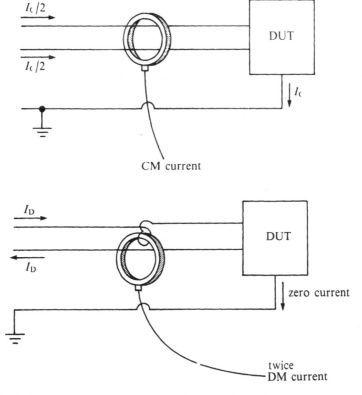

Fig. 23-13. *Measurement of common- and differential-mode currents with a current transformer.*

practical situations, because other fault currents might pass through the ground leads back to their source. When common-mode currents are to be measured, all of the wires through the current transformer are arranged so that positive sense of I_C enters on the same side of the current transformer. When differential-mode currents are to be measured, the two wires through the current transformer are arranged so that positive I_D enters on the same side of the current transformer.

If the output voltage of a current transformer is too small and a more sensitive current transformer is not available, one can pass the same wire(s) through the core of the transformer more than once. If the wires are passed through the core N times, the indicated current will be N times the actual current in the wires.

There is another type of current transformer that has an output current instead of an output voltage. This type of current transformer must have its secondary properly terminated in an external load resistor. Omission of the load resistor while passing a current through the primary wire may damage the current transformer by producing insulation breakdown on the secondary coil or by permanently magnetizing the core. This type of current transformer is used for measuring steady-state current on the mains and is not suitable for measurement of surge currents.

2. Current Viewing Resistors

A current viewing resistor (CVR) is a simple resistor that is used to convert current to voltage, as shown in Fig. 23-14. At low frequencies (below about 10 kHz) and modest magnitudes of currents (e.g., $1\,\mu A$ to $100\,mA$),

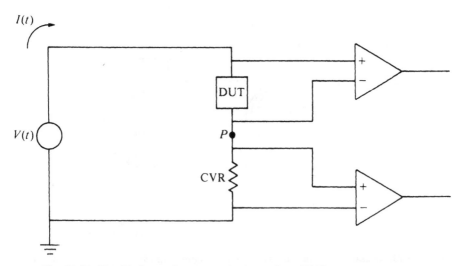

Fig. 23-14. *Simplified use of a current-viewing resistor (CVR) to measure current.*

ordinary resistors used in electronic circuits can be used as a CVR. As measurements are made of larger values of currents, the resistance value becomes smaller, perhaps as small as 0.001 Ω. When wide bandwidth is required, the CVR must be mounted in a coaxial housing and shielded from ambient electromagnetic fields. Work with transient overvoltages requires both large currents *and* wide bandwidth.

As shown in Fig. 23-14, the straightforward use of a CVR requires differential amplifiers to measure voltage. This may be a problem, because differential amplifiers are expensive and have limited bandwidth. One might avoid the requirement for differential amplifiers by grounding point *P* in Fig. 23-14 and floating the surge generator. This practice is unsafe! Further, the large parasitic capacitance between the frame of the surge generator and ground can cause appreciable errors.

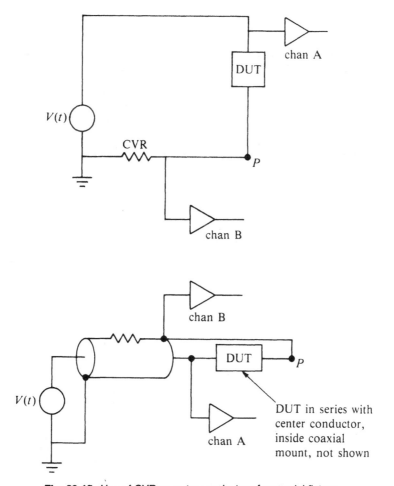

Fig. 23-15. *Use of CVR on outer conductor of a coaxial fixture.*

There is a simple way to avoid differential amplifiers, by placing the CVR in series with the coaxial shield near the ground at the surge generator, as shown in Fig. 23-15. If the resistance of the CVR is suitably small, then the potential of point P will be only a few volts above ground. If the voltage across the device under test (DUT) is at least a few hundred volts, then the voltage drop across the CVR can be neglected. This permits single-ended amplifiers to be used to measure both the voltage across the CVR and the voltage across the DUT.

E. SURGE GENERATORS

There are three basic types of surge generators: (1) capacitor discharge into a RLC pulse shaping network, (2) charged transmission lines, and (3)

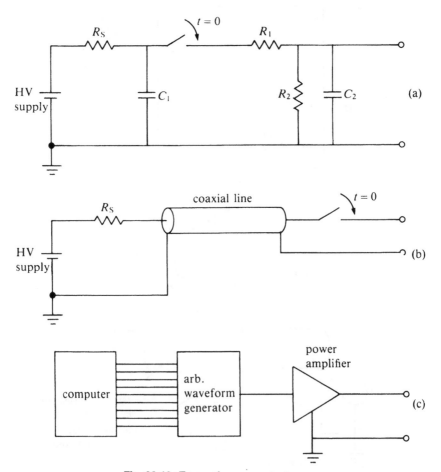

Fig. 23-16. Types of surge generators.

waveform generators that feed high-power amplifiers. These three types are illustrated in Fig. 23-16 and discussed below.

In Fig. 23-16a, the power supply charges capacitor C_1. When the switch is closed, the charge on C_1 flows into the network composed of R_1, R_2, and C_2 and forms a double-exponential voltage waveform across the output port.* The presence of a low-impedance load across the output port will change the voltage waveform significantly from the open-circuit waveform. The high-voltage supply, energy-storage capacitor C_1 and switch can be built into a mainframe. Various RLC pulse-shaping networks and back filters can be built into modules that plug into the mainframe of the surge generator. This approach allows many different waveforms to be obtained without duplicating the most expensive parts of the surge generator. This method is commonly used to produce all of the standard test waveforms such as the $1.2/50\,\mu s$ impulse, $8/20\,\mu s$ current waveform, and the $100\,kHz$ ring wave.

In Fig. 23-16b, when the switch is closed, charge flows out of the transmission line and forms a rectangular voltage waveform across a resistive load at the output port. This method is commonly used to produce brief-duration pulses with a very short rise time. The short-circuit output current is limited because the output impedance of this generator is relatively large; typical values are $50\,\Omega$.

The switches in Fig. 23-16a and b are usually mercury-wetted relays for small peak currents. At larger currents, the switches are usually a triggered gas-filled spark gap. Resistor R_S prevents damage to the high-voltage supplies when the switches are closed and the output is a short circuit.

The last general method uses electronic circuits to generate a waveform with an amplitude of only a few volts, which is fed to a high-power amplifier to drive the load. This method is the most versatile of the three general methods, but it is restricted to modest peak voltage levels owing to restrictions on amplifier technology. In the form shown in Fig. 23-16c, a waveform is provided in digital format by a computer. The computer could have obtained the waveform as the result of a voltage measured by a digitizer, or the computer could have obtained the waveform by calculating points from a mathematical relation. These points are fed into an *arbitrary waveform generator* which is a laboratory instrument that contains a digital-to-analog (D-to-A) converter, a clock, and a computer interface. This generator then creates an analog waveform from the equally spaced digital data from the computer. The output of this generator is typically restricted to $\pm 5\,V$. The voltage and power levels, are increased by a high-power amplifier to values that are useful for overstress testing.

Surge generators with RLC pulse-shaping networks are the most common type of surge generator. They can supply large peak currents and a

---

* The circuit in Fig. 23-16a is illustrative of a generic design but is not recommended for actual use.

variety of waveforms. The major disadvantage of generators with pulse-shaping networks is that the load affects the waveshape. The design of the pulse-shaping network for non-oscillatory voltage generators is discussed by Thomason (1934, 1937), Eaton and Gebelein (1940), Ellesworth (1957), and Creed and Collins (1971). The design of a pulse-shaping network for the combination 1.2/50 μs voltage wave and 8/20 μs current wave is given by Wiesinger (1983). The schematic of a circuit that generates the 0.5 μs–100 kHz ring wave is shown in ANSI C62.41-1980.

Surge generators with charged transmission lines are inexpensive and can provide a very short rise time. However, the waveform is not selectable, and the peak current is relatively small.

Surge generators that use a high-power amplifier tend to be the most expensive type of surge generator, owing to the cost of the amplifier. They also have a relatively small peak output voltage (typically less than 1 kV) with technology available when this book was written. However, the ability to reproduce a wide variety of waveforms with no additional hardware is attractive.

A very important part of a high-voltage surge generator is various safety features. Interlocks should be provided to prevent the surge generator from being operated without a ground connection. The pushbutton switches that cause the generator to fire should be spaced so that the operator needs two hands to fire the generator. This prevents the operator from having one hand somewhere else that is unsafe and getting an electric shock. Crowbars should be installed inside the surge generator so that when the power is turned off, all of the energy storage capacitors are discharged, and no electric shock hazard is present. Some of the general principles are discussed in the next chapter, but this book does not contain all of the information necessary to design a safe surge generator.

It is difficult to design a good surge generator or back filter. However, the addition of appropriate safety features that do not compromise the performance of the surge generator is even more difficult. Therefore, readers are encouraged to purchase and use commercially available surge generators and back filters rather than design and build their own generators and couplers.

24

Safety

A. INTRODUCTION

This chapter contains some hints for the safe use of high-voltage laboratory apparatus and ends with a sketch of first aid and medical treatment for electrical shock. There is no assurance that the following discussion of safety is sufficient. However, the author believes that the following discussion of safety, even though it may be incomplete, is still more helpful than vague warnings to "work safely" and "don't get shocked." People who work with hazardous equipment or materials are responsible for devising and following safe operating procedures.

The treatment throughout this book, and especially in Chapters 23 and 24, is oriented toward systems that are exposed to a maximum overvoltage between about 6 and 10 kV. Systems that are exposed to larger voltages, such as overhead conductors for transmission and distribution of electric power, need to be tested with greater voltages. Such testing is beyond the scope of this book.

References to the medical literature in this chapter are compiled at the end of this chapter, rather than mixed with the references to the engineering literature that are listed at the end of this book.

B. TRADITIONAL RULES

1. One Hand In Pocket

One of the oldest rules is, "Keep one hand in your pocket". The intent of this rule is to prevent accidental electric currents from passing from one hand to the other, by way of the heart. Such a pathway is more dangerous than a pathway from one hand down the side of the torso to the foot, because electrical current can induce cardiac failure. A better rule would be, "Keep your *left* hand in your pocket," since the heart is on the left side of the thorax. This old rule was appropriate for use with vacuum tube circuits, where the maximum potential was often no more than 300 V. However, the author believes that is inappropriate to work on energized high-voltage circuits, as explained later in this chapter.

2. Two People in Room

Another traditional rule is to require that two people always be present in the laboratory where high voltage is being used. The intent of this rule is that if one person is injured, the other can call for help and render first aid. Preferably, both people should be knowledgeable so that they can check each other's work. There is no telling how many accidents have been avoided by a partner calling out, "Don't touch that!"

3. Ground Stick

Every high-voltage laboratory should have a "ground stick"* in a convenient location. A common type of ground stick consists of a rigid, insulating rod (e.g., fiberglass) with a copper braid permanently attached to one end. The other end of the copper braid is securely connected to earth ground. The ground stick should be touched to apparatus before people are allowed to touch it, in order to make certain that the apparatus is not at a differential potential from earth. Apparatus that contains capacitors or devices with large parasitic capacitances can store charge and be especially dangerous.

4. Check Grounds

Common oscilloscopes measure voltage with respect to ground. To measure voltages with respect to some other potential, some engineers "float" the oscilloscope by disconnecting the green wire in the oscilloscope's power cord and placing the oscilloscope on an insulating platform. This is a very dangerous practice. The oscilloscope chassis might be, for example, 1 kV above earth ground. Then someone who touches one of the knobs or switches on the front panel of the oscilloscope can be shocked or even electrocuted (killed). Floating an oscilloscope should be forbidden for safety reasons alone. However, it is also bad measurement technique, since the parasitic capaticance between the chassis and earth will defeat any benefit from floating the oscilloscope during brief pulses. Better ways to make measurements are presented in Chapter 23.

* It is common to call it a *hot stick,* but that usage makes no sense.

5. Coaxial Cable

All high-voltage conductors should be the center conductor in coaxial cable (e.g., type RG-8 or RG-11). The braided shield of the coaxial cable should be securely connected to the ground of the high-voltage power supply or surge generator. By using coaxial cable in this way, there is a good ground between the high-voltage conductor and the part of the cable that people touch.

6. Safety Resistors

Most surge generators contain a high-voltage power supply that is used to charge a capacitor bank that stores energy. The high-voltage power supply often has a small maximum rated current, perhaps only a few milliamperes. In custom-built surge generators one should consider permanently adding a series resistor between the high-voltage power supply and the load. The value of this resistor might be the maximum output voltage of the power supply divided by the lesser of 5 mA or the maximum output current of the power supply. This resistor has several desirable properties. It protects the power supply from damage when its output is shorted to ground, perhaps by a faulty capacitor. The resistor also makes the power supply somewhat safer by limiting the output current, in case someone touches the output. (Note: this resistor does not make it safe to touch the energized capacitor bank!)

A bleeder resistor, which is shown as R_B in Fig. 24-1, should be connected across *each* capacitor that can be charged to a high voltage, so that the charge is drained automatically when the high-voltage power supply is turned off. The value of the resistance is usually chosen to give an RC time constant of a few seconds. This resistor should have appropriate steady-state voltage and power ratings for this application.

Energy storage capacitors in surge generators are obvious candidates for a bleeder resistor. However, there are other capacitors in a typical high-voltage test laboratory that should have bleeder resistors: capacitors in

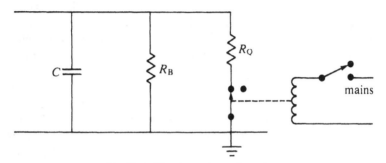

Fig. 24-1. *Bleeder resistors for safety.*

coupling devices between the surge generator and the EUT, and capacitors in the back filter.

It is also desirable to have a relay and small resistance, shown as R_Q in Fig. 24-1, that will quickly discharge the capacitor bank whenever the equipment is turned off. The resistance R_Q is included to avoid welding the contacts of the relay. This resistor should have appropriate pulse current and power ratings so that it does not fail open and provide a false sense of security. The value of R_Q should be selected so that the voltage across capacitor C will decay from its maximum possible value to no more than 34 V 1 second after the mains power is switched off. (The criterion of less than 34 V at 1 second is stated in VDE Standard 0730.)

Resistors R_B and R_Q have the same purpose: to discharge the capacitor bank when the charging current is turned off. Requiring both in the same circuit provides redundant protection. Resistor R_Q is part of a "crowbar" circuit that quickly discharges the capacitor bank and would appear to give more desirable protection to people. However, the relay, with its moving parts, is subject to failure. The bleeder resistor, R_B, is simple and less likely to fail. However, to avoid interfering with normal operation of the capacitor bank, the value of R_B must be large enough to prevent rapid discharge of the capacitor bank.

7. Shorting Strap

A ground stick should be used to discharge an energy storage capacitor before anyone approaches the capacitor. No one should work near or touch a high-voltage capacitor without first connecting a shorting strap across it. The strap is usually a wide piece of flat copper braid with an appropriate clamp or bolt terminal on the end. First it is verified that one end of the shorting strap is connected securely to ground (this end may be permanently connected). Then the free end can be tossed gently at the high-voltage terminal, so that the strap is not being touched when it gets near the high-voltage terminal. Once it has been verified that the capacitor is fully discharged, the free end of the shorting strap should be connected securely to the high-voltage terminal.

One should not assume that a capacitor that was discharged a few minutes ago is still discharged unless the shorting strap is still connected. A high-voltage capacitor can be discharged and yet have sufficient charge to kill someone a few minutes later. The reason for this apparent paradox is somewhat complicated. In some dielectric materials, of which oils are a notorious example, charge can migrate into the dielectric. If the two terminals of the capacitor are short-circuited for a few seconds, all of the charge on the plates of the capacitor *at that time* will be neutralized. When the shorting strap is disconnected, charge in the dielectric can migrate slowly back to the plates of the capacitor. Depending on the amount of charge that migrates back, a lethal shock may be given to a person who touches the terminals of this "discharged" capacitor. The conclusion is simple: Leave the shorting strap securely connected across high-voltage capacitors whenever people are near these capacitors.

C. BARRIERS

1. Explosive Failures

On occasion a component will fail in an explosive manner during surge testing. Fragments of the ruptured case and the component may cause injury to personnel in the vicinity. A transparent barrier should be placed between the device under test and the observers. A suitable barrier may be a sheet of transparent polycarbonate plastic with a thickness of at least 3 mm.

When testing small components (e.g., spark gaps, varistors, diodes), it may be convenient to place the components in a transparent plastic box that is enclosed on four sides and the bottom. The open top should be pointed away from observers. In addition, personnel in the room should wear industrial safety glasses, face shields, or other appropriate protection.

2. Fence Around Apparatus

It is desirable to have an area dedicated to high-voltage testing. This area should be kept free of clutter and distractions.

All high-voltage apparatus, including the object under test, should be confined either inside a grounded metal enclosure or behind a fence of grounded screen wire. This discourages careless people from touching high-voltage connections. Moreover, it helps to contain fragments of any high-voltage capacitor that may explode. Capacitors have a large internal mechanical stress when they are operated at high voltages. When the stress is suddenly relieved by quick discharge (e.g., to produce a transient overvoltage for testing), the material in the capacitor may become fatigued after many surges. It may fail when subsequently charged, and, owing to the large mechanical stress, it is likely to explode. Many high-voltage capacitors are filled with a dielectric fluid to eliminate corona discharge and possibly to provide cooling. Failure of the capacitor's internal insulation will create an arc inside the capacitor that may vaporize enough fluid to rupture the capacitor's case.

If the capacitor bank is securely enclosed inside a cabinet, such as the case of a commercial surge generator, it may be unnecessary to put a fence around the surge generator. Certainly for work at potentials above 10 kV, a fence would be a good idea.

Special precautions are desirable when automatic, repetitive surge generators are used. Such generators are like machine guns, except that someone must pull the trigger on a machine gun whereas repetitive surge generators fire away without human attention. The author believes that such automatic surge generators should only be used inside a fenced area that is dedicated to surge testing. People should not be inside the fenced area when automatic surge testing is being done. Interlocks should be provided so that opening any door to the fenced area will shut off the surge generator.

D. DO NOT WORK WITH ENERGIZED HIGH-VOLTAGE CIRCUITS

1. Redundant Safety Rules

It is common practice for engineers and technicians to work on live *electronic* circuits. They move oscilloscope probes and clip leads from one point to another without turning off the equipment under test. The maximum voltage encountered in electronic equipment is often no more than 20 V (except for power supplies and high-power amplifiers), so working on these energized electronic circuits is not dangerous.

However, the practice of working on live high-voltage circuits is hazardous and should be forbidden. In particular, all personnel should keep their hands far away from *all* high-voltage cables, oscilloscope probes, current probes, and the device or equipment under test until *several redundant* steps have been taken to remove all dangerous voltages. The exact steps depend on the particular situation, but often include the following:

1. Switch off mains power to high-voltage supplies and surge generators.
2. Connect the output of the high-voltage supply or surge generator to earth by using a ground stick.
3. Disconnect the cable that connects the high-voltage or surge generator to the test fixture (disconnect it at the source end of the cable).
4. Discharge all exposed energy storage capacitors, and then connect a shorting strap across them.

In a properly designed system, switching off the high-voltage supply allows energy storage capacitors to discharge slowly through bleeder resistors and more quickly through crowbar circuits, as explained earlier in this chapter.

Commercial cables and probes for high-voltage work are well-insulated but are not fool-proof. For example, it is possible to be shocked while holding the connector of a voltage probe that is connected to a high-voltage circuit, as shown in Fig. 24-2. Commercial high-voltage probes tend to be bulky and awkward to handle, which makes it easy for them to wiggle or slip during handling. To manipulate the probe accurately, you may need to get your hands or face close to the high-voltage test fixture. There are many

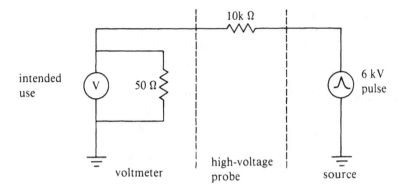

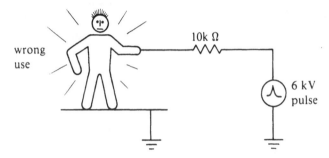

Fig. 24-2. *Misuse of high-voltage probe.*

possible ways to get injured. Rather than analyze all of the possibilities (and eventually overlook one), it is simple and safe to decide not to work on energized high-voltage circuits. The consequences of an accident are too serious to take chances or learn by experience.

2. Test Mains

High-voltage surges are often coupled to the mains for testing equipment that is operating or for examining possible follow current, so people in a high-voltage laboratory are also exposed to shock hazards from low-voltage supply mains. It is suggested that electric shock from the mains *may* be more dangerous to people than shocks from brief high-voltage pulses, as explained later in this chapter. Although it is common for engineers and technicians to be complacent about the low-voltage mains, there is a definite possibility of serious injury or even death from electric shock from the ac supply mains.

It is appropriate to have at least two branches of the ac mains in the high-voltage test area: one for powering laboratory instruments, and one for powering the EUT—the so-called *test mains*. There are several advantages to having two branches. The laboratory instruments are on a dedicated branch, so tripping a circuit breaker or blowing a fuse on the test mains does not interrupt power to the instruments. This is particularly important when digital oscilloscopes are interfaced to computers in the laboratory. Placing the instruments on a separate branch increases the overvoltage protection for them, since any surge currents that get out of the test area will be divided at the distribution panel.

Before approaching the test fixture, one should extinguish power to the EUT by several redundant steps. The exact steps depend on the particular situation, but often include the following: (1) remove power on the test mains at a wall switch; (2) disconnect the cable that connects the back filter to the test mains; or (3) disconnect the cable that connects the EUT to either the back filter or the test mains.

In addition to these safety precautions, it is wise to have warning lamps that are illuminated whenever power is available at the test mains in the laboratory. Such lamps might be a simple neon lamp that is plugged into one of the two receptacles at each duplex outlet (suitable neon lamps are sold in hardware stores as "night lights"). Neon lamps are also available built into wall outlets. A bright tungsten lamp could be mounted on the wall and connected to the test mains.

E. X-RAYS

Still another hazard of high-voltage systems is possible generation of X-rays. This begins to be a serious concern as potentials get above 10 kV, since X-rays with an energy less than 10 keV are rapidly attenuated by traveling in air. X-rays are a serious hazard if some of the surge current travels in a region of low gas pressure.

F. CORONA AND OZONE

The electric field at the surface of a conducting sphere is V/a where a is the radius of the surface and V is the potential at the spherical surface relative to the potential at infinity. The potential at infinity may be defined to be zero; the surface at "infinity" can be approximated by the walls, floor, and ceiling when they are at a distance of at least $100a$ from the sphere. The important result is that a large magnitude of electric field can be obtained by a large potential difference *or* by a small radius of curvature. When the magnitude of the electric field exceeds the dielectric strength (and the maximum potential difference exceeds about 35 V), corona discharge can occur. The dielectric strength of air at sea level pressure and room temperature is about 3 MV/m. In corona discharge new ions can be formed by collisions between neutral molecules and older ions that are accelerated in the intense electric field. Collisions of charged and neutral ions that transfer more than about 5.5 eV of energy can produce ultraviolet light. Ultraviolet light can produce ozone, O_3, from oxygen, O_2, in the air. Ozone is a very nasty substance, because it is a powerful oxidizing agent. A concentration of one ozone molecule per million air molecules is quite irritating to the eyes, nose, and throat and should be avoided.

In addition to being a health hazard, corona discharge can produce broadband electromagnetic noise which interferes with measurements.

There are two techniques that can be used to reduce or eliminate corona discharge, without reducing the operating voltage:

1. reduce the electric field by avoiding conductors with small radii of curvature, and
2. cover high-voltage conductors with an insulation that has a greater dielectric field strength and excludes air from the surface of the conductor.

The radius of curvature is increased by using large spherical electrodes on all high-voltage terminals. One should be careful not to leave sharp edges on high-voltage conductors that have been cut, nor should fine strands of wire be exposed.

There are several ways to apply insulation. One is to use room temperature vulcanizing silicon rubber caulking compound. Another is to use an insulating putty similar to molding clay. Insulating paints, which are marketed to television repair personnel, can be applied to exposed conductors. The entire high-voltage apparatus could be immersed in transformer oil or silicone oil.

G. FIRST AID FOR ELECTRICAL SHOCK

Accidents can happen even to careful workers. Proper safety planning includes having contingency plans for accidents. Therefore it is important to know how to treat injuries from electric shock.

The first-aid treatment is rather simple: (1) remove the victim from the electric circuit; (2) check for breathing and heartbeat; (3) administer cardiopulmonary resuscitation (CPR) as necessary; (4) call for an ambulance. CPR is a good first-aid technique, but it is no substitute for more sophisticated medical treatment. Therefore, do not delay calling for an ambulance.

If the victim was shocked by the output of a high-voltage pulse generator, the victim may no longer be in contact with an energized electric circuit. The victim may have discharged the capacitor bank, or a violent muscle contraction may have thrown the victim free from the circuit.

It is particularly bad when the victim touches a continuously energized conductor with the palm of the hand. An electrical current greater than about 10 ± 5 mA rms (at 60 Hz) will cause the muscles to contract the fingers around the energized conductor, and the victim will be unable to let go of the conductor (Dalziel and Massoglia, 1956). If the electrical current can pass continuously through the victim, as when grasping a mains conductor, the victim can be electrocuted (killed). The victim's resistance is probably great enough that the magnitude of the current is insufficient to open a fuse or circuit breaker in the mains distribution panel. Lethal situations are particularly common when the victim is partly immersed in a grounded, conducting fluid, as in a bathtub or swimming pool, because the victim's resistance is decreased by a large contact area and fluid absorbed in the skin.

The first rescuer on the scene must consider the situation carefully before taking any action. If the rescuer simply grabs the victim, the rescuer may also become part of the electric circuit, with the result that there will be an additional victim.

Depending on the situation, there are various actions that can be taken to free the victim from the circuit.

1. A ground stick can be touched to the offending conductor, which may result in the opening of the appropriate circuit breaker or fuse.
2. The circuit breakers in the electrical panel may be switched off, but one should be careful not to also switch off the illumination. It would be helpful in this regard to clearly label circuit breakers that go to the test mains in the high-voltage laboratory prior to any accident.
3. If banana plugs or other friction connectors are used to complete the circuit, the current may be easily interrupted by pulling on the conductors with a ground stick or insulated rod.

All personnel who work in a high-voltage laboratory should be trained in CPR. The rule of having at least two people present during any high-voltage work tends to ensure that someone will promptly begin rescue efforts. In addition, it is also helpful to have CPR training for personnel who work in nearby offices. CPR training of secretaries is often overlooked but is highly desirable, since they tend to stay in their offices to answer telephone calls, so they are readily available. Appropriate training in CPR is conducted by local groups such as the American Red Cross, Heart Association, industrial safety offices, fire departments, hospitals, etc.

Incidentally, closed chest cardiac massage, which is part of CPR, was developed by an electrical engineer (Kouwenhoven et al., 1960).

H. MEDICAL TREATMENT OF ELECTRICAL SHOCK

Knowledge of the physiology of electrical shock helps one understand how to treat it. Most physicians have little experience with treatment of electrical shock, because it is relatively rare compared to many other diseases and injuries. There have been a number of recent reviews of electrical shock in the medical literature (DiVincenti et al., 1969; Morely and Carter, 1972; Solem et al., 1977; Butler and Gant, 1977; Kobernick, 1982).

There are four major effects of electrical current on man: (1) cardiac arrest, (2) respiratory arrest, (3) loss of consciousness (grand mal seizure), and (4) destruction of tissue by thermal energy.

If the victim is to survive, treatment of the first two effects must be initiated promptly by rescuers. Without circulation of blood and reoxygenation of the blood by the lungs, death will ensure within a few minutes.

Cardiac arrest following electric shock often involves fibrillation, a rapid twitching of the heart muscle, which results in negligible blood flow. Fibrillation due to electrical shock may occur in either the atrial or ventricular chambers of the heart, but ventricular fibrillation appears to be much more common. Death will ensue within a few minutes unless adequate circulation can be maintained. The only first-aid technique for cardiac arrest is closed-chest cardiac massage (part of CPR). In some cases normal heart action can be restored by the use of CPR. If CPR does not restore normal heart action, it may at least preserve life until paramedics arrive or the victim can be transported to a hospital. Paramedics or physicians may be able to restore the normal heartbeat by using electrical defibrillation.

Respiratory arrest is often caused by passage of electric current through the brainstem or across the thorax. It can be treated immediately by mouth-to-mouth artificial respiration (part of CPR). Positive pressure ventilation with compressed oxygen, which is administered by paramedics, is even more effective at providing adequate oxygenation.

The loss of consciousness (grand mal seizure) usually resolves spontaneously, provided any cardiac or respiratory failures are reversed. The seizure is caused by passage of electric current through the brain. Such an effect is often induced deliberately as electroconvulsive treatment of psychiatric patients.

The passage of a current, I, through a resistance, R, liberates heat at the rate of I^2R. (Although conduction in tissues and fluid is not ohmic, the concept of resistance is still well-defined.) If the value of $\int I^2R\, dt$ is sufficiently large, serious biological injury can occur. In graphic terms, the victim becomes "cooked meat." Most of the current will pass through the parts of the body that are good conductors—nerves, blood vessels, and the muscle mass. Subsequent to the electrical injury, perivascular hemorrhage (bleeding around blood vessels,) progressive ischemia (reduced blood supply to tissues), and gangrene may develop. There is no way for medical staff to repair this kind of damage. Fasciotomy or debridement is often required. Fasciotomy is surgical separation of muscle planes; debridement is surgical removal of dead tissue. In severe cases, amputation of affected limbs may be necessary to prevent spread of gangrene. Tissue necrosis may lead to renal failure. Renal failure may be avoided by replacement of fluids to maintain adequate urine output. Oliguria (inadequate urine output relative to fluid input) can be treated with diuretics, such as mannitol. It is a sad fact that, though many victims of electrical shock can be resuscitated, a few die later from the effects of massive thermal injury.

There are two incidental biological effects of electrical injury: muscle contraction and cataracts. The severe muscular contraction that occurs during electric shock can fracture bones. Bilateral dislocation of the shoulders can occur in victims of electrical shock (Brown, 1984). If the victim is standing on a ladder or near a railing at the edge of an elevated structure, the victim may fall as a result of either the muscular contraction or loss of consciousness. Such falls may result in serious additional trauma. Optical cataracts occur in between 5% and 20% of patients with electrical injury to the head (Saffle et al., 1985). These cataracts usually develop within 1 year after the electrical injury.

Most of the reports of electrical injuries in the medical literature result from the victim's contact with the mains or with high-voltage (more than 1 kV rms) distribution lines. Accidental discharge of a capacitor bank through a victim in a high-voltage laboratory may be somewhat different. Accidents with the mains can have the victim connected to the electrical circuit for tens of seconds or longer. During this time, a relatively large amount of electrical energy can produce massive thermal destruction of tissue in the victim. Accidental discharge of a capacitor bank will result in one brief pulse of current in the victim, which may produce less thermal damage in the victim. There are three types of situations reported in the medical literature that may be similar to this type of high-voltage laboratory accident: (1) victims who stand near an object struck by lightning, and who receive a small part of the lightning current; (2) patients who receive electroconvulsive therapy for psychiatric disorders; and (3) patients who receive electrical defibrillation for cardiac arrest.

Since everyone has heard of people killed by contact with the low-voltage mains, one might assume that contact with high-voltage surge generators would be invariably fatal to people. This is not so, as the following examples illustrate. Dalziel (1953) collected reports of 14 accidental shocks that occurred in high-voltage laboratories. There were no fatalities among the 13 unipolar discharges, including two 1.2/50 μs waveforms with peak voltages of 230 and 750 kV. The energy dissipated in the victim ranged from about 9 J to 25 kJ. None of these 13 victims had respiratory or cardiac failure. The single oscillatory discharge, which had a frequency of 5 Hz, resulted in a fatality. The current pathway in this fatal accident was across the chest from one hand to another, which is the worst possible path.

Soon after the invention of closed-chest cardiac massage, Taussig (1968) urged that victims "killed" by lightning receive CPR, since many of them could be revived. This same advice applies to victims of electrical accidents who appear to be dead.

Just because it is possible to survive shock in a high-voltage laboratory is not a good reason to become complacent about such accidents. It is the author's earnest hope that the reader will consider the possibility of electrical shock seriously and observe good engineering practices to avoid accidents.

MEDICAL REFERENCES

Brown, R.J., "Bilateral Dislocation of the Shoulders," *Injury*, 15: 267–273, Jan 1984.

Butler, E.D., and T.D. Gant, "Electrical Injuries, with Special Reference to the Upper Extremities," *American Journal of Surgery*, 134:95–101, July 1977.

Dalziel, C.F., "A Study of the Hazards of Impulse Currents," *Transactions AIEE*, 72:1032–1043, Oct 1953.

Dalziel, C.F., and F.P. Massoglia, "Let-Go Currents and Voltages," *Transactions AIEE*, pt. II, 75:49–56, May 1956.

DiVincenti, F.C., J.A. Moncrief, and B.A. Pruitt, "Electrical Injuries: A Review of 65 Cases," *Journal of Trauma*, 9:497–507, 1969.

Kobernick, M., "Electrical Injuries: Pathophysiology and Emergency Management," *Annals of Emergency Medicine*, 11:633–638, Nov 1982.

Kouwenhoven, W.B., J.R. Jude, and G.G. Knickenbocker, "Closed-Chest Cardiac Massage," *Journal of the American Medical Association*, 173:1064–1067, July 1960.

Morley, R., and A.O. Carter, "First Aid Treatment of Electric Shock," *Archives of Environmental Health*, 25:276–285, Oct 1972.

Saffle, J.R., A. Crandall, and G.D. Warden, "Cataracts: A Long-Term Complication of Electrical Injury," *Journal of Trauma*, 25:17–21, Jan 1985.

Solem, L., R.P. Fischer, and R.G. Strate, "The Natural History of Electrical Injury," *Journal of Trauma*, 17:487–92, July 1977.

Taussig, H.B., "'Death' from Lightning—and the Possibility of Living Again," *Annals of Internal Medicine*, 68:1345–1353, June 1968.

Appendix A

Glossary

The following words are defined as they are used in the context of describing overvoltages, disturbances on the mains, and overvoltage protective circuits. Many of these terms also have other valid usages in other areas of engineering, science, and mathematics. The definitions that the author prefers are listed. Other common usages of these terms in the overvoltage community are also listed, with appropriate notation.

ac 1. A time-varying voltage or current, what remains after the time-averaged (dc) component has been subtracted from the instantaneous value. Because the value of an ac voltage or current changes with time, the ac value is usually expressed as a root mean square (rms) value. 2. A type of source or circuit where the polarity of voltage changes periodically with time, usually as a sinusoidal function of time, as in "ac supply mains."

arrester 1. A surge protective device that is intended either for connection to high-voltage transmission and distribution equipment, or for connection to the mains at the point of entry of electric power into a building. 2. An arrester is supposed to arrest or stop the surge. However, it actually diverts surge current, as explained in Chapters 6 and 19. Also misspelled "arrestor," by false analogy with resistor, capacitor, transistor, varistor, etc.

asymmetrical Common-mode.

avalanche diode A type of semiconductor diode that is intended for operation in the reverse breakdown region. See Chapter 9.

bifilar choke A common-mode choke.

bipolar Pertaining to something that changes sign with time. 1. A bipolar protective device can provide protection from both polarities of an overvoltage (e.g., a bipolar avalanche diode). 2. A bipolar waveform has a voltage and current that change sign with time, more commonly called oscillatory.

breakdown Dielectric breakdown or insulation breakdown (e.g., the change in properties of a spark gap as it goes from insulator to conductor is called *breakdown*). In the overvoltage community "breakdown" is *not* a synonym for damage or malfunction.

brownout A deliberate reduction in rms voltage on the mains, often in response to an excessive demand for power.

carbon block A spark gap formed by two carbon electrodes and air at atmospheric pressure, commonly used to protect telephone equipment.

choke An inductor. Often refers to an inductor that carries a large steady-state current (e.g., more than 1 A rms).

clamping voltage The voltage across a nonlinear shunt protective device during the passage of a specified surge current.

common-mode current In a system of two or more conductors, common-mode current divides into equal parts that pass in the same direction on each conductor and returns through the grounding conductor. See discussion in Chapter 4.

common-mode voltage The voltage that appears between each of two (or more) conductors and ground. See discussion in Chapter 4.

common-mode choke A choke that offers a relatively large impedance to common-mode signals and a much smaller impedance to differential-mode signals. See Chapter 12.

coordination The art of using two or more different nonlinear shunt protective devices. If the two devices are poorly coordinated, either (1) one device absorbs nearly all of the energy and provides a relatively large clamping voltage for the protected load, or (2) the more robust device conducts little current while the more fragile device conducts most of the surge current and may be damaged.

damage Failure of hardware that requires replacement of defective components or modules.

dc 1. A time-independent voltage or current, the average value of a voltage or current. 2. A type of source or circuit where the voltage is essentially constant, independent of time and load current—e.g., a 5 V dc supply. 3. An abbreviation for "direct coupled" (e.g., an input switch on the front panel of an oscilloscope). A direct-coupled signal includes both the time-independent and the time-dependent parts.

differential-mode current In a system of two conductors, differential-mode current passes in opposite directions on each conductor. See Chapter 4.

differential-mode voltage The voltage that appears between two conductors, neither of which is at ground potential. See Chapter 4.

diode A semiconductor component that has a different *V-I* characteristic depending on the polarity of *V* or *I*. 1. Useful as a rectifier (see Chapter 10). 2. Also used to refer to an "avalanche diode" or "zener diode" (see Chapter 9).

disturbance An abnormal and undesirable voltage, usually on the mains.

downstream Toward the protected system, away from the surge protective device.

duration In surge testing with unipolar waveforms, *duration* is either (1) the time between the virtual zero and the 50% point on the decaying tail of the waveform or (2) the full-width at half-maximum. See Chapter 5.

DUT Abbreviation for "device under test," the component being tested.

electrical surge arrestor (ESA) See Arrester.

electromagnetic interference (EMI) Noise. EMI may be either conducted or radiated. Conducted EMI is usually understood as having a frequency between about 10 kHz and 20 MHz and an amplitude of less than 10 V; radiated EMI is usually understood as having a frequency greater than 100 kHz. EMI may be composed of an isolated pulse, a pulse that has a definite repetition rate, a burst of pulses, or a continuous phenomenon that persists from milliseconds to hours. EMI may cause upset, but damage is rare.

electromagnetic pulse (EMP) A brief pulse of electromagnetic energy from an unspecified source, most commonly detonation of a nuclear weapon. See nuclear electromagnetic pulse.

EUT Abbreviation for equipment under test. The equipment may (should) be operating during the test.

flashover An electric discharge over the surface of an insulator as a result of an overvoltage. The discharge path can be either in a fluid (such as air) or along the surface of a solid insulator. As a result of flashover, current can pass to places unintended by engineers who designed the system. In this way, significant safety hazards can be created by flashover. Flashover currents can shunt resistors, inductors, and other current-limiting components and cause failure of devices downstream.

flicker Momentary outage or reduction in rms mains voltage, so called because the symptom is the perceptible flickering of electric illumination.

follow current The current from a steady-state source (e.g., dc power supply or ac mains) that passes through a spark gap after a transient overvoltage has caused the spark gap to conduct. See Chapter 7.

front time A measure of the time required for the initial increase of voltage or current in a standard overstress test waveform. See Chapter 5.

full-width at half-maximum (FWHM) A measure of the duration of a standard overstress test waveform. See Chapter 5.

gas tube A spark gap that is enclosed and filled with a gas other than air, see Chapter 7.

grounded conductor 1. In the special case of the ac mains, "grounded" refers to the neutral conductor, which has insulation that is color-coded blue in international practice and white in the United States. 2. In other cases, the grounded conductor is the signal return path that is connected to earth (e.g., the shield on coaxial cable).

grounding conductor In the special case of the ac mains, refers to the protective conductor that is connected to earth. The insulation of the grounding conductor is color-coded green with a yellow stripe (or green in the United States).

hardened The ability of a circuit or system to avoid damage from transient overvoltages.

harmonic distortion Distortion of the steady-state voltage which is characterized by the superposition of sinusoids of various frequencies, amplitudes, and phases. The frequencies are integer multiples of the fundamental frequency of the steady-state waveform.

high-altitude electromagnetic pulse (HEMP) A disturbance caused on or near the earth's surface by the radiated electromagnetic field from detonation of a nuclear weapon above the atmosphere. See Chapter 2.

high-voltage 1. An ac system whose difference in potential is normally at least 1 kV rms. 2. A dc system that operates at a potential difference of at least 1.4 kV.

hybrid SPD A surge protective device that contains two or more stages of nonlinear devices, such as a spark gap and avalanche diode, that are separated by an impedance. See Chapter 6.

immunity The ability of a system to avoid being upset by disturbances.

impulse* 1. A unipolar transient overvoltage test waveform that is not found in the environment. 2. Defined by IEC 664 as the $1.2/50\,\mu s$ overvoltage test waveform (see Chapter 5), which is more restrictive than conventional usage. 3. In general electrical engineering, impulse is the Dirac delta function that has a value of infinity at one point and is zero elswhere. 4. In physics, impulse is $\int F\,dt$, where F is any force.

isolation Interruption of a conducting path between input and output ports; commonly done with either (1) a transformer that has an electrostatic shield between the primary and secondary coils, or (2) an optically coupled device (e.g., a light-emitting diode and a phototransistor). The purpose of isolation is to block common-mode voltages and break ground loops. See Chapter 14.

let-through The part of a transient overvoltage that propagates downstream from a surge protective device before the device conducts. The let-through has essentially the same voltage and current as the incident surge. A more general term is *remnant*.

---

* It may be wise to avoid using a word with so many different meanings.

lightning electromagnetic pulse (LEMP) A disturbance caused by the electromagnetic field from lightning, not direct injection of current by lightning.

line conditioner A device used to remove some disturbances from the mains; it provides all of the following three functions: (1) regulation of the rms voltage, (2) isolation, (3) differential-mode low-pass filtering. See Chapter 20.

longitudinal Common-mode.

mains The conductors that carry low-voltage ac power inside buildings (e.g., 120/240 V rms single-phase, 60 Hz in the United States).

metallic Differential-mode.

noise An unwanted or undesirable voltage or current, usually high-frequency (e.g., greater than 10 kHz) and small-amplitude (e.g., peak voltage less than 10 V).

nonlinear device A device that has a voltage that is not proportional to current: for example, $I = kV^\alpha$ where $\alpha \neq 1$. Note: superposition is not valid for circuit analysis involving nonlinear devices.

nonsymmetrical voltage A voltage that appears between one conductor and ground in a multiconductor system, also called *unsymmetrical*. See Chapter 4.

normal mode Differential mode.

notch A disturbance on the mains in which the magnitude of the voltage decreases abruptly for a few milliseconds or less. See Chapter 3.

nuclear electromagnetic pulse (NEMP) The electromagnetic field from the detonation of a nuclear weapon. See Chapter 2.

outage Absence of usable power on the mains.

overcurrent An abnormally large current. Fuses and circuit breakers are designed to protect against overcurrents. The word "overcurrent" should not be used by analogy with "overvoltage."

overvoltage Significantly greater magnitude of voltage than that expected during normal operation. Overvoltages often cause damage or upset of electronic systems. See Chapter 1.

positive temperature coefficient (PTC) resistor A type of resistor, whose resistance increases at larger values of temperature (or current, due to self-heating). See Chapter 12.

power follow See follow current.

radio frequency interference (RFI) Old term for what is now called electromagnetic interference (EMI).

regulation (of voltage) A measure of the ability of a voltage regulator to maintain a constant output voltage when either (1) the input voltage, (2) the output current, or (3) the ambient temperature changes, or (4) as time increases. When voltage regulation of an ac waveform is specified, the rms voltage is used, since the average voltage is zero.

remnant   The part of a transient overvoltage that propagates downstream from a surge protective device.

rise time   The time required for the initial edge of a voltage or current waveform to go from 10% to 90% of the peak value. See Chapter 5.

sag   A momentary reduction in the rms voltage on the mains. See Chapter 3.

secondary arrester   A surge protective device intended for connection to the secondary of a distribution transformer. See Chapter 19.

source-region electromagnetic pulse (SREMP)   EMP from detonation of a nuclear weapon at or near the surface of the earth. See Chapter 2.

spark gap   A device that has two or more electrodes separated by a gas. During an overvoltage the gas may change from an insulator to a conductor. See Chapter 7.

spike   Brief transient overvoltage (colloquial). "Brief" is dependent on the scale of the display. A spike appears as a vertical line when the time axis is horizontal.

steady state   1. The condition where the amplitude and frequency of voltage and current do not change appreciably over a long period of time so that circuit analysis, rather than transmission line theory, is adequate to describe the steady state. 2. In mathematical analysis the steady state is reached when the amplitude of the transient has decayed to zero. 3. The amplitude of a steady-state oscillation is usually characterized by an rms voltage, rms current, or time-averaged power.

surge   A type of electrical overstress. In the absence of protective devices, the magnitude of the peak voltage of a surge is usually understood as at least twice the normal system voltage, and the duration of the overvoltage is less than a few milliseconds. (The word "surge" is also used by some engineers and technicians to indicate what should properly be called a *swell.*)

surge protective device (SPD)   A device that attenuates transient overvoltages or diverts surge currents. May be a single nonlinear component or a circuit with two or more nonlinear components. (Engineers who work for the U.S. Department of Defense use "TPD"; U.S. engineers in commercial applications use "SPD".)

surge withstand capability (SWC)   The ability of a system to avoid damage and upset due to transient overvoltages.

susceptibility   The ability of a system to be upset by disturbances. See Chapter 1. It is the antonym of immunity.

swell   A momentary increase in rms voltage on the mains to an abnormally large value. The duration of a swell is longer than one-half cycle and usually less than a few minutes.

symmetrical mode Differential mode.

system-generated electromagnetic pulse (SGEMP) The disturbance created in the interior of a satellite or space vehicle due to detonation of a nuclear weapon in space. See Chapter 2.

terminal protection device (TPD) A device that attenuates transient overvoltages or diverts surge currents. (Engineers who work for the U.S. Department of Defense use "TPD"; U.S. engineers in commercial applications use "SPD".)

transient 1. A brief event, usually lasting less than a few milliseconds. In many situations transmission line theory, rather than circuit analysis, must be used to describe the propagation of a transient voltage or current. 2. In mathematical analysis the transient is the part of the system's behavior before the steady state is reached. 3. The word "transient" is often used to indicate a "transient overvoltage".

transient control level (TCL) 1. The largest peak voltage that will produce neither upset nor damage in an electronic circuit or system. 2. The largest peak voltage that appears across the protected port of a surge protective device when a specified overstress is applied across the input port. 3. A simple concept for specifying vulnerability and susceptibility of electronic systems and for specifying performance of SPDs. See Chapter 22.

transient radiation effects on electronics (TREE) Damage or upset of electronic circuits caused by illumination with photons or subatomic particles. See Chapter 2.

transverse mode Differential mode.

uninterruptible power supply (UPS) An apparatus that supplies continuous power to a load, despite disturbances and outages in the mains. A UPS contains a bank of rechargeable batteries that supply power in the absence of acceptable supply voltage. See Chapter 20.

unipolar Pertaining to something that does not change sign with time. 1. Unipolar protective devices can provide protection from only one polarity of an overvoltage. 2. A unipolar waveform has a voltage and current that do not change sign. More commonly called a *nonoscillatory waveform*.

unsymmetrical voltage A voltage that appears between one conductor and ground in a multiconductor system. Also called *nonsymmetrical*. See Chapter 4.

upset Temporary malfunction of a circuit or system. Recovery does *not* require replacement of defective components but may require an operator's intervention.

upstream Toward the source of the overvoltage, away from the surge protective device.

varistor A nonlinear device with a large value of V/I during normal operation and a very small value of V/I during passage of a surge current.

A varistor is bipolar and dissipates energy in a bulk material rather than in a thin semiconductor junction. See Chapter 8.

vulnerability   The ability of a system to be damaged by disturbances. See Chapter 1.

zener diode   1. A type of semiconductor diode that is intended for applications in the reverse breakdown region. 2. An avalanche diode (colloquial). See Chapter 9.

List of Manufacturers

Because some of the components and test equipment described in this book are unusual items that are unfamiliar to most electrical engineers, a list of manufacturers may be of assistance to the practicing engineer who is beginning to work with overvoltage protection techniques. Within each category the companies are listed in alphabetical order to avoid showing preference for any particular company. Foreign companies are identified; most of them have representatives in the United States. Telephone numbers are not given because these change with time.

Spark Gaps

Cerberus (Switzerland)

C.P. Clare

Cook Electric, division of Northern Telecom

E.G. & G., Inc.

Elevam (Japan)

English Electric Valve Co. (England)

Ericsson (Sweden)

Joslyn Electronic Systems Division

Lightning Protection Corporation

Mitsubishi (Japan)

M–O Valve Company (England)

Siemens (German)

Wickmann (German)

Coaxial Spark Gaps

Cerberus (Switzerland)

Cushcraft

English Electric Valve Co. (England)

Huber + Suhner (Switzerland)

M–O Valve (England)

Reliable Electric, division of Reliance Comm/Tec

Metal Oxide Varistors for Low-Voltage Applications

CKE, Inc.

Fuji Semiconductor (Japan)

General Electric

Iskra (Yugoslavia)

Marcon America Corp.

Mepco/Electra

Panasonic (Matsushita Electric Corp. of America)

Siemens (Germany)

Thomson (France)

Victory Engineering Corp.

Avalanche Diodes for Overvoltage Suppression

CKE, Inc.

General Semiconductor Industries

Ishizuka (Japan)

Lucas (England)

Microsemi

Motorola

Semicon

Thomson (France)

Unitrode

Thyristors for Transient Overvoltage Suppression

General Semiconductor Industries

Motorola

RCA

Teccor

Texas Instruments

Thomson (France)

PTC Resistors

Keystone Carbon Company, Thermistor Div.

Mepco/Electra (N. V. Philips components)

Midwest Components Inc.

Murata-Erie

Raychem Corp., Polyswitch Div.

Siemens

Test Equipment

High-power amplifiers
  Amplifier Research (Souderton, PA)

Wide-band voltage attenuators
  Barth Electronics, Inc. (Boulder City, NV)

Voltage attenuators
  Bird Electronic Corp. (Cleveland, OH)

Capacitors
  Condenser Products Corp. (Brooksville, FL)

Current transformers
  EG&G Washington Analytical Services (Albuquerque, NM)

Current transformers
  Ion Physics Corp. (Wilmington, MA)

Capacitors, spark gap switches, pulse generators
  Maxwell Laboratories (San Diego, CA)

Current transformers, voltage dividers
  Pearson Electronics, Inc. (Palo Alto, CA)

Capacitors, high-voltage dc power supplies
  Plastic Capacitors, Inc. (Chicago)

Electrostatic voltmeters
  Rawson-Lush Instrument Co. (Acton, MA)

Electrostatic voltmeters
  Sensitive Research Instruments (Mount Vernon, NY)

Current viewing resistors, voltage dividers
  T&M Research Products, Inc. (Albuquerque, NM)

Voltage probes models P6015 and P6009, differential amplifier model 7A13
  Tektronix (Beaverton, OR)

Surge Generators

Elgal Electronic Industries, Ltd. (Carmiel, Israel)

EG&G Washington Analytical Services (Albuquerque, NM)

Haefely Test Systems (Basel, Switzerland)

KeyTek Instrument Corp. (Wilmington, MA)

Schaffner Electronik AG (Luterbach, Switzerland)

Velonex (Santa Clara, CA)

Appendix C

Bibliography

The reference librarian at a local technical library (e.g., the science or engineering library at a university), can be helpful in obtaining copies of documents that are listed in this bibliography. The following additional sources should be helpful in obtaining items that are not available at a local technical library.

1. Reports sponsored by U.S. military are available from the Defense Documentation Center (DDC) in Alexandria, Va. Only employees at U.S. military installations can request copies of these documents. However, engineers working on a current military contract can get documents through their contract monitor. This bibliography lists the DDC document number of items available from this source. Some of the documents available from the DDC are also available from NTIS (see next item).

2. All reports sponsored by agencies of the U.S. Government that have been cleared for release to the public, including citizens of other countries, are available from:

> National Technical Information Service (NTIS)
> 5285 Port Royal Road
> Springfield, VA 22161

This bibliography lists the NTIS document number of items known to be available from this source.

3. Copies of patents issued by the U.S. Government are available from:

> Patent and Trademark Office
> Washington, DC 20231

4. Standards issued by the American National Standards Institute (ANSI) and the International Electrotechnical Commission (IEC) are available in the United States from:

> American National Standards Institute
> 1430 Broadway
> New York City, NY 10018

5. The following two libraries have extensive holdings in technology and can furnish photocopies by mail:

> Engineering Societies Library
> 345 East 47th Street
> New York, NY 10017

> Linda Hall Library
> 5109 Cherry Street
> Kansas City, MO 64110

6. Copies of U.S. military handbooks and standards can be obtained from:

> Naval Publications and Forms Center
> 5801 Tabor Ave.
> Philadelphia, PA19120

BIBLIOGRAPHY

Abrahams, P., "Specifications and Illusions," *Communications of the ACM* [Association for Computing Machinery], 31:480–481, May 1988.

Abramson, P., R. Beastrom, B. Hobgood, and G. Ligon, "Lightning Surge Suppressor," *IBM Technical Disclosure Bulletin*, 26:6304–6305, May 1984.

Albertson, V.D., and J.M. Thorson, "Power System Disturbances During a K-8 Geomagnetic Storm: August 4, 1972," *IEEE Trans. Power Apparatus and Systems*, pp. 1025–1030, 1973.

Albertson, V.D., J.M. Thorson, and S.A. Miske, "The Effects of Geomagnetic Storms on Electrical Power Systems," *IEEE Trans. Power Apparatus and Systems*, pp. 1031–1044, 1973.

Alexander, D.R., et al., "EMP Susceptibility of Semiconductor Components," U.S. Air Force Weapons Laboratory AFWL-TR-74-280, Sept 1975.

Allibone, T.E., and F.R. Perry, "Standardization of Impulse-Voltage Testing," *IEE Journal*, 78:257–284, Mar 1936.

Allen, G.W., "Design of Power-Line Monitoring Equipment," *IEEE Transactions on Power Apparatus and Systems*, 90:2604–2609, Dec 1971.

Allen, G.W., and D. Segall, "Monitoring of Computer Installations for Power Line

Disturbances," paper C-74-199-6 presented at 1974 IEEE Power Engineering Society Winter Meeting.

American National Standard 4-1978, *IEEE Standard Techniques for High-Voltage testing*, 125 pp.

American National Standard C37.90a-1974, *IEEE Guide for Surge Withstand Capability Tests* (formerly IEEE Standard 472).

American National Standard C62.1-1984, *IEEE Standard for Surge Arresters for AC Power Circuits* (formerly IEEE Standard 28).

American National Standard C62.41-1980, *IEEE Guide for Surge Voltages in Low-Voltage AC Power Circuits* (formerly IEEE Standard 587).

American National Standard C62.45-1987, *IEEE Guide on Surge Testing for Equipment Connected to Low-Voltage AC Power Circuits.*

American National Standard C63.4-1981, *Methods of Measurement of Radio-Noise Emissions from Low-Voltage Electrical and Electronic Equipment in the Range of 10 kHz to 1 GHz.*

American National Standard C84.1-1982, *American National Standard for Electric Power Systems and Equipment—Voltage Ratings* (60 Hz).

Andrich, E., "Properties and Applications of PTC Thermistors," *Electronics Applications Bulletin* (Netherlands), 26:123–144, 1966.

Armstrong, H.L., "On the Switching Transient in the Forward Conduction of Semiconductor Diodes," *IRE Trans. Electron Devices,* 4:111–113, Apr 1957.

Armstrong, H.R., E. Beck, and G.F. Lincks, "The International Standardization of Lightning Arresters," *AIEE Trans.* Part IIIb, 78:1561–1565, Feb 1960.

Bachman, L., "Investigation of Power Line Transient Suppression Components and Devices," U.S. Naval Electronic Systems Engineering Activity 82-02-26, DDC-B067542, 7 Sept 82. (Distribution limited to U.S. Government agencies only: test and evaluation. Unclassified.)

Bachman, L., M. Gullberg, F. Stricker, and H. Sachs, "An Assessment of Shipboard Power Line Transients," *IEEE International Symposium on Electromagnetic Compatibility,* pp. 218–223, 1981.

Baishiki, R.S., and D.W. Deno, "Interference from 60 Hz Electric and Magnetic Fields on Personal Computers," *IEEE Transactions on Power Delivery,* 2:558–563, Apr 1987.

Bazarian, A., "Gas Tube Surge Arresters for Control of Transient Voltages," *Rome Air Development Center 1980 Electrostatic Discharge Symposium,* pp. 44–53, 1980.

Bellaschi, P.L., "Lightning Surges Transferred from One Circuit to Another Through Transformers," *AIEE Trans.,* 62:731–738, Dec 1943.

Bellcore, "Lightning and 60 Hz Disturbances at the Bell Operating Company Network Interface," Bell Communications Research TR-EOP-000 001, June 1984.

Bennison, E., A.J. Ghazi, and P. Ferland, "Lightning Surges in Open Wire, Coaxial, and Paired Cable, *IEEE Trans. Communications,* 21:1136–1143, Oct 1973

Bewley, L.V., "Traveling Waves Due to Lightning," *AIEE Trans.,* 48:1050–1064, July 1929.

Bodle, D.W., and P.A. Gresh, "Lightning Surges in Paired Telephone Cable Facilities," *Bell System Tech. J.*, 40:547–576, Mar 1961.

Bodle, D.W., and J.B. Hayes, "Lightning Protection Circuit," U.S. Patent 2,789,254, 16 Apr 1957.

Bodle, D.W., and J.B. Hayes, "Electrical Protection for Transistorized Equipment, *Trans. AIEE,* 78:232–237, July 1959.

Bodle, D.W., A.J. Ghazi, M. Syed, and R.L. Woodside, *Characterization of the Electrical Environment,* Toronto: University of Toronto Press, 321 pp., 1976.

Bowling, E.L., "Electrolytic Capacitors," U.S. Patent 4,141,070, 20 Feb 1979.

Brainard, J.P., and L.A. Andrews, "Dielectric Stimulated Arcs in Lightning-Arrestor Connectors," *IEEE Trans. Components, Hybrids, and Mfg. Technology,* 2:309–316, Sep 1979.

Brainard, J.P., L.A. Andrews, and R.A. Anderson, "Varistor-Initiated Arcs in Lightning Arrestor Connectors," *IEEE 1981 Conference on Electronic Components,* Atlanta, pp. 308–312, May 1981.

Bridges, J.E., et al., "DNA EMP Preferred Test Procedures," Defense Nuclear Agency 3286H, DDC AD-787-482, Sept 1976. (Distribution limited to U.S. Government agencies only: test and evaluation. Unclassified.)

Broad, W.J., "Nuclear Pulse (III): Playing a Wild Card," *Science,* 212:1248–1251, 12 June 1981.

Brook, M., N. Kitagawa, and E.J. Workman, "Quantitative Study of Strokes and Continuing Currents in Lighting Discharges to Ground," *J. Geophysical Research.,* 67:649–659, 1962.

Brown, R.M., et al., "EMP Electronic Design Handbook," U.S. Air Force Weapons Laboratory AFWL-TR-74-58, DDC-AD-918277, April 1973 (cleared for public release).

Brown, R.W., J.R. Miletta, and R.E. Parsons, "Spark Gap Devices for EMP Protection," paper 5-2B-4, Joint EMP Technical Meeting Proceedings, Sept 1973. (Distributed limited to U.S. Government agencies only: test and evaluation. Unclassified.)

Bull, J.H., "Voltage Spikes in Low Voltage Distribution Systems and Their Effects on the Use of Electronic Control Equipment," Electrical Research Association Report Nr. 5254, Leatherhead, Surrey, U.K., May 1968.

Bull, J.H., and W. Nethercot, "The Frequency of Occurrence and the Magnitude of Short Duration Transients in Low-Voltage Supply Mains," *The Radio and Electronic Engineer,* 28:185–190, Sept 1964.

Burger, J.R., "Protection Against Junction Burnout by Current Limiting," *IEEE Trans. Nuclear Science,* 21:28–30, Oct 1974.

Buurma, Gerald, "CMOS Schmitt Trigger—A Uniquely Versatile Component," National Semiconductor Application Note 140, reprinted in *National Semiconductor CMOS Databook,* 1978.

Campi, Morris, "EMP Line Filter using MOV Devices," U.S. Patent 4,021,759, 3 May 1977.

Cannova, S.F., "Short-time Voltage Transients in Shipboard Electrical Systems," *IEEE Transactions Industry Applications,* 9:533–538, 1973.

Carroll, R.L., "Loop Transient Measurement in Cleveland, South Carolina," *Bell System Technical Journal*, 59:1645–1680, Nov 1980.

Carroll, R.L., and P.S. Miller, "Loop Transients at the Customer Station," *Bell System Technical Journal*, 59:1609–1643, Nov 1980.

CCITT (International Telegraph and Telephone Consultative Committee), *Protection Against Interference*, Vol. 9, Recommendation K17. Original text Geneva, 1976; modified Malaga-Torremolinos, 1984.

Cergel, L., "General CMOS Characteristics," chapter in *Motorola McMos Handbook*, 1974.

Chowdhuri, P., "Breakdown of P-N Junctions by Transient Voltages," *Direct Current*, 10:131–139, Aug 1965.

Chowdhuri, P., "Transient-Voltage Characteristics of Silicon Power Rectifiers," *IEEE Trans. Industry Applications*, 9:582–592, Sept 1973.

Chowdhuri, P., "Circuit for Protecting Semiconductors Against Transient Voltages," U.S. Patent 3,793,535, 19 Feb 1974.

Clark, O.M., "Suppression of Fast Rise-Time Transients," *IEEE EMC Conference Proceedings*, pp. 66–71, 1975.

Clark, O.M., "Power Surge Protection System," U.S. Patent 3,934,175, 20 Jan 1976.

Clark, O.M., "Lightning Protection for Computer Data Lines," *Electrical Overstress/ESD Symposium*, pp. 212–218, 22 Sep 1981.

Clark, O.M., "Four Terminal Pulse Suppressor," U.S. Patent 4,325,097, 13 Apr 1982.

Clark, O.M., and J.J. Pizzicaroli, "Effect of Lead Wire Lengths on Protector Clamping Voltages," 1979 Federal Aviation Administration Workshop on Grounding and Lightning Technology, Report FAA-RD-79-6, pp. 69–73, 1979.

Clark, O.M., and R.D. Winters, "Feasibility Study for EMP Terminal Protection," General Semiconductor Industries Report TPD003 for U.S. Army Material Command, DDC-AD-909267, Mar 1973.

Cohen, E.J., J.B. Eppes, and E.L. Fisher, "Gas Tube Arresters," *IEEE 1972 International Communications Conference Proceedings*, paper 43, *1972*.

Cooper, H.K., "Forward Transient Response of Silicon Diffused P–N Junctions," *Proc. National Electronics Conference*, 18:107–113, 1962.

Cooper, J.A., and L.J. Allen, "The Lightning Arrester-Connector Concept: Description and Data," *IEEE Trans. Electromagnetic Compatibility*, EMC 15:104–110, Aug 1973.

Cornelius, H.A., "Use of 10×20 Current Waves for Lightning-Arrester Tests," *AIEE Trans.* Part III, 72:895–896, Oct 1953.

Creed, F.C., and M.M.C. Collins, "Shaping Circuits for High Voltage Impulses," *IEEE Trans. Power Apparatus and Systems*, 90:2239–2246, 1971.

Cuffman, J.D., J. Linders, and M.A. Zucker, "Power Factor Correction Capacitors and Their Side Effects," *IEEE 28th Annual Conference of Electrical Engineering Problems in the Rubber and Plastics Industry*, pp. 37–49, 1976.

Cushman, L.A., "Lightning Arrester," U.S. Patent 2,922,913, 26 Jan 1960.

Damljanovic, D., and V. Arandjelovic, "Input Protection of Low Current DC

Amplifiers by GaAsP Diodes," *J. Phys. E (Scientific Instruments)*, 14:414–417, 1981.

De Souza, A.A., "Surge Protection Circuit," U.S. Patent 3,353,066, 14 Nov 1967.

Doljack, F.A., "Polyswitch PTC Devices," *IEEE Trans. Components, Hybrids, and Mfg. Technology*, 4:372–378, Dec 1981.

Domingos, H., and R. Raghavan, "Circuit Design for EOS/ESD Protection," *Rome Air Development Center 1982 Electrostatic Discharge Symposium*, pp. 169–174, 1982.

Domingos, H., and D.C. Wunsch, "High Pulse Power Failure of Discrete Resistors," *IEEE Trans. Parts, Hybrids, and Packaging*, 11:225–229, Sept 1975.

Dowling, R.C., "Lightning Protection on the Stevens Point–Wisconsin Rapids Intercity Telephone Cable," *AIEE Trans.*, 75:697–701, Jan 1957.

Duell, A.H., and W.V. Roland, "Power Line Disturbances and Their Effect on Computer Design and Performance," *Hewlett-Packard Journal*, 32:25–32, Aug 1981.

Durgin, D.L., C.R. Jenkins, and G.J. Rimbert, "Methods, Devices, and Circuits for the EMP Hardening of Army Electronics," U.S. Army Electronics Command TR-ECOM-0275-F, July 1972. (Distribution limited to U.S. Government agencies only: test and evaluation. Unclassified.)

Eaton, J.R., and J.P. Gebelein, "Circuit Constants for the Production of Impulse Test Waves," *General Electric Review*, 43:322–332, Aug 1940.

Edwards, F.S., A.S. Husbands, and F.R. Perry, "The Development and Design of High-Voltage Impulse Generators," *Proc. IEE* Part I, 98:155–180, 1951.

EIA Standard RS-232-C, "Interface Between Data Terminal Equipment and Data Communication Equipment Employing Serial Binary Data Interchange," Washington, D.C.: Electronic Industries Association, 1981.

Ellesworth, G., "Some Characteristics of Double Exponential Pulse-Shaping Networks in High-Voltage Impulse Generators," *Proc. IEE*, 104C:403–410, Apr 1957.

Enlow, E.W., "Determining an Emitter-Base Failure Threshold Distribution of NPN Transistors," *Rome Air Development Center 1981 Electrostatic Discharge Symposium*, pp. 145–150, 1981.

Erickson, J., "Lightning and High Voltage Surge Protection for Balanced Digital Transmission Devices," U.S. Army Electronics Command ECOM-4027, NTIS-AD-752448, Sept 1972.

Ferrieu, J., and Y. Rochard, *Comptes Rendu*, 252:2931, 1961.

Fisher, F.A., "Overshoot: A Lead Effect in Varistor Characteristics," General Electric Company, Schenectady, NY, Corporate Research and Development Report 78CRD201, Sept 1978.

Fisher, F.A., and F.D. Martzloff, "Transient Control Levels," *IEEE Trans. Power Apparatus and Systems*, 95:120–129, Jan 1976.

Florig, H.K., "The Future Battlefield: A Blast of Gigawatts?", *IEEE Spectrum*, 25:50–54, Mar 1988.

Fuquay, D.M., A.R. Taylor, R.G. Hawe, and C.W. Schmid, "Lighting Discharges That Caused Forest Fires," *J. Geophys. Research*, 77:2156–2158, Apr 1972.

Gilber, J.L., and C.L. Longmire, "Theory of Ground Burst Source Region EMP (U)," *Journal of Defense Research*, special issue 84-1, pp. 120–126, May 1985.

Glasstone, S., and P.J. Dolan, *The Effects of Nuclear Weapons*, 3rd Ed., U.S. Government Printing Office: Washington, D.C., 653 pp., 1977.

Goedbloed, J.J., "Transient in Low-Voltage Supply Networks," *IEEE Trans. Electromagnetic Compatibility*, 29:104–115, May 1987.

Golde, R.H. (ed.), *Lightning*, Academic Press, London, 2 vols., 849 pp., 1977.

Goldstein, M., and P. Speranza, "The Quality of U.S. Commercial AC Power," *IEEE Intelec Conference*, pp. 28–33, 1982.

Greenwood, A., *Electrical Transients in Power Systems*, Wiley, New York, 544 pp., 1971.

Grossner, N.R., *Transformers for Electronic Circuits*, 2nd Ed., McGraw-Hill, New York, 467 pp., 1983.

Gutzwiller, F.W., "Protective Control Circuits," U.S. Patent 3,213,349, 19 Oct 1965.

Harnden, J.D., F.D. Martzlof, W.G. Morris, and F.G. Golden, "Metal-Oxide Varistor: A New Way to Suppress Transients," *Electronics*, pp. 91–95, 9 Oct 1972.

Hays, J.B., and D.W. Bodle, "Electrical Protection of Tactical Communication Systems," Bell Laboratories Technical Report, Aug 1958. (Available from National Technical Information Service, catalog Nr. AD-693300.)

Hayter, H.D., "High Voltage Nanosecond Duration Power Line Transients," *Tenth Tri-Service Conference on EMC*, pp. 729–778, Nov 1964.

Herbig, H.F., and J.D. Winters, "Investigation of the Selenium Rectifier for Contact Protection," *AIEE Trans.* Part II, 70:1919–1923, 1951.

Higgins, D.F., K.S.H. Lee, and L. Marin, "System-Generated EMP," *IEEE Trans. Electromagnetic Compatibility*, 20:14–22, Feb 1978.

Howatt, J.R., "Optically Isolated Interface Circuits," U.S. Patent 4,079,272, 14 Mar 1978.

Howell, E.K., "How Switches Produce Electrical Noise," *IEEE Trans. Electromagnetic Compatibility*, Vol. EMC-21, Nr. 3, pp. 162–170, Aug 1979.

Huddleston, G.K., and G.C. Bush, "Lightning Protection for Status and Control Lines of the Mark III Instrument Landing System," *IEEE 1975 EMC Symposium*, paper 3AIb, 1975.

Hyltén-Cavallius, N., "Discussion of: Definition of Switching Surge Waveshapes," *IEEE Trans. Power Apparatus and Systems*, 86:1407, Nov 1967.

International Electrotechnical Commission Publication 60-2, "High-Voltage Test Techniques: Test Procedures," 1973.

International Electrotechnical Commission Publication 801-4, "Electromagnetic Compatibility for Industrial-Process Measurement and Control Equipment—Part 4: Electrical Fast Transient/Burst Requirements," 1987.

Johnson, I.B., "Switching Surges," *AIEE Trans. Power Apparatus and Systems*, 80:240–261, June 1961.

Johnson, W.C., *Transmission Lines and Networks*, McGraw-Hill, New York, 361 pp., 1950.

Kalab, B.M., "Design-Dependent Variability of Pulse Hardness of Types of Discrete Semiconductor Devices (Intervendor Variations)," U.S. Army Harry Diamond Laboratories Technical Report, HDL-TR-1999, 44 pp., Dec 1982.

Kania, M.J., R.F. Piasecki, D.R. Sewart, and S. Danis, "Protected Power for Computer Systems," *Western Electric Engineer*, 24:40–47, 1980.

Kappenman, J.G., V.D. Albertson, and N. Mohan, "Current Transformer and Relay Performance in the Presence of Geomagnetically Induced Currents," *IEEE Trans. Power Apparatus and Systems*, 100:1078–1088, Mar 1983.

Kawiecki, C.J., "Surge Protector," U.S. Patent 3,564,473, 16 Feb 1971a.

Kawiecki, C.J., "Spark Gap Device Having a Thin Conductive Layer for Stabilizing Operation," U.S. Patent 3,588,576, 28 June 1971b.

Kawiecki, C.J., "Spark Gap Device," U.S. Patent 3,811,064, 14 May 1974.

Key, T.S., "Diagnosing Power Quality–Related Computer Problems," *IEEE Transactions on Industry Applications*, 15:381–393, July 1979.

Kimball, J.D., "Static-Magnetic Regulators: Part I," *Electronic Products*, pp. 58–63, Dec 1966.

Kimball, J.D., "Static-Magnetic Regulators: Part II," *Electronic Products*, pp. 74–75, Jan 1967.

Klein, K.W., P.R. Barnes, and H.W. Zaininger, "Electromagnetic Pulse and the Electric Power Network," *IEEE Trans. Power Apparatus and Systems*, 104:1571–1577, June 1985.

Knight, S., "Circuit Protection Apparatus Utilizing Optical Transmission Path," U.S. Patent 3,648,110, 7 Mar 1972.

Knox, K.A.T., "Semiconductor Devices in Hostile Electrical Environments," *Electronics and Power*, 19:557–560, Dec 1973.

Koval, D.O., J.R. Beristain, and D.H. Bent, "Evaluating the Reliability Cost of Computer System Interruptions Due to Power System Disturbances," *Proceedings IEEE Industry Applications Society Annual Meeting*, pp. 1061–1069, Sept 1986.

Kreck, J.A., "Actual Operating Characteristics of Three Terminal Protection Devices Applied to the AN/PRC-77 Field Radio," U.S. Army Harry Diamond Laboratories HDL-TR-1826, DDC-AD-B027-699, Dec 1977. (Distribution limited to U.S. Government agencies only: test and evaluation. Unclassified.)

Kreider, E.P., C.D. Weidman, and R.C. Noggle, "The Electric Field Produced by Lightning Stepped Leaders," *Journal of Geophysical Research*, 82:951–960, 20 Feb 1977.

V. Kübel, "Characteristics and Application of Radio Interference Suppression Filters with Current Compensated Chokes," *Components Report*, 10:108–111, Oct 1975.

Kulkarni, V., "Electrostatic Discharge Prevention—Input Protection Circuits and Handling Guide for CMOS Devices," *National Semiconductor Application Note 248*, June 1980.

Lambert, K.N., and R.C. Peterson, "Electrolytic Capacitor and Filter Network," U.S. Patent 3,439,230, 15 Apr 1969.

Lasitter, H.A., and D.B. Clark, "Nuclear Electromagnetic Pulse Protective

Measures Applied to a Typical Communications Shelter," Naval Civil Engineering Laboratory TN-1091, DDC AD-707-696, Apr 1970.

Lee, K.S.H. (Ed.), *EMP Interaction: Principles, Techniques, and Reference Data,* Hemisphere Publishing Corp., Washington, D.C., 744 pp., 1986.

Legro, J.R., N.C. Abi-Samra, J.C. Crouse, and F.M. Tesche, "A Methodology to Assess the Effects of Magnetohydrodynamic Electromagnetic Pulse (MHD-EMP) on Power Systems," *IEEE Trans. Power Delivery,* 1:203–210, July 1986.

Lennox, C.R., "Experimental Results of Testing Resistors Under Pulse Conditions," Sandia Laboratories SC-TM-67-559, Nov 1967.

Leonard, S.L., "Basic Macroscopic Measurements," chapter in R.H. Huddlestone and S.L. Leonard (eds.), *Plasma Diagnostic Techniques,* Academic Press, New York, 627 pp., 1965.

Lesinski, L.C., "Electronic Surge Arrestor," U.S. Patent 4,390,919, 28 June 1983.

Levinson, L.M., and H.R. Philipp, "ZnO Varistors for Transient Protection," *IEEE Trans. Parts, Hybrids, and Packaging,* 13:338–343, Dec 1977.

Levinson, S.J., and E.E. Kunhardt, "Statistical Investigation of Overvoltage Breakdown," *Proceedings IEEE Third International Pulsed Power Conference,* pp. 226–229, 1981.

Liao, T.-W., and T.H. Lee, "Surge Suppressors for the Protection of Solid-State Devices," *IEEE Trans. Industry General Applications,* 2:44–52, Jan 1966.

Longacre, A., "Optoisolators Couple CRT Terminals to Printer Lines," *Electronics,* p. 118, 2 Oct 1975.

Longmire, C.L., "On the Electromagnetic Pulse Produced by Nuclear Explosions," *IEEE Trans. Electromagnetic Compatibility,* 20:3–13, Feb 1978.

Longmire, C.L., "Introduction to EMP Generation Theory (U)," *Journal of Defense Research,* special issue 84-1, pp. 91–95, May 1985a.

Longmire, C.L., "EMP on Honolulu from the Starfish Event," U.S. Air Force Weapons Laboratory Theoretical Note 353, 12 pp., Mar 1985b.

Longmire, C.L., R.M. Hamilton, and J.M. Hahn "A Nominal Set of High-Altitude EMP Environments," U.S. Air Force Weapons Laboratory Theoretical Note 354, 83 pp., Jan 1987.

Lopez, C., A. Garcia, and E. Munoz, "Deep-Level Changes Associated with the Degradation of Ga $As_{0.6}P_{0.4}$ LEDs," *Electronic Letters,* 13:459–461, 4 Aug 1977.

Lord, H.W., "High-Frequency Transient Voltage Measuring Technique," *IEEE Transactions on Communications and Electronics,* 82:602–605, Nov 1963.

McEachron, K.B., "Thyrite: A New Material for Lightning Arresters," *Journal Amer. Institute Elect. Engineers,* 49:350–353, May 1930.

McEachron, K.B., "Discussion of: Recommendations for Impulse Voltage Testing," *AIEE Trans.,* 52:473, June 1933.

McNeill, R., "Lightning Arrester," U.S. Patent 1,382,795, 28 June 1921.

McMorris, W.A., "Discussion of: Use of 10×20 Current Waves for Lightning-Arrester Tests," *AIEE Trans.* Part III, 72:898–890, Oct 1953.

Malack, J.A., and J.R. Engstrom, "RF Impedance of United States and European Power Lines," *IEEE Trans. Electromagnetic Compatibility,* 18:36–38, Feb 1976.

Martzloff, F.D., "Semiconductor Performance Under Reverse Transient Overvoltages," General Electric Technical Information Series, publication 64GL173, Schenectady, NY, Dec 1964 (cleared for public release).

Martzloff, F.D., and G.J. Hahn, "Surge Voltages in Residential and Industrial Power Circuits," *IEEE Trans. Power Apparatus and Systems*, 89:1049–1056, July 1970.

Martzloff, F.D., "Transient Control Levels," *Telephone Engineer and Management*, 81:55–57, Sept 1977.

Martzloff, F.D., "Coordination of Surge Protectors in Low-Voltage AC Power Circuits," *IEEE Trans. Power Apparatus and Systems*, 99:129–133, Jan 1980.

Martzloff, F.D., "The Coordination of Transient Protection for Solid-State Power Conversion Equipment," 1982 IEEE/Industry Applications Society International Semiconductor Power Conference, pp. 97–105, 24 May 1982.

Martzloff, F.D., "The Propagation and Attenuation of Surge Voltages and Surge Currents in Low-Voltage AC Circuits," *IEEE Trans. Power Apparatus and Systems*, 102:1163–1170, May 1983.

Martzloff, F.D., "Matching Surge Protective Devices to Their Environment (Varistors and Fuses)," *IEEE Trans. Industry Applications*, 21:99–106, Jan 1985a.

Martzloff, F.D., "The Protection of Computer and Electronic Systems Against Power Supply and Data Line Disturbances," General Electric Corporate Research and Development Report 85CRD084, July 1985b.

Martzloff, F.D., "Discussion of: Measurements of Voltage and Current Surges on the AC Power Line in Computer and Industrial Environments, by Odenberg and Braskich," *IEEE Trans. Power Apparatus and Systems*, 104:2689–2691, Oct 1985c.

Martzloff, F.D., "Varistor vs. Environment: Winning the Rematch," *IEEE Trans. Power Systems*, 1:59–66, Apr 1986.

Martzloff, F.D., and H.A. Gauper, "Surge and High-Frequency Propagation in Industrial Power Lines," *IEEE Trans. Industry Applications*, 22:634–640, July 1986.

Martzloff, F.D., and T.M. Gruzs, "Power Quality Site Surveys: Facts, Fiction, and Fallacies," *IEEE Industry Applications Society Industrial and Commercial Power Conference*, pp. 21–33, May 1987 (scheduled for publication in *IEEE Trans. IAS*, late 1988).

Martzloff, F.D., and P.F. Wilson, "Fast Transient Tests—Trivial or Terminal Pursuit?," *International EMC Symposium*, Zürich, pp. 283–288, 1987.

Matsuoka, M., T. Masuyama, and Y. Iida, "Non-linear Resistors," U.S. Patent 3,496,512, 17 Feb 1970.

Mayer, F., "Absorptive Low-Pass Cables: State of the Art and an Outlook to the Future," *IEEE Trans. Electromagnetic Compatibility*, 28:7–17, Feb 1986.

Maylandt, H., and H. Sträb, U.S. Patent 2,741,730, 10 Apr 1956.

Meissen, W., "Überspannungen in Niederspannungsnetzen," *Elektrotech. Z.*, 104:343–346, 1983.

Meissen, W., "Transiente Netzüberspannungen," *Elektrotech. Z.*, 107:50–55, 1986.

Mellitt, B., "Transient Voltages Generated by Inductive Switching in Control Circuits," *Proc. IEE*, 121:668–676, July 1974.

Miletta, J.R., R.J. Chase, M. Campi, and G.L. Roffman, "Defense Switched Network Design Practices for Protection Against High-Altitude Electromagnetic Pulse," U.S. Army Harry Diamond Laboratories HDL-SR-82-2, DDC-B066470, July 1982. (Distribution limited to U.S. Government agencies only: test and evaluation. Unclassified.)

Mills, G. W., "The Mechanisms of the Showering Arc," *IEEE Trans. Parts, Material, and Packaging,* 5:47–55, Mar 1969.

Mitchel, G.R., and C. Melançon, "Subnanosecond Protection Circuits for Oscilloscope Inputs," *Rev. Sci. Instrum.,* 56:1804–1808, Sept 1985.

Morrison, R., *Grounding and Shielding Techniques in Instrumentation,* 2nd Ed., Wiley, New York, 146 pp., 1977.

Nash, H.O., and F.M. Wells, "Power Systems Disturbances and Considerations for Power Conditioning," *IEEE Trans. Industry Applications,* 21:1472–1481, Nov 1985.

Netherwood, P.H., "Electrical Capacitors," U.S. Patent 2,938,153, 24 May 1960.

Netherwood, P.H., "Self-Healing Capacitor," U.S. Patent 3,202,892, 24 Aug 1965.

Nicholson, J.R., and J.A. Malack, "RF Impedance of Power Lines and Line Impedance Stabilization Networks in Conducted Interference Measurements," *IEEE Trans. EMC,* 15:84–86, May 1973.

Odenberg, R., and B.J. Braskich, "Measurements of Voltage and Current Surges on the AC Power Line in Computer and Industrial Environments," *IEEE Transactions on Power Apparatus and Systems,* 104:2681–2688, Oct 1985.

Ott, H.W., *Noise Reduction Techniques in Electronic Systems,* Wiley, New York, 294 pp., 1976.

Palmer, J.R., "Power and Lightning Surges in Coaxial Distribution Systems," *IEEE 1977 Cable Television Reliability Symposium,* pp. 27–31, 1977.

Palueff, K.K., and J.H. Hagenguth, "Effect of Transient Voltages on Power Transformer Design—IV: Transition of Lightning Waves from One Circuit to Another Through Transformers," *AIEE Trans.,* 51:601–620, Sept 1932.

Philipp, H.R., and L.M. Levinson, "ZnO Varistors for Protection Against Nuclear Electromagnetic Pulses," *J. Appl. Phys.,* 52:1083–1090, Feb 1981.

Pierce, J.R., *Evaluation of Methodologies for Estimating Vulnerability to Electromagnetic Pulse Effects,* National Academy Press, Washington, D.C., 101 pp., 1984.

Piety, R.A., "Intrabuilding Data Transmission Using Power-Line Wiring."*Hewlett-Packard Journal,* 38:35–40, May 1987.

Popp, E., "Lightning Protection of Line Repeaters," *Entwicklungs Berichte der Siemens Halske Werke,* 31:25–28, Sept 1968.

Pujol, H.L., "COS/MOS Electrostatic Discharge Protection Networks," *RCA Application Note ICAN-6572,* reprinted in *RCA COS/MOS Integrated Circuit Data Book,* 1977.

Reich, B., "Protection of Semiconductor Devices, Circuits, and Equipment from Voltage Transients," *Proc. IEEE,* 55:1355–1361, Aug 1967.

Reynolds, J.E., "Low Cost Transient Suppressors: Exploratory Development," U.S. Army Electronics Command ECOM-0325-F, DDC-AD-743996, May 1972.

Rhoades, W.T., "Development of Power Main Transient Protection for Commercial Equipment," *IEEE 1980 Electromagnetic Compatibility Symposium*, pp. 235–244, 1980.

Richman, P., "Single Output, Voltage and Current Surge Generation for Testing Electronic Systems," *IEEE 1983 Electromagnetic Compatibility Symposium*, pp. 47–51, 1983.

Richman, P., "Changes to Classic Surge Test Waves Required by Back Filters Used for Testing Powered Equipment," *Proc. Sixth EMC Symposium*, Zürich, pp. 413–418, May 1985a.

Richman, P., "Comments on Measurements of Voltage and Current Surges on the AC Power Line in Computer and Industrial Environments, by Odenberg and Braskich," *IEEE Transactions on Power Apparatus and Systems*, 104:2689, Oct 1985b.

Richmann, P., and A. Tasker, "ESD Testing: The Interface Between Simulator and Equipment Under Test," *Proc. of the Sixth EMC Symposium*, Zürich, pp. 25–30, 1985.

Richman, P., "Classification of ESD Hand/Metal Current Waves Versus Approach Speed, Voltage, Electrode Geometry and Humidity," *Proc. IEEE International Symposium on EMC*, pp. 451–460, 1986.

Ricketts, L.W., J.E. Bridges, and J. Milletta, *EMP Radiation and Protective Techniques*, Wiley, New York, 380 pp., 1976.

Roehr, B., "Hardening Power Supplies to Line Voltage Transients," *Power Electronics Design Conference*, Anaheim, CA, 1985.

Roehr, B., "An Effective Transient and Noise Barrier for Switching Power Supplies," *Power Conversion International Conference*, Munich, pp. 189–194, June 1986.

Rosch, W.L., "The Hair-Trigger Zenith Power Supply," *PC Magazine*, p. 183, 16 Sept 1986.

Rovere, L.H., P.H. Estes, and J.W. Milnor, "Electrical Protective Device," U.S. Patent 1,971,146, 21 Aug 1934.

Saburi, O., and K. Wakino, "Processing Techniques and Applications of Positive Temperature Coefficient Thermistors," *IEEE Trans. Component Parts*, 10:53–67, June 1963.

Sandia Laboratories, "Electromagnetic Pulse Handbook for Missiles and Aircraft in Flight," Sandia Labs SC-M-71-0346, 520 pp., Sept 1972.

Schlicke, H.M., "High Voltage Suppressor for Transmission Lines," U.S. Patent 3,824,431, 16 July 1974.

Schlicke, H.M., and H. Weidmann, "Compatible EMI Filters," *IEEE Spectrum*, 4:59–68, Oct 1967.

Schwab, A.J., *High-Voltage Measurement Techniques*, MIT Press, Cambridge, MA, 1972.

Shaw, R.N., and R.D. Enoch, "Electrostatic Discharge Testing of Integrated Circuits Using Step-Stress Transients or Multiple Transients," *Electronics Letters*, 22:813–815, 17 July 1986.

Sherman, R., *EMP Engineering and Design Principles*, Bell Telephone Laboratories, Whippany, NJ, 151 pp., 1975.

Sherwood, R.A., "Protecting CATV Transmission Equipment from Surges," *IEEE* 1977 *Cable Television Reliability Symposium,* pp. 35–38, 1977.

Shi, J., and R.M. Showers, "Switching Transients in Low Current DC Circuits," *IEEE International EMC Symposium,* pp. 273–279, Apr 1984.

Singer, H., J.L. ter Haseborg, F. Weitze, and H. Garbe, "Response of Arresters and Spark Gaps at Different Impulse Steepnesses," *Fifth International Symposium on High Voltage Engineering,* Braunschweig, Federal Republic of Germany, paper 81.07, Aug 1987.

Singletary, J.B., and J.A. Hasdal, "Methods, Devices, and Circuits for the EMP Hardening of Army Electronics," U.S. Army Electronics Command ECOM-0085-F, DDC-AD-885224, June 1971.

Skilling, H.H., *Electric Transmission Lines,* McGraw-Hill, New York, 442 pp., 1951.

Smith, M.N., "Practical Application and Effectiveness of Commercially Available Pulse Voltage Transient Suppressors," U.S. Naval Civil Engineering Laboratory TN-1312, DDC AD-773-074, Dec 1973.

Smith, M.W., and M.D. McCormick (ed.), *Transient Voltage Suppression,* 3rd Ed., General Electric, Auburn, NY, 162 pp., 1982.

Smith, R.S., G.E. Thomas, D.C. Coleman, and T.J. Pappalardo, "Gas-Breakdown Transmit-Receive Tube Turn-On Times," *IEEE Trans. Plasma Science,* 14:63–66, Feb 1986.

Smithson, A.K., "Lightning Protection for W.R.E. Computer Cable Network," Weapons Research Establishment, Adelaide, Australia, WRE-TM-1796, DDC-AD-A047080, May 1977.

Sola, J.G., "Transformer Having Constant and Harmonic-Free Output Voltage," U.S. Patent 2,694,177, 9 Nov 1984.

Speranza, P.D., "A Look at the Quality of AC Power Serving the Bell System," *Bell Laboratories Record,* 60:148–152, July 1982.

Sporn, P., "Recommendations for Impulse Voltage Testing," *Electrical Engineering,* 52:17–22, Jan 1933. (Reprinted in *AIEE Trans.* 52:466–471, June 1933.)

Standler, R.B., *Transient Protection of Electronic Circuits,* U.S. Air Force Weapons Laboratory Technical Report, AFWL-TR-85-34, available from NTIS: AD-A178 945, 222 pp., Aug 1984.

Standler, R.B., "Passive Overvoltage Protection Devices, Especially for Protection of Computer Equipment Connected to Data Lines," U.S. Patent 4,586,104, 29 Apr 1986.

Standler, R.B., "An Experiment to Monitor Disturbances on the Mains," *IEEE Industry Applications Society Conference Proceedings,* pp. 1325–1330, Oct 1987.

Standler, R.B., "Equations for Some Transient Overvoltage Test Waveforms," *IEEE Trans. Electromagnetic Compatibility,* 30:69–71, Feb 1988a.

Standler, R.B., "Use of Low-Pass Filters to Protect Equipment from Transient Overvoltages on the Mains," *IEEE Industry Applications Society Industrial and Commercial Power Systems Conference Proceedings,* pp. 66–73, May 1988b.

Standler, R.B., "Technology of Fast Spark Gaps," U.S. Army Harry Diamond Laboratory Report, 56 pp., 1988c.

Standler, R.B., "Protection of Small Computers from Disturbances on the Mains," *IEEE Industry Applications Society Annual Meeting Conference Record*, Oct 1988d.

Standler, R.B., "Transients on Mains in a Residential Area," scheduled for publication in *IEEE Trans. Electromagnetic Compatibility*, May 1989.

Standler, R.B., and A.C. Canike, *Mitigation of Mains Disturbances*, U.S. Air Force Weapons Laboratory Technical Report TR-86-80, 188 pp., Aug 1986.

Stansberry, C.L., "EMP, Lightning, and Power Transients: Their Threat and Relationship to Future EMP Standards," National Communications Systems TN-22-77, DDC-AD-A058404, Nov 1977.

Steinbruner, J., "Launch Under Attack," *Scientific American*, 250:37–47, Jan 1984.

Stoelting, H.O., "Discussion of: New Lightning Arrester Standard," *AIEE Trans.* Part III, 75:963–964, Oct 1956.

Tasca, D.M., "Pulse Power Response and Damage Characterization of Capacitors," *Rome Air Development Center 1981 Electrostatic Discharge Symposium*, pp. 174–191, 1981.

Tasker, A., "ESD Discharge Waveform Measurement, the First Step in Human ESD Simulation," *Proc. of the IEEE International Symposium on EMC*, pp. 246–250, Aug 1985.

Tetreault, M., and F.D. Martzloff, "Characterization of Disturbing Transient Waveforms on Computer Data Communication Lines," *Proceedings Sixth EMC Symposium*, Zürich, pp. 423–428, Mar 1985.

Thomas, M.E., and F.L. Pitts, "Direct Strike Lightning Data," NASA-TM-84-626, Mar 1983.

Thomason, J.L., "Impulse Generator Circuit Formulas," *Electrical Engineering*, 53:169–176, Jan 1934.

Thomason, J.L., "Impulse Generator Voltage Charts for Selecting Circuit Constants," *Electrical Engineering*, 56:183–188, Jan 1937.

Todd, C.D., *Zener and Avalanche Diodes*, Wiley-Interscience, New York, 266 pp., 1970. (Reprinted by University Microfilms, Ann Arbor, MI, 1980.)

Tront, J.G., "Predicting URF Upset of MOSFET Digital ICs" *IEEE Trans. Electromagnetic Compatibility*, 27:64–69, May 1985.

Trybus, P.R., A.M. Chodorow, D.L. Endsley, and J.E. Bridge, "Spark Gap Breakdown at EMP Threat Level Rates of Rise," *IEEE Trans. Nuclear Science*, 26:4959–4963, Dec 1979.

Tucker, T.J., "Spark Initiation Requirements of a Secondary Explosive," *Annals of the New York Academy of Science*, 152:643–653, Oct 1968.

Uman, M., *Lightning*, McGraw-Hill, New York, 164 pp., 1969. (Enlarged and corrected republication by Dover Publications, New York, 1984.)

Uman, M., *The Lightning Discharge*, Academic Press, 377 pp., 1987.

Uman, M., and E.P. Kreider, "A Review of Natural Lightning," *IEEE Transactions on Electromagnetic Compatibility*, 24:79–112, May 1982.

Uman, M., D.F. Seacord, G.H. Price, and E.T. Pierce, "Lightning Induced by Thermonuclear Detonations," *J. Geophysical Research*, 77:1591–1596, 20 Mar 1972.

Uman, M., and R.B. Standler, "Lightning Activated Relay," U.S. Patent 4,276,576, 30 June 1981.

Underwriters Laboratory Standard 1283, "Electromagnetic Interference Filters," Revised 2nd Ed., Feb 1986.

Underwriters Laboratory Standard 1414, "Across the Line, Antenna Coupling, and Line Bypass Capacitors for Radio and Television-Type Appliances," 3rd Ed., Apr 1982.

U.S. Department of Defense, Defense Civil Preparedness Agency TR-61A, "EMP Protection for Emergency Operating Centers," July 1972. (Reprinted as Lawrence Livermore Laboratory Report PEM-8.)

U.S. Department of Defense Military Standard 188-124, "Grounding, Bonding, and Shielding for Common Long Haul/Tactical Communication Systems," 14 June 1978.

U.S. Department of Defense Military Handbook 419, "Grounding, Bonding, and Shielding for Electronic Equipments and Facilities," 21 Jan 1982.

Van Keuren, E., "Effects of EMP Induced Transients on Integrated Circuits," *IEEE 1975 Electromagnetic Compatibility Symposium,* paper 3AIIe, 1975.

Vance, E.F., "EMP-Induced Transients in Long Cables," *IEEE 1975 Electromagnetic Compatibility Symposium,* paper 3AIIa, 1975.

Vance, E.F., and M.A. Uman, "Differences Between Lightning and Nuclear Electromagnetic Pulse Interactions," *IEEE Trans. Electromagnetic Compatibility,* 30:54–62, Feb 1988.

Vicaud, A., "AC Voltage Ageing of Zinc Oxide Ceramics," *IEEE Trans. Power Systems,* 1:49–58, Apr 1986.

Vines, R.M., H.J. Trussell, L.J. Gale, and J.B. O'Neal, "Noise on Residential Power Distribution Circuits," *IEEE Trans. Electromagnetic Compatibility,* 26:161–168, Nov 1984.

Vines, R.M., H.J. Trussell, K.C. Shuey, and J.B. O'Neal, "Impedance of the Residential Power-Distribution Circuit," *IEEE Trans. Electromagnetic Compatibility,* 27:6–12, Feb 1985.

Voorhoeve, E.W., "Power-Limiting Electrical Barrier Device," U.S. Patent 3,878,434, 15 Apr 1975.

Wernström, H., M. Bröms, and S. Boberg, *Transient Overvoltages on AC Power Supply Systems in Swedish Industry,* Försvarets Forskningsanstalt E30002-E2, Apr 1984.

Wiesinger, J., "Hybrid Generator for the Coordination of Insulation," *Mess- und Proftechnik,* 104:1102–1105, 1983.

Williams, D.P., and M. Brook, "Magnetic Measurements of Thunderstorm Currents," *J. Geophys. Research,* 68:3243–3247, May 1963.

Williams, R.L., "Component Evaluation for Terminal Protection," paper 5-2B-2, Joint EMP Technical Meeting Proceedings, Sept 1973. (Distribution limited to U.S. Government agencies only: test and evaluation. Unclassified.)

Williams, R.L., "Test Procedures for Evaluating Terminal Protection Devices Used in EMP Applications," U.S. Army Harry Diamond Laboratories HDL-TR-1709, DDC-AD-A019098, June 1975.

Woody, J.A., "Modeling of Parasitic Effects in Discrete Passive Components," Rome Air Development Center, RADC-TR-83-32, Feb 1983.

Wooten, S.J., "Testing of Transient Suppression Networks," U.S. Naval Civil Engineering Laboratory TN-725, 30 Aug 1965. (Distribution limited to U.S. Government agencies only: test and evaluation. Unclassified.)

Wroblewski, T., "Voltage Regulating Transformer," U.S. Patent 4,075,547, 21 Feb 1978.

Wunsch, D.C., and R.R. Bell, "Determination of Threshold Failure Levels of Semiconductor Diodes and Transistors Due to Pulse Voltages," *IEEE Trans. Nuclear Science*, 15:244–259, Dec 1968.

Wynn-Williams, C.E., "An Investigation into the Theory of the Three Point Gap," *Philosophical Magazine*, 1:353–378, Feb 1926.

Index

CPSIA information can be obtained
at www.ICGtesting.com
Printed in the USA
BVHW040215010219
539241BV00012B/124/P

9 780486 425528